Preface

This book is intended for a one-year course in Mathematics for business and the Social Sciences. It should also be satisfactory for a similar service course in the biological sciences.

Since students in these fields are seldom interested in proofs of mathematical theorems, only proofs that are both simple and instructive are included. The exposition, therefore, relies heavily on the student's intuition and his willingness to accept the truth of many of the theorems needed in the logical development of the subject.

This book is divided into two parts. The first half consists of topics in finite mathematics; the second half is a brief treatment of elementary calculus. There is considerably more material on finite mathematics than can be covered in a one-semester course. It is assumed that the instructor will select the topics that best suit his needs. There is also too much material on elementary calculus to be covered in one semester. It will be necessary for him to omit a number of the more difficult sections and to refrain from assigning too many problems.

Since the book contains a large number of exercises, some of which are fairly difficult, it is possible, by varying the number of exercises covered, to construct a three-quarter course in which one quarter is devoted to finite mathematics and two quarters are devoted to calculus. Other course combinations are also possible through the proper selection of material and exercises.

Chapter 1, on functions, is essentially a review chapter and may be omitted if the students have an adequate foundation in high school mathematics. Chapters 2 and 3 deal with linear systems, and Chapter 4 applies this knowledge to linear programming. Chapters 5 and 6 deal with probability and are independent of the preceding chapters. Chapter 7, however, requires some knowledge of both probability and linear programming and thus should not be included unless the related chapters are studied. The chapters on calculus, which are independent of the material on finite mathematics, could well begin with the review of functions in Chapter 1 before the instructor proceeds to Chapter 8.

The emphasis in the course for which this book is designed should be on understanding the basic ideas and seeing how they can be applied to real life problems. It is therefore important to spend a considerable amount of time on problem solving in class. Some of the more difficult but interesting problems are intended for class discussion rather than for homework assignments. Answers to the odd numbered exercises are given in the Appendix. Instructors, however, may obtain a booklet with a complete list of answers from the publisher.

Paul G. Hoel

Contents

Finite Mathematics and Calculus
with Applications to Business

1 | Functions and Graphs

1. DEFINITIONS

This chapter is devoted to a review of material concerning functions and their geometrical representation that is usually learned in high school algebra. The concept of a function is fundamental in mathematics for both theoretical and practical reasons. It is easy to discuss the word function from an intuitive point of view. Thus it is customary to say that the consumer demand for a given product is a function of its price, or that the quality of a French wine is a function of the year in which it was produced. For a mathematical theory, however, we must be more precise. Toward this end we begin with the concept of a set, which may be described as follows:

Definition. A set is any well defined collection of objects.

The phrase well defined is inserted to make certain that there is no difficulty in determining whether an object is a member of the collection. It would be improper, for example, to talk about the set of happy people living in a given city unless a satisfactory definition of happy were available. However, the individuals in that city who voted at the last election would form a set because it is possible to determine from the records which individuals voted.

Any object in a set of objects is called an *element* of the set. Sets are usually denoted by capital letters, and elements are denoted by corresponding lower case letters. Thus we might let A denote the set of individuals in a given city who voted in the last election and let *a* denote any one

1

of those individuals. Similarly, we might let B denote the set of all real numbers and let b denote any one of those real numbers. With this notation and background, we are ready for the definition of a function.

Definition. Let A and B be any two nonempty sets. Then a function f, from A to B, is a rule that to each element, x, in A associates a unique element in B.

The set A is called the *domain* of the function f and the set of elements in B that are associated with elements of A is called the *range* of the function f. The element in B that is associated with the element x in A is denoted by the symbol $f(x)$, and is read "f of x."

The most familiar functions are those for which A and B are sets of real numbers. Thus for the function f defined by the formula $f(x) = x^2$ we might choose the domain A to be the set of all real numbers. For this function each real number x is associated with its square x^2; consequently, the range would consist of the set of all nonnegative numbers.

To illustrate functional notation, let us consider the function f defined by the formula $f(x) = x^2$ and the following particular set of x values: 2, 1, $\frac{1}{2}$, 0, -1. The value of the function corresponding to each of these x values is given by the respective numbers: $f(2) = 4$, $f(1) = 1$, $f\left(\frac{1}{2}\right) = \frac{1}{4}$, $f(0) = 0$, $f(-1) = 1$.

For another illustration, let the function g be defined by the formula $g(x) = \sqrt{x+1}$. Since the square root of a negative number is not a real number, we restrict the domain of g to be the set of numbers larger than -1. Then, corresponding to the x values 1, 0, and $-\frac{3}{4}$, for example, we have the functional values $g(1) = \sqrt{2}$, $g(0) = 1$, and $g\left(-\frac{3}{4}\right) = \frac{1}{2}$.

When a function f of a variable x is defined by a formula, such as $f(x) = x^2$, it is customary to abbreviate this statement and speak of the function $f(x) = x^2$. Thus this symbolism is used both to describe the function and also to represent its numerical value at a specified point x. We shall use this abbreviated language frequently in the material that follows when it creates no ambiguity.

Although the functions that we study in the next few chapters are the familiar kind in which both the domain and range are sets of real numbers, the preceding more general definition will be needed later when we study probability. There the domain may consist of things such as a set of letters, or a set of points in a plane.

As an example of this more general type of function, consider a function that gives the number of children in the family of a governor of any one of the 50 states. It might consist of a table of the following type:

Alabama	2
Alaska	0
.	.
.	.
.	.
Wyoming	1

If f were used to denote this function, we could write f(Alabama) = 2, f(Alaska) = 0, . . . , f(Wyoming) = 1. Here the domain A consists of the 50 state names and the range consists of the nonnegative integers that correspond to the number of children in at least one governor's family.

2. RECTANGULAR COORDINATES

A function for which both the domain and range are sets of numbers has the advantage of possessing a simple geometrical representation. We begin our study of functions by studying functions of this type. In doing so, we need to introduce a rectangular coordinate system in a plane. We first draw a horizontal line and a vertical line in the plane. These lines will intersect in a point, labeled O, called the origin. The two lines are called the coordinate axes. Then we select a unit length and mark the horizontal axis, starting at the origin, into segments of lengths 1,2,3, . . . , both to the left and to the right of O. We do the same for the vertical axis. The marked points on the horizontal axis to the right of O are associated with the positive integers, and those to the left are associated with the negative integers. Similarly, the marked points on the vertical axis above O are associated with the positive integers, and those below are associated with the negative integers. This construction is shown in Fig. 1.

Now consider an ordered pair of numbers, such as (3,2). We associate this pair of numbers with the point P shown in Fig. 2. The first number of the ordered pair tells us to go to the right of O along the horizontal axis a distance of 3 units and the second number tells us to go up vertically a distance of 2 units. The pair $(-2,1)$ would tell us to go to the left of O a distance of 2 units and up 1 unit. This pair corresponds to the point in Fig. 2 labeled Q. We operate in the same manner for any ordered pair of real numbers (a,b). Conversely, given any point in the coordinate plane, we associate it with an ordered pair of numbers (a,b) where a gives the horizontal distance, measured along the horizontal axis, and b gives the vertical distance, measured along the vertical axis, that the point is from the origin O. The sign of a determines whether to measure to the right or to the left of the origin and the sign of b determines whether to measure above or below the origin. This correspondence between ordered pairs of numbers and points in the plane enables us to set up a correspondence between the algebra and the geometry of functions. This is a very useful correspondence because we

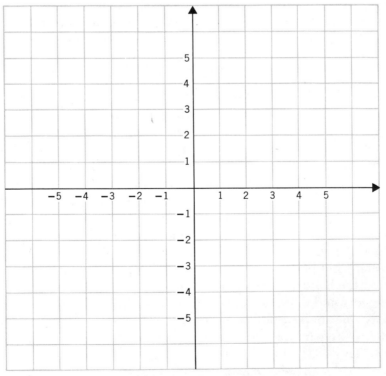

FIGURE 1 A rectangular coordinate system.

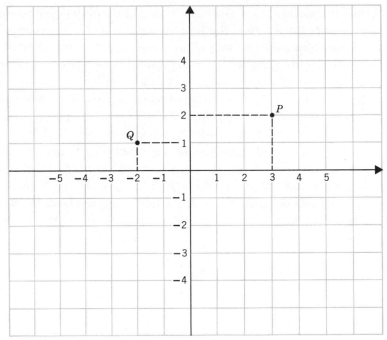

FIGURE 2   Points in a rectangular coordinate system.

can use our geometrical intuition to help solve problems involving numbers and functions, and conversely we can use algebraic techniques to help solve problems that arise from geometrical considerations.

It is customary in coordinate geometry to call the horizontal axis the x axis and the vertical axis the y axis and to designate an arbitrary point in the plane by the symbol (x,y). Thus the point P in Fig. 2 which is given by $(3,2)$ would be said to have an x coordinate of 3 and a y coordinate of 2.

3. GRAPHS

The purpose of introducing a coordinate system in the plane is to enable us to visualize a functional relationship geometrically when the domain and range of the function are sets of real numbers. Let f be any such function. Then the graph of f is defined to be the collection of all pairs of numbers $(x,f(x))$ where x is any number in the domain of f and $f(x)$ is the value of the function at x. Since each such pair of numbers corresponds to a point in the plane, the graph of the function yields a collection of points in the plane. If the domain is an interval on the x axis, this collection of points when sketched yields a curve that represents the function geometrically. The word graph is customarily used for both the collection of number pairs $(x,f(x))$ and the curve that results from considering these number pairs as the coordinates of points in the plane. We shall use the geometrical curve interpretation because it is more descriptive than the algebraic version.

As an illustration, consider once more the function given by the formula $f(x) = x^2$. Its graph consists of the collection of points with coordinates (x,x^2). Thus the points $(0,0)$, $(1,1)$, $(2,4)$, $(3,9)$, $(-1,1)$, $(-2,4)$, and $(-3,9)$ would be part of this collection. These particular points have been plotted in Fig. 3, together with a partial sketch of the graph of the function. A

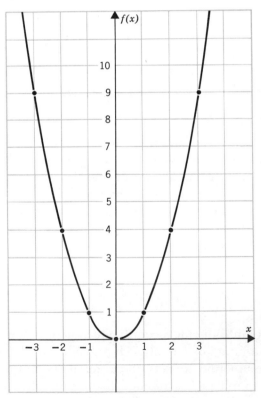

FIGURE 3   A graph of $f(x) = x^2$.

sketch of a graph, such as that in Fig. 3, is necessarily only an approximation since it is obtained by finding only a few points of the graph and then drawing a smooth curve between them.

It should be observed that in graphing a function, the functional value $f(x)$ corresponding to x is the y coordinate of the plotted point. Thus the y axis is the axis of functional values. Because of this relationship, it is often convenient to write $y = f(x)$ and then let (x,y) represent the coordinates of points on the graph of the function.

As additional illustrations of simple functions and their graphs, consider the following examples.

Example 1. Let $f(x) = 2x - 1$. Here the graph of f is the curve whose coordinates are given by $(x, 2x - 1)$, where x is any number. A few such points are the following: $(0,-1)$, $(-1,-3)$, $(1,1)$, $(2,3)$, $\left(\frac{1}{2}, 0\right)$, $\left(-\frac{1}{2}, -2\right)$. These points have been plotted in Fig. 4. It appears from these few points that the collection of all such points will determine a straight line. A sketch of this

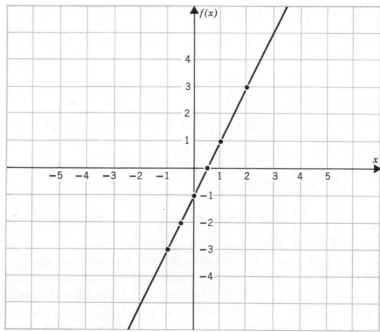

FIGURE 4   A graph of $f(x) = 2x - 1$.

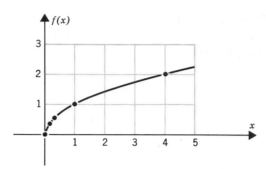

FIGURE 5   A graph of $f(x) = \sqrt{x}$, $x \geq 0$.

line is also shown in Fig. 4. In the next section we demonstrate that the graph must be a straight line.

Example 2. Let $f(x) = \sqrt{x}$. Here it is necessary to restrict the domain of f to be the set of all nonnegative numbers. Some of the points belonging to the collection $(x, \sqrt{x})$ are: $(0,0)$, $\left(\frac{1}{9}, \frac{1}{3}\right)$, $\left(\frac{1}{4}, \frac{1}{2}\right)$, $(1,1)$, $(4,2)$. The plot of these points, together with a sketch of a smooth curve joining them, is shown in Fig. 5.

Example 3. Let $f(x) = \frac{1}{x}$. This function is not defined for $x = 0$; therefore, we choose as its domain the set of all real numbers except 0. A few of the points whose coordinates are given by $\left(x, \frac{1}{x}\right)$ are the following: $\left(\frac{1}{3}, 3\right)$, $\left(\frac{1}{2}, 2\right)$, $(1,1)$, $\left(2, \frac{1}{2}\right)$, $\left(3, \frac{1}{3}\right)$, $\left(-\frac{1}{3}, -3\right)$, $\left(-\frac{1}{2}, -2\right)$, $(-1,-1)$, $\left(-2, -\frac{1}{2}\right)$, $\left(-3, -\frac{1}{3}\right)$. The plot of these points, together with a smooth curve passing through them, is shown in Fig. 6.

If a function is defined by a formula, which was true for the preceding three illustrations, we assume without stating it each time that the domain of the function is the set of real numbers for which the formula yields a real number value. This is the basis for the domains chosen in the foregoing illustrations. Occasionally, however, we may wish to restrict the domain further because of practical considerations.

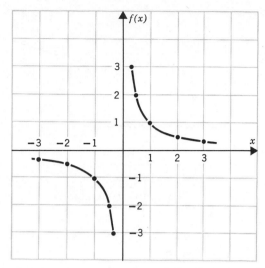

FIGURE 6   A graph of $f(x) = \dfrac{1}{x}$, $x \neq 0$.

4. LINEAR FUNCTIONS

The function given by $f(x) = 2x - 1$ that was discussed in the preceding section appears to yield a straight-line graph. A more general function of this type is the function given by the formula $f(x) = ax + b$, where a and b are two constants. We wish to show that the graph of this function is a straight line. For convenience of notation we let $y = f(x)$ and consider the graph of the relation $y = ax + b$.

A straight line in the plane is characterized by the property that the slope of any segment of the line is the same as that for any other segment. The slope of a straight line is a measure of how much the line rises vertically for a unit change to the right on the x axis. More precisely, let (x_1, y_1) and (x_2, y_2) be two points lying on the line, as shown in Fig. 7. Then $y_2 - y_1$ measures the amount the line has risen vertically in traversing the horizontal distance $x_2 - x_1$. The slope, denoted by m, is therefore given by the formula

(1)
$$m = \frac{y_2 - y_1}{x_2 - x_1}$$

Although this formula was obtained by inspecting Fig. 7 where the line is rising, we easily see that it also holds true for a line that is falling. In the

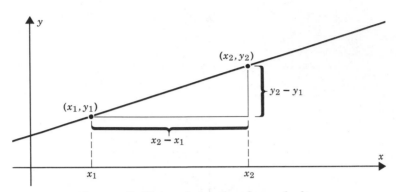

FIGURE 7   Determining the slope of a line.

latter case the slope is negative. This formula is valid provided $x_1 \neq x_2$, that is, provided the line is not vertical.

To determine whether a curve in the plane is a straight line, it suffices to choose three arbitrary points P_1, P_2, and P_3 on the curve and then determine whether the slope of the line segment connecting P_1 and P_2 is the same as that for the line segment connecting P_2 and P_3. Let these three arbitrary points lying on the graph of $y = ax + b$ be denoted by (x_1,y_1), (x_2,y_2), and (x_3,y_3). Then, from formula (1), the slopes of these two line segments are, respectively

$$\frac{y_2 - y_1}{x_2 - x_1}$$

and

$$\frac{y_3 - y_2}{x_3 - x_2}$$

But the y coordinates of these points are given by the functional formula $y = f(x) = ax + b$; consequently, these slopes assume the values

$$\frac{ax_2 + b - (ax_1 + b)}{x_2 - x_1} = \frac{ax_2 - ax_1}{x_2 - x_1} = \frac{a(x_2 - x_1)}{x_2 - x_1} = a$$

and

$$\frac{ax_3 + b - (ax_2 + b)}{x_3 - x_2} = \frac{ax_3 - ax_2}{x_3 - x_2} = \frac{a(x_3 - x_2)}{x_3 - x_2} = a$$

Since the slopes of these two line segments are the same and this holds true for all such line segments, the graph of $f(x) = ax + b$ must be a straight line.

Because of this fact the function given by $f(x) = ax + b$ is called a *linear function* of x.

The preceding demonstration shows that the coefficient of x in the linear equation $y = ax + b$ is the slope of the line. The number b is the value of y when $x = 0$ and, hence, is the y coordinate of the point on the y axis where the line crosses that axis. The number b is called the y *intercept* of the line.

Linear functions are often used to express the relationship between certain variables occurring in the study of business problems. For example, the total cost, C, of producing x units of a manufactured article during one week's production might be expressed in the form:

$$C(x) = 2x + 200$$

In this formula, 200 represents the fixed setup cost in dollars and 2 represents the cost in dollars per unit produced for labor and materials. From the nature of the problem, x must be a nonnegative integer here; however, for convenience in studying this function, we might choose the domain to consist of the set of all positive numbers.

As another illustration, a study of the demand for a certain product over a number of years yielded the formula

$$D(p) = 32 - .3p$$

Here p denotes the price per unit of the product and D denotes the demand in units as a function of the price. The negative coefficient of p determines how rapidly the demand decreases as the price increases. A practical formula like this is usually valid only for small changes in p from some average value. It obviously would not be applicable for, say, $p = 0$ or $p = 200$.

The preceding illustrations were concerned with the geometry of a linear function when the function was given by an algebraic formula. Now let us consider the reverse problem, that is, the problem of finding the algebraic form of a linear function when the function is described geometrically.

Since a straight line is uniquely determined geometrically by knowing two points through which it passes, or by knowing its slope and one point through which it passes, the problem here is to find the equation of a straight line for which this information is given. Assume, therefore, that we are given two points P_1 and P_2 with coordinates (x_1, y_1) and (x_2, y_2), respectively, through which the line passes, and that x_2 is not equal to x_1. Let P with coordinates (x, y) denote an arbitrary point on this line, with the exception of the point (x_1, y_1). From the earlier characterization of a straight line, it then follows that the slope of the line connecting the points P and P_1 must be the same as the slope of the line connecting the points P_1 and P_2. From formula (1), this requires that

$$\frac{y - y_1}{x - x_1} = \frac{y_2 - y_1}{x_2 - x_1}$$

Or, using m in place of the expression on the right, we obtain

$$\frac{y - y_1}{x - x_1} = m$$

Multiplying through by $x - x_1$, we obtain

(2) $$y - y_1 = m(x - x_1)$$

Since (x,y) was any point on the line, with the exception of the point (x_1, y_1), this equation must hold for every point on the line except (x_1, y_1). It obviously holds for (x_1, y_1) also; therefore, that earlier restriction disappears here. If (x,y) is any point off the line, the two preceding slopes will not be equal and, therefore, equation (2) will not be satisfied by such a point. Thus equation (2) is satisfied by every point on the line and by no other points; therefore it must be the equation of the line. This formula applies to any line that is not vertical. The equation of a vertical line is of the form $x = c$, where c denotes the point where the line cuts the x axis.

Formula (2) can be used to find the equation of a straight line either when two points are given or when the slope and one point are given. In the first case, we calculate the slope by means of formula (1) and then apply formula (2). In the second case, we apply formula (2) directly.

As an illustration, suppose that a line passes through the two points $(3,1)$ and $(2,-1)$. From formula (1), its slope is

$$m = \frac{-1 - 1}{2 - 3} = 2$$

From formula (2), its equation is

$$y - 1 = 2(x - 3)$$

Or

$$y = 2x - 5$$

Problems of the preceding type often arise in practical situations. For example, suppose a manufacturing firm has found that its total production costs were $20,000 during a month when it produced 500 units and were

$15,000 during a month when it produced 300 units. If the relationship between units produced, x, and total production costs, y, is linear, what is the equation of this relationship?

We may choose $(x_1,y_1) = (300, 15{,}000)$ and $(x_2,y_2) = (500, 20{,}000)$. Then by formula (1),

$$m = \frac{20{,}000 - 15{,}000}{500 - 300} = 25$$

Hence, by formula (2), the equation of the line is

$$y - 15{,}000 = 25(x - 300)$$

Or

$$y = 25x + 7500$$

This relationship can now be used to estimate production costs for various production levels. Its accuracy will, of course, depend on the degree to which the relationship is strictly linear. A graph of the records of production costs for many months in the past would undoubtedly show whether a linear relationship between these two variables is justified.

Since linear functions are used extensively in the social sciences, it is well to become familiar with their construction. The following examples illustrate how the equation of a straight line is determined when various kinds of geometrical information are given.

Example 1. Find the equation of the line that has slope -2 and that passes through the point $(2,-3)$.

Formula (2) may be applied directly to give

$$y + 3 = -2(x - 2) \qquad \text{or} \qquad y = -2x + 1$$

Example 2. Find the equation of the line that is parallel to the line $4x - 3y = 7$ and that passes through the origin.

We may rewrite the equation of the given line in the form $y = \dfrac{4}{3}x - \dfrac{7}{3}$. This shows that the slope is $\dfrac{4}{3}$. Parallel lines have the same slope; hence, the desired line has the equation

$$y = \frac{4}{3}x$$

Example 3. Find the equation of the line that cuts the x axis at $x = 5$ and the y axis at $y = -3$.

This information is equivalent to being told that the line passes through the two points $(5,0)$ and $(0,-3)$; hence, its slope is

$$m = \frac{0+3}{5-0} = \frac{3}{5}$$

and its equation is

$$y + 3 = \frac{3}{5}(x - 0) \qquad \text{or} \qquad y = \frac{3}{5}x - 3$$

Example 4. Find the equation of the line that passes through the point $(-3,2)$ and that has equal nonzero intercepts on the two axes.

If $(a,0)$ and $(0,a)$ are the points where the line crosses the two axes; the slope will be given by

$$m = \frac{a-0}{0-a} = -1$$

Its equation is, therefore,

$$y - 2 = -1(x + 3) \qquad \text{or} \qquad y = -x - 1$$

5. QUADRATIC FUNCTIONS

Many functions that arise in business problems are the type in which the function increases to a maximum value and then decreases, or decreases to a minimum value and then increases. A highly useful model for such problems is the quadratic function

$$f(x) = ax^2 + bx + c$$

where a, b, and c are constants and $a \neq 0$. The function $f(x) = x^2$ that was graphed in Fig. 3 is a special case of this general quadratic. The graph of a quadratic function is a curve called a *parabola*. It is easily shown that a quadratic function has a single maximum, or minimum, value and that this value occurs when $x = -\dfrac{b}{2a}$. This is accomplished by carrying out the following algebraic operations:

$$f(x) = ax^2 + bx + c$$

$$= a\left(x^2 + \frac{b}{a}x + \frac{c}{a}\right)$$

$$= a\left(x^2 + \frac{b}{a}x + \frac{b^2}{4a^2} + \frac{c}{a} - \frac{b^2}{4a^2}\right)$$

$$= a\left(x^2 + \frac{b}{a}x + \frac{b^2}{4a^2}\right) + c - \frac{b^2}{4a}$$

$$= a\left(x + \frac{b}{2a}\right)^2 + \frac{4ac - b^2}{4a}$$

If a is a positive number, $f(x)$ will be minimized by choosing x to make the term in parentheses equal zero, that is, by choosing $x = -\frac{b}{2a}$. If a is a negative number, $f(x)$ will be maximized by this same choice of x. The point where this maximum, or minimum, occurs is called the *vertex* of the parabola. The vertex is very useful in graphing the function because a parabola is symmetric about a vertical line that passes through the vertex.

We illustrate this property of a parabola on the function $f(x) = 2x^2 - 4x + 5$. Here

$$-\frac{b}{2a} = -\frac{-4}{2 \cdot 2} = 1$$

and $f(1) = 3$; hence, the vertex of the parabola is at the point $(1,3)$. Calculations give $f(2) = 5$, $f(3) = 11$, $f(0) = 5$, and $f(-1) = 11$, which shows that $(1,3)$ must be a minimizing point. A plot of these points and a sketch of the parabola is shown in Fig. 8.

As an illustration of an application of quadratic functions, let us assume that the demand for a product is a linear function of the price, as was done in the second illustration of the preceding section. Hence, we have a relation of the form $x = \alpha p + \beta$, where α and β are constants, where p is the price per unit, and where x is the number of units demanded. If we solve this equation for p as a function of x, we obtain a relation of the form

$$p = ax + b$$

In terms of the original constants, we readily observe that $a = 1/\alpha$ and $b = -\beta/\alpha$. The total revenue, denoted by R, and obtained by selling x units at this price is then given by the function

$$R(x) = xp = ax^2 + bx$$

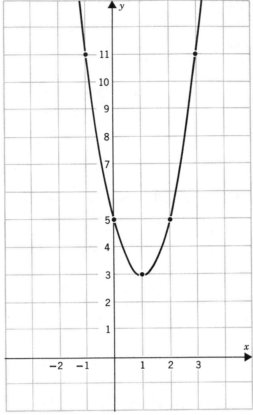

FIGURE 8   A graph of $f(x) = 2x^2 - 4x + 5$.

Hence, the total revenue function is quadratic if the demand function is linear.

Quadratic functions are also frequently used as models for representing the cost per unit of a product when producing x units of it because the cost per unit is usually at a minimum under an ideal set of production conditions and rises as the number of units produced becomes smaller or larger than this ideal number.

6. SOME ADDITIONAL FUNCTIONS

If we want a function that has greater flexibility than that possessed by linear and quadratic functions, we can add higher degree terms in x to ob-

the relationship is given by

$$x = 300 - 2p$$

We.may wish to study the price per unit to charge for the product as a function of the demand, in which case we must solve for p in terms of x to obtain

$$p = 150 - \frac{1}{2}x$$

In the first equation, x is a function of p, whereas in the second equation the roles have been interchanged. The second relation defines what is called the *inverse function* of the first functional relationship.

The graph of the first function is the straight line shown in Fig. 10. Since the second functional relationship is identical with the first, except in the manner of expressing it, the graph of this relationship must be the same as for the first relationship. It is customary, however, in graphing a function of a variable always to have the variable axis horizontal and the functional axis vertical. If we follow this tradition, then in the second graph our horizontal axis should be the x axis and our vertical axis should be the p axis. To accomplish this rearrangement of axes, it will suffice to rotate the original graph counterclockwise one fourth of a complete turn until the original x axis becomes horizontal and the p axis becomes vertical. The only difficulty with this technique is that then the positive x axis will be positive to the left instead of to the right on the horizontal axis. Therefore, we must flip the

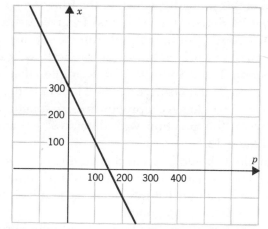

FIGURE 10   A graph of the function $x = 300 - 2p$.

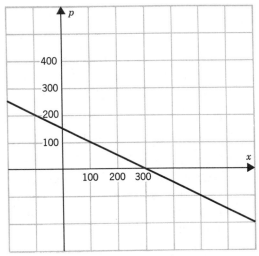

FIGURE 11   A graph of the function $p = 150 - \dfrac{1}{2}x$.

picture over, or make it out of transparent paper and look at it from the back side.

The graphs of the original function and its inverse, properly oriented, are shown in Figs. 10 and 11.

Not every function possesses an inverse function because it is not always possible to solve the original functional relationship, say, $y = f(x)$, for x in terms of y and obtain a unique value of x corresponding to each value of y. As an illustration of this difficulty, consider the function $f(x) = x^2$. We write this in the form $y = x^2$. Solving for x in terms of y gives $x = \pm\sqrt{y}$. Here there are two values of x corresponding to each value of y; therefore, unless we restrict the domain of our original function properly we cannot talk about the inverse function here. Suppose, for example, that we choose the domain of $f(x) = x^2$ to be the positive x axis. Then only the positive root in $x = \pm\sqrt{y}$ applies, and we do have a proper inverse function. The graphical relationship between this function with domain $x > 0$ and its inverse is shown in Figs. 12 and 13.

From this example it should be clear that an inverse function will exist if to each value of y there exists a unique value of x. Geometrically, this will be true if every horizontal line that intersects the graph of the original function intersects it in precisely one point.

One of the simplest methods for determining whether a function pos-

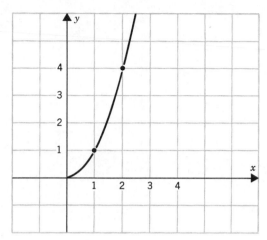

FIGURE 12   A graph of the function $y = x^2$ for $x \geqslant 0$.

sesses a unique inverse is to graph the function and observe whether this property is satisfied. As an illustration, consider the function $f(x) = \dfrac{x}{x + 1}$. By choosing convenient values of x and by calculating the corresponding values of $f(x)$, the graph shown in Fig. 14 was constructed. From that graph it is clear that this function does possess a unique inverse, except at $y = 1$ where the inverse is not defined.

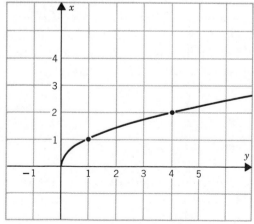

FIGURE 13   A graph of the function $x = \sqrt{y}$.

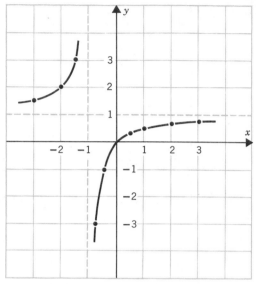

FIGURE 14   A graph of the function $f(x) = \dfrac{x}{x+1}$.

COMMENTS

This chapter has been concerned with the mathematical concept of function, and in particular with the relationship between the algebra and the geometry of functions. It studied some of the simpler functions that are frequently used as mathematical models in the social sciences. This material which is usually presented in high school mathematics courses is included here for the benefit of those students who may have forgotten it. It is essential for an understanding of later chapters.

EXERCISES

Section 3

1. Graph each of the following functions over a suitable domain by plotting a few points and drawing a smooth curve through those points.
   (a) $f(x) = -x + 3$ (b) $f(x) = -2$
   (c) $f(x) = x^2 - 1$ (d) $f(x) = x^3$
   (e) $f(x) = \sqrt{x-1}$ (f) $f(x) = -\sqrt{2x}$

2. Graph each of the following functions over a suitable domain by plotting a few points and drawing a smooth curve through the points that represent a piece of the entire graph. The entire graph may consist of several parts.

(a) $f(x) = \begin{Bmatrix} x \,, x \geqslant 0 \\ -x \,, x < 0 \end{Bmatrix}$

(b) $f(x) = \begin{Bmatrix} 1 \,, x \leqslant 1 \\ x^2, x > 1 \end{Bmatrix}$

(c) $f(x) = \dfrac{1}{x - 2}$

(d) $f(x) = \dfrac{1}{x^2}$

(e) $f(x) = \begin{Bmatrix} x \quad\;\; , x \leqslant 1 \\ x + 2, x > 1 \end{Bmatrix}$

(f) $f(x) = \begin{Bmatrix} 2, x = 1 \\ x, x \neq 1 \end{Bmatrix}$

3. For what values of x are the following functions defined?

(a) $f(x) = \dfrac{x}{x - 1} + \sqrt{2x - 1}$

(b) $f(x) = \dfrac{\sqrt{4 - x}}{\sqrt{x - 2}}$

4. Graph each of the following functions.

(a) $f(x) = \begin{Bmatrix} x + 1, x \leqslant 0 \\ x^2 + 1, x > 0 \end{Bmatrix}$

(b) $f(x) = \begin{Bmatrix} x + 1, x \leqslant 0 \\ x^2 \quad\;\;, x > 0 \end{Bmatrix}$

5. Experiments with individuals show that they underestimate heavy weights and overestimate light weights when asked to estimate the weights of a set of objects. It has also been found that the relationship between the actual weight, x, of an object and its estimated weight, y, is approximated well by a formula of the type

$$y = ax^b$$

where a and b are constants. An experiment based on eight objects whose weights ranged from 20 to 200 grams gave the values $a = \dfrac{1}{5}$ and $b = \dfrac{3}{2}$. Graph this relationship.

6. According to a theory of economics, called Pareto's law of the distribution of incomes, the number of individuals, y, of a given population whose income exceeds x units of money is given by the formula

$$y = \dfrac{a}{x^b}$$

where a and b are constants. This relationship is assumed to hold true for large incomes only. For a certain population and for each unit of x corresponding to $1000, it was found that $a = 50$ and $b = \dfrac{3}{2}$. Graph this relationship for $x = 10, 20, 30, 40$, and 50, and draw a smooth curve through the resulting points.

7. Graph the postage function that gives the cost of postage corresponding to various weights of letters if a letter weighing 1 ounce or less costs 8 cents, a letter weighing more than 1 ounce but not more than 2 ounces costs 16 cents, and each additional ounce, or fraction thereof, costs 8 cents.

8. The monthly bank charges for a checking account are \$2 for 20 checks or less and 5 cents for each check cashed after the first 20. Let x denote the number of checks cashed in a month and let $C(x)$ denote the monthly charges. Find an algebraic expression for $C(x)$ for $0 \leqslant x \leqslant 50$ and graph it. Although x must be a nonnegative integer here, ignore this fact and draw the graph for all values of x in this interval.

9. Let x denote the number of hours a week that a laborer works. Let r denote his hourly wage rate in dollars and let $W(x)$ denote his wages for one week. He gets paid time and a half for all hours worked over 40 hours. Find an algebraic expression for $W(x)$ for $0 \leqslant x \leqslant 60$. Graph this function for $r = 3$. Choose different scales on the two axes to keep the graph from climbing too high.

Section 4

1. In each case find the slope of the line that passes through the given pair of points.
   (a) (4,2) and (2,$-$3)   (b) (5,1) and ($-$2,2)
   (c) $\left(\frac{1}{2}, -3\right)$ and $\left(\frac{1}{4}, 2\right)$ (d) $\left(-\frac{3}{2}, -1\right)$ and $\left(\frac{3}{2}, -1\right)$

2. Determine which of the following points lie on the line whose equation is $y = 2x - 6$.
   (a) (4,2)   (b) ($-$1,$-$4)   (c) $\left(\frac{1}{2}, -5\right)$ (d) ($-$3,0)

3. Given the equation of a line in the form $ax + by + c = 0$:
   (a) What is its slope?
   (b) What is its y intercept?
   (c) Where does it cut the x axis?

4. Graph the two lines whose equations are $2x - y = 1$ and $2x - y = 2$. What can you say about the graph of $2x - y = c$ where c is any real number?

5. Describe the graph of the line whose equation is $\dfrac{x}{a} + \dfrac{y}{b} = 1$.

6. What is the equation of (a) the y axis, (b) the x axis?

7. Find the equation of the line that has the given slope and that passes through the given point:
   (a) $m = 3$, $(2,1)$ (b) $m = -2$, $(-1,1)$
   (c) $m = \dfrac{1}{2}$, $(-2,-3)$ (d) $m = -\dfrac{1}{3}$, $\left(-\dfrac{1}{3}, \dfrac{2}{3}\right)$

8. Find the equation of the line that passes through the two given points:
   (a) $(2,-1)$ and $(3,2)$ (b) $(-1,4)$ and $(-3,1)$
   (c) $(-2,-3)$ and $(-5,-6)$ (d) $\left(\dfrac{1}{2}, -2\right)$ and $\left(\dfrac{1}{4}, 1\right)$

9. Find the equation of the line that:
   (a) Is parallel to the y axis and passes through the point $(3,4)$.
   (b) Is perpendicular to the y axis and passes through the point $(-3,2)$.
   (c) Has slope -3 and y intercept of 2.

10. Find the equation of the line that:
    (a) Is parallel to the line $2x - 3y + 4 = 0$ and passes through the point $(2,-1)$.
    (b) Has slope -2 and cuts the x axis at $x = 4$.
    (c) Passes through the point $(3,2)$ and cuts the positive x and y axes the same distance from the origin.

11. Find the equation of the line that cuts the x axis at $x = -1$ and the y axis at $y = 3$.

12. The line $y = ax + b$ passes through the points $(1,2)$ and $(3,4)$. Find the values of a and b.

13. Find the equation of the line that passes through the origin and that is parallel to the line $4x + 3y = 2$.

14. What is the equation of a line whose x and y intercepts are equal, assuming that it does not pass through the origin?

15. Let x denote the number of units produced in a factory in one day and let $C(x)$ denote the cost function. Two processes can be used. In one, the setup cost is $100 and the unit cost for labor and materials is $5. In the other, the setup cost is $200 and the unit cost for labor and materials is $4; however, this process can be used only if at least 100 units are produced.

   (a) Express $C(x)$ as a function of x for both processes and graph those functions for $0 < x \leqslant 200$.

   (b) What would you do to minimize costs for this entire interval?

Section 5

1. Graph each of the following parabolas by first locating its vertex and then finding a few points that lie on it.
   (a) $y = 4x^2$ (b) $y = 3x^2 - 3x + 3$
   (c) $y = x^2 - 4x + 2$ (d) $y = 5x^2 + 7x - 2$

2. Graph each of the following parabolas by first locating its vertex and then finding a few points that lie on it.
   (a) $y = -x^2 - x + 4$ (b) $y = -4x^2 + 8x + 2$

   (c) $y = -\dfrac{x^2}{6} - \dfrac{5x}{3} - \dfrac{19}{6}$ (d) $2x^2 + y - 4x + 6 = 0$

3. Find the equation of the parabola $y = ax^2 + bx + c$ if it passes through the origin and its vertex is located at the point $(2,-1)$.

4. Find the equation of the parabola $y = ax^2 + bx + c$ if its vertex is located at the origin and it passes through the point $(1,3)$.

5. Find the equation of the parabola $y = ax^2 + bx + c$ if it passes through the point $(1,1)$ and its vertex is located at the point $(2,2)$.

6. A parabolic arch has a height of 20 feet and a width of 36 feet where the ends of the arch meet the ground. Let a coordinate system be introduced with the origin chosen at the midpoint of the base of the arch and the y axis passing through the top of the arch. Find the equation of the arch in this coordinate system.

7. A new apartment has 50 units. They will all be occupied if the rent is only $150 a month. For each $5 per month added on to this basic amount, one apartment on the average will be vacant. A vacant apartment costs management $5 a month to maintain, whereas an occupied apartment costs management $15 a month to maintain. Let x denote the

number of apartments that are rented and let $P(x)$ denote the revenue minus maintenance costs. Find an algebraic expression for $P(x)$ and graph it. What rent would you charge if you were the manager?

Section 6

1. Graph each of the following functions over a suitable domain.

   (a) $f(x) = \dfrac{2}{x-4}$ (b) $f(x) = \dfrac{1}{\sqrt{x}}$

   (c) $f(x) = \dfrac{x}{x^2+1}$ (d) $f(x) = \dfrac{x^2}{x^2+1}$

2. Graph each of the following absolute value functions.

   (a) $f(x) = |x|$ (b) $f(x) = |x-2|$

3. Graph each of the following functions over a suitable domain.

   (a) $f(x) = \sqrt{x-x^2}$ (b) $f(x) = \dfrac{1}{\sqrt{x+x^2}}$

   (c) $f(x) = \dfrac{x+1}{x-1}$ (d) $f(x) = (x-1)(x-2)(x-3)$

4. An author is paid by his publishing firm according to the following royalty schedule: \$1.40 for each copy sold until 10,000 copies have been sold, and thereafter \$1.70 for each additional copy sold. Let $R(x)$ denote the author's royalties in dollars where x denotes the number of copies sold. Find an algebraic expression for $R(x)$ and graph it.

5. An electric company charges its customers the following amounts for service: \$3 for the first 20 kilowatt hours or less, 10 cents per kilowatt hour for the next 80 kilowatt hours, and 5 cents per kilowatt hour for any hours above 100 kilowatt hours. Let x denote the number of kilowatt hours consumed by a customer and let $E(x)$ denote his electricity bill. Find an algebraic expression for $E(x)$ and graph it.

6. Let x denote the number of units produced in a hosiery mill and let $C(x)$ denote the cost of producing this number of units. Experience over a period of time yielded the following cost function, which was fairly accurate over the range of x values encountered in this mill.

$$C(x) = 10,000 + 7.20x + .0004x^2$$

Let $A(x) = C(x)/x$ denote the average cost per unit produced. Write down the algebraic expression for $A(x)$ and graph it for $0 < x \leqslant 15,000$. From this graph, guess the production level that will minimize the average cost per unit.

7. Let x denote the number of units produced and let $A(x)$ denote the average cost per unit when x units are produced. For a certain type of manufactured product, experience has shown that for the range of values of x that are normally encountered, $A(x)$ is given by the function

$$A(x) = \frac{x^2}{100} - 10x + 3000, \qquad 100 \leqslant x \leqslant 800$$

(a) For what production level is the average cost per unit a minimum?

(b) Let $C(x) = xA(x)$ denote the total cost of producing x units. Write down the algebraic expression for $C(x)$ and graph it over the indicated domain.

Section 7

1. By means of a graph, or otherwise, determine which of the following functions possesses an inverse.

   (a) $f(x) = \sqrt{x-4}$ (b) $f(x) = x^3$

   (c) $f(x) = \dfrac{1}{x^2+1}$ (d) $f(x) = \dfrac{1}{x^3-1}, \quad x \neq 1$

2. By means of a graph, or otherwise, determine which of the following functions possesses an inverse.

   (a) $f(x) = x^4$ (b) $f(x) = (x+1)^3$

   (c) $f(x) = x^3 + x$ (d) $f(x) = x^2 - \dfrac{1}{x}, \quad x > 1$

   (e) $f(x) = \dfrac{x^2}{x+1}, \quad x > -1$ (f) $f(x) = \dfrac{x^3}{x-1}, \quad x > 1$

3. Given $f(x) = \dfrac{x+2}{x}$, $x \neq 0$, find the inverse function and graph both the function and its inverse.

4. Given $f(x) = x^3 + 1$, find the inverse function and graph both the function and its inverse.

5. Given $f(x) = \dfrac{1}{x^3}$, $x \neq 0$, find the inverse function and graph both the function and its inverse.

6. Given $f(x) = \dfrac{1-x^3}{x^3}$, $x \neq 0$, find the inverse function and graph both the function and its inverse.

2 | Systems of Linear Equations

Thus far we have considered functions of a single variable only. Many social science problems, however, involve several variables and their relationships. Since linear functions are the simplest to manipulate and are also satisfactory models in many problems, we restrict ourselves to these functions in this chapter.

1. TWO EQUATIONS IN TWO UNKNOWNS

Suppose that we have a linear function, such as the total cost function discussed in Section 4, Chapter 1. If we let y denote the total cost, the relationship between cost and production, x, becomes

$$y = 2x + 200$$

Suppose a second method of producing this product is being considered in which the setup cost is 300 dollars and the cost for labor and materials per unit produced is 1.5 dollars. The relationship between total cost and production is then given by

$$y = 1.5x + 300$$

For the purpose of comparing these two manufacturing processes we graph the two functions on the same set of axes. Both graphs will, of course, be straight lines. Since the ordinates on the two lines, corresponding to a given value of x, give the respective costs for the two processes for that amount of production, we will be able to compare the two processes visually

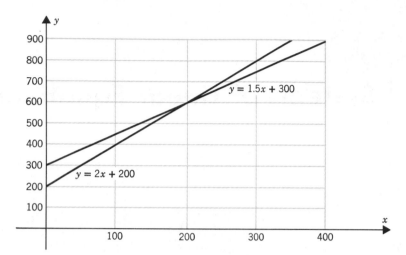

FIGURE 1　The graphs of two linear functions.

by means of these graphs. Let us assume that the two cost formulas hold true only for $100 \leqslant x \leqslant 300$.

Although two points are sufficient to determine a line, three points were used for each line to guard against errors. The first line was determined by the pairs of values (100,400), (200,600), and (300,800), whereas the second line was determined by the pairs (100,450), (200,600), and (300,750). The graphs of the two lines are shown in Fig. 1.

From these graphs we see that the first process is cheaper if production is expected to be less than 200 units, that the costs are equal when production is exactly 200 units, and that the second process is cheaper for production greater than 200 units.

If we were interested only in determining for what production level the costs are the same, it would suffice to find the value of x that makes the two values of y equal. This is accomplished by solving the pair of equations

$$y = 2x + 200$$
$$y = 1.5x + 300$$

Equating the two values of y gives

$$2x + 200 = 1.5x + 300$$

The solution is obtained by carrying out the following familiar algebraic operations:

$$2x - 1.5x = 300 - 200$$
$$0.5x = 100$$
$$x = \frac{100}{0.5} = 200$$

This value was obtained earlier in the process of graphing the two functions; however, we cannot expect to be so lucky in general. The algebraic approach requires no luck.

Another interesting application of the need to solve a pair of linear equations arises in the following manner. Suppose we have a total cost function and a revenue function that are both functions of the amount produced. In particular, suppose the cost function is the first of the two functions used in the preceding problem. Thus

$$C(x) = 2x + 200$$

If each unit produced can be sold for, say, $4 and if the fixed sales costs for this product are $100, then the total revenue obtained by selling x units is given by

$$R(x) = 4x - 100$$

The following question should now be answered: How many units must be produced before the manufacturer will break even? This is equivalent to asking for the value of x that will make $R(x) = C(x)$. The answer is obtained by finding the solution of this pair of linear equations:

$$y = 2x + 200$$
$$y = 4x - 100$$

Equating the y values, we obtain

$$4x - 100 = 2x + 200$$

Here the solution is $x = 150$. This value of x is called the *break-even point*. It takes this much production before the manufacturer can expect to begin to make a profit.

Another useful application of solving a pair of linear equations occurs in economic theory when the demand and supply functions for a given product are both linear functions of the price. If the demand for a product, x, is a linear function of the price, p, then, as p decreases, x will increase because consumers usually increase their purchases when the price falls. The inverse relationship in which p is a function of x will therefore be linear and its graph will have a negative slope. Similarly, if the number of units produced, y, is a linear function of the price, p, then, as p increases, y will increase because the producer will be happy to produce more if the price is raised. This implies that the inverse relationship in which p is a function of y will also be linear and its graph will have a positive slope. The first of these two inverse functions is called the *price-demand function*, and the second is called the *price-supply function*. The following two formulas are illustrations of such functions.

$$\text{Price-demand:} \qquad p = -\frac{2}{3}x + 200$$

$$\text{Price-supply:} \qquad p = \frac{3}{4}y + 30$$

Graphs of these two functions using the same set of axes are shown in Fig. 2. It is understood therein that x plays the role of y in the price-supply graph.

When the price of a product is such that the number of units demanded at that price as given by the price-demand curve is equal to the number of units that will be produced at that price as given by the price-supply curve, we have *market equilibrium* for the product. That price is called the *equilibrium price* and the corresponding quantity is called the *equilibrium quantity*. It may be obtained by replacing y by x in the price-supply function and then solving the resulting two linear equations in the variables

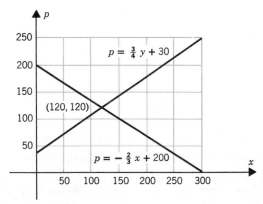

FIGURE 2   Graphs of a price-demand and a price-supply function.

p and x. For this illustration, we may equate the p's and obtain

$$-\frac{2}{3}x + 200 = \frac{3}{4}x + 30$$

This gives $x = 120$, which in turn gives $p = 120$; hence, these are the equilibrium quantity and price, respectively. Geometrically, these values give the coordinates of the point of intersection of the price-demand and price-supply graphs.

Since problems of the preceding type occur so often in the social sciences, we shall solve several sets of two equations in two unknowns for the purpose of observing the nature of the solutions.

First, consider the problem of solving the following pair of equations.

$$2x + y = 1$$
$$3x + 2y = 3$$

Multiplying the first equation by 2, we obtain the following equivalent pair of equations.

$$4x + 2y = 2$$
$$3x + 2y = 3$$

Subtracting the second equation from the first gives $x = -1$. Substituting $x = -1$ into the first equation and solving for y gives $y = 3$. Hence, the desired solution is $x = -1$, $y = 3$.

Next, consider the following pair of equations.

$$2x + y = 1$$
$$4x + 2y = 2$$

If we proceed as before by multiplying the first equation by 2, we obtain the equivalent pair

$$4x + 2y = 2$$
$$4x + 2y = 2$$

Any pair of numbers (x,y) that satisfies one of the equations obviously satisfies the other equation. Since there are an infinite number of pairs (x,y) that satisfy a single linear equation, it follows that the original pair of equations possesses an infinite number of solutions. Thus, if we choose $x = k$, where k is any number, the corresponding value of y is given by $y = 1 - 2k$.

The infinite number of possible solutions may then be represented by the infinite number of pairs of values $(k, 1 - 2k)$.

Finally, consider the following pair of equations.

$$2x + y = 1$$
$$4x + 2y = 3$$

As previously, we multiply the first equation by 2. This gives the equivalent pair

$$4x + 2y = 2$$
$$4x + 2y = 3$$

Now any pair of numbers (x,y) that satisfies the first equation cannot satisfy the second equation. Thus, there cannot exist a solution to this pair of equations. Equations like these are said to be *inconsistent*.

The geometrical interpretation of what is occurring in the three preceding problems is shown in Figs. 3, 4, and 5. In the first problem the graphs of the two linear equations intersect in a single point, $(-1,3)$, whose coordinates yield the unique solution of the two equations. In the second problem the graphs of the two equations are identical; consequently, the two lines intersect at all points on this common line. Thus every point on the line has coordinates that satisfy both equations. In the third problem the graphs are

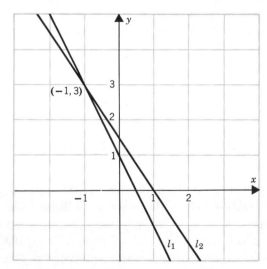

FIGURE 3   Two intersecting lines.

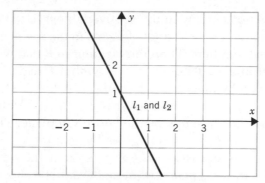

FIGURE 4   Two identical lines.

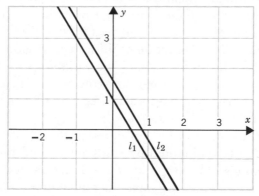

FIGURE 5   Two parallel lines.

those of two parallel lines; therefore, the lines never intersect and no solution is possible.

The three preceding examples illustrate the three possibilities that can arise. A pair of linear equations may have a unique solution, or an infinite number of solutions, or no solution. These possibilities correspond geometrically to the straight-line graphs of the two equations intersecting in a single point, coinciding, or being parallel.

2. LINEAR FUNCTIONS OF SEVERAL VARIABLES

In this section and the next we study functions of more than one variable. Our definition of a function, which is given in Section 1, Chapter 1, is applicable to functions of any number of variables. In particular, suppose that we choose the set A to be the set of points in the x, y plane and the set B to be the set of real numbers. Then the element in B that we associate with the element (x,y) in A is denoted by the symbol $f(x,y)$. Functions of this type are often given by means of a formula.

As an illustration, the formula $f(x,y) = x + 2y + 3$ defines a function whose domain may be chosen to be the entire x, y plane. This function can take on any real value. The function given by $g(x,y) = \sqrt{x - y}$, however, is defined only when x is at least as large as y; consequently, its domain could not consist of the entire x, y plane. Its domain could be chosen, for example, to be that part of the x, y plane lying below the line whose equation is $y = x$.

Often we conveniently denote a function by a single letter. Thus we might denote the function $f(x,y) = x + 2y + 3$ by the letter z and write $z = x + 2y + 3$. A function of the form $f(x,y) = ax + by + c$, where a, b, and c are constants, is called a *linear function* of those two variables. It is an extension of the name for a linear function of one variable. Thus $z = x + 2y + 3$ is a linear function of x and y. Furthermore, the relationship $z = x + 2y + 3$, which may also be written in the form $x + 2y - z + 3 = 0$, is called a *linear equation* in the three variables x, y, and z.

In the next section we will be concerned with sets of linear equations. These equations will be obtained by studying several linear functions of the variables x and y and by assigning those functions letter symbols, such as z. Then a basic problem to be solved is that of determining what values of x and y will yield the same value to all the functions involved. Thus we are interested in a type of break-even analysis for functions of two variables similar to that encountered in the preceding section for functions of one variable.

3. THREE EQUATIONS IN THREE UNKNOWNS

Suppose that we have three linear functions of the two variables x and y and we wish to find the values of x and y that make the three functional values equal. Since we wish the functional values to be equal, we may denote them all by the letter z. The three relationships then become three linear equations in three unknowns and the problem reduces to one of solving three linear equations in the three unknowns x, y, and z. For the purpose of developing a systematic method for doing this, consider the following particular set of equations:

$$x - 2y + 3z = 4$$
$$2x + y - 3z = 5$$
$$-x + y + 2z = 3$$

The method that will be used to solve these equations is commonly called the *Gauss elimination method*. It is also called the *pivotal reduction method*. It hinges on reducing the given three equations to three equivalent equations of the form:

(1)
$$x + b_1 y + c_1 z = d_1$$
$$y + c_2 z = d_2$$
$$z = d_3$$

The technique for accomplishing this reduction proceeds as follows.

Since the first equation is already in the desired form, it is copied down again to form the first of the new set of equations:

$$x - 2y + 3z = 4$$

Next, we multiply the first equation by -2 and add it to the second to obtain an equation that has no x term in it. This gives

$$5y - 9z = -3$$

This equation is divided by 5 to yield the desired new second equation. Thus, we obtain

$$y - \frac{9}{5} z = -\frac{3}{5}$$

The first equation is then added to the third equation to obtain another equation without an x term:

$$-y + 5z = 7$$

The original set of equations is now replaced by this equivalent set of equations:

(2)
$$x - 2y + 3z = 4$$
$$y - \frac{9}{5} z = -\frac{3}{5}$$
$$-y + 5z = 7$$

The name pivotal reduction arises from the technique of "pivoting" on the x term of the first equation to work for zero coefficients for the x terms in the remaining equations.

We now ignore the first equation in (2) and work on the last two equations by pivoting on the y term of the second equation to work for a zero coefficient of the y term in the third equation. Thus we add the second equation to the third to obtain a new third equation which has no y term in it:

$$\frac{16}{5} z = \frac{32}{5}$$

Or

$$z = 2$$

The desired new set of equations of the form shown in (1) is, therefore, the set

$$x - 2y + 3z = 4$$
$$y - \frac{9}{5} z = -\frac{3}{5}$$
$$z = 2$$

It is now a simple matter to substitute the value of z from the third equation into the second and then solve that equation for y. This gives

$$y = -\frac{3}{5} + \frac{9}{5}(2) = 3$$

Finally, we substitute the values of z and y already found into the first equation to obtain the value of x:

$$x = 4 + 2(3) - 3(2) = 4$$

The solution for this set of equations is, therefore, $x = 4$, $y = 3$, and $z = 2$.

It is easy to show that the pivotal reduction method produces a new set of equations that is equivalent to the original set in the sense that the two sets possess the same solutions. Thus we do not lose or gain any solutions by our pivotal reduction operations.

The set of equations that was solved at the beginning of this section is now modified to illustrate the fact that just as is the case for two equations in two unknowns, there exists the possibility that a set of equations may possess a unique solution, an infinite number of solutions, or no solution.

With this in mind, consider the following set:

$$x - 2y + 3z = 4$$
$$2x + y - 3z = 5$$
$$x + 3y - 6z = 1$$

Since the first two equations are the same as before, the only change in the pivotal reduction method will occur in the third equation in the first set of operations. Thus, proceeding as before, we obtain

(3)
$$x - 2y + 3z = 4$$
$$y - \frac{9}{5}z = -\frac{3}{5}$$
$$5y - 9z = -3$$

Multiplying the second equation by -5 and adding it to the third equation to obtain a new third equation that does not possess a y term, we obtain the equation:

$$0z = 0$$

This equation is certainly satisfied for all values of z. The third equation in (3) is seen to be equivalent to the second equation in (3), differing from it only by a factor of 5. Thus any set of numbers (x,y,z) that satisfies the first two equations in (3) must satisfy the third equation as well and also must satisfy the original set. But by assigning z some numerical value, the first two equations become two equations in the two unknowns x and y and will, as is easily seen, possess a unique solution. This will be true for every value of z assigned; therefore, since z can be assigned an infinite number of values, this set of equations must possess an infinite number of solutions.

The solution of the preceding equations can be expressed in the following way. Let $z = k$, where k is any real number. Then the second equation in (3) gives $y = -\frac{3}{5} + \frac{9}{5}k$. These values, when substituted into the first equation, give $x = 4 - 3k + 2\left(-\frac{3}{5} + \frac{9}{5}k\right) = \frac{14}{5} + \frac{3}{5}k$. Hence, the solutions are expressible in the form:

$$x = \frac{14}{5} + \frac{3}{5}k, \qquad y = -\frac{3}{5} + \frac{9}{5}k, \qquad \text{and} \qquad z = k$$

where k is any real number.

To illustrate the fact that a set of equations may possess no solution, consider the following set:

$$x - 2y + 3z = 4$$
$$2x + y - 3z = 5$$
$$x + 3y - 6z = 2$$

Proceeding as before, we obtain

$$x - 2y + 3z = 4$$

(4)
$$y - \frac{9}{5}z = -\frac{3}{5}$$

$$5y - 9z = -2$$

If we multiply the second equation by -5 and add it to the third equation, we obtain a new third equation of the form

$$0z = 1$$

But it is impossible to find a value of z that will satisfy this equation. The second and third equations in (4) are inconsistent and, therefore, the original set of three equations in three unknowns is inconsistent.

The three preceding possibilities can be illustrated geometrically by looking at the graphs of the three equations in three dimensions. The graph of a single linear equation in three variables is a plane in three dimensions; hence, three equations in three variables will yield three planes in three dimensions. The three planes may intersect in a unique point, intersect in a common line or plane, or may have two or more of the planes parallel. Since it is difficult to sketch planes in three dimensions and observe these possibilities with any accuracy, we do not pursue this geometrical interpretation further.

The following problem illustrates how sets of linear equations can arise in actual problems. Consider a small manufacturing plant that turns out three types of toys, which we label types A, B, and C. Because of limited facilities and employees, this plant, when running at full capacity, has 800 man-hours a week available in its machining department, 900 man-hours in its assembly department, and 700 man-hours in its painting department. The production of one toy of each of these types requires the following amounts of man-hours in the three departments.

	Machining	Assembly	Painting
A	.3	.3	.3
B	.4	.3	.2
C	.2	.4	.3

Let the number of toys of types A, B, and C that can be produced with the total amounts of times available be denoted by x, y, and z, respectively. The problem is to determine the values of these variables. Because of the preceding requirements, x, y, and z must satisfy the equations:

$$.3x + .4y + .2z = 800$$
$$.3x + .3y + .4z = 900$$
$$.3x + .2y + .3z = 700$$

To eliminate the decimals, each equation is multiplied through by 10 to yield the equivalent set:

(5)
$$3x + 4y + 2z = 8000$$
$$3x + 3y + 4z = 9000$$
$$3x + 2y + 3z = 7000$$

To solve these equations, we use the pivotal reduction method as before. Here, however, it is easier to eliminate the x terms from the second and third equations by subtracting the first equation from each of the other two equations instead of first dividing through by 3 in the first equation to obtain a coefficient of 1 for x. Performing that operation, we get

$$3x + 4y + 2z = 8000$$
$$-y + 2z = 1000$$
$$-2y + z = -1000$$

Now dividing the first equation through by 3, multiplying the second equation by -1, and then eliminating the y term in the third equation, we obtain

$$x + \frac{4}{3}y + \frac{2}{3}z = \frac{8000}{3}$$
$$y - 2z = -1000$$
$$- 3z = -3000$$

Finally, dividing the third equation by -3 and solving backward for y and x, we obtain the desired solution:

$$x = 666\,\frac{2}{3}, \qquad y = 1000, \qquad z = 1000$$

Since these values are obtained under the optimistic assumption of full plant capacity and no problems arising, the manufacturer would undoubtedly scale these numbers down slightly to integers somewhat below these numbers. In applied problems, such as this one, it often suffices to find approximate solutions to a set of equations. This is discussed further in a later chapter.

4. UNEQUAL NUMBER OF EQUATIONS AND UNKNOWNS

The preceding linear problems involved the same number of unknowns and equations. Problems arise, however, in which this is not true. If, for example, the manufacturing plant of the preceding section manufactured only two types of toys, say, A and B, the equations given in (5) would not contain any z terms. As a result we would be asked to solve three equations in two unknowns. However, if the firm were manufacturing four types of toys, it would be necessary to introduce a fourth variable and then solve three equations in four unknowns.

One of the striking advantages of the pivotal reduction method for solving sets of linear equations is that it works equally well when the number of unknowns is not equal to the number of equations. Furthermore, in carrying out the solution, it becomes clear whether there exists a unique solution, an infinite number of solutions, or no solution.

For the purpose of illustrating the preceding properties of the pivotal reduction method, we solve the following set of equations in which there are more unknowns than equations.

$$\begin{aligned}
x + 2y - z + t &= -3 \\
-x + 3y + z + 4t &= -2 \\
2x + y + 2z - 4t &= 9.
\end{aligned}$$

Pivoting on the x term in the first equation, we obtain

$$\begin{aligned}
x + 2y - z + t &= -3 \\
5y + 5t &= -5 \\
-3y + 4z - 6t &= 15.
\end{aligned}$$

Dividing the second equation by 5 gives

$$
\begin{aligned}
x + 2y - z + t &= -3 \\
y + t &= -1 \\
-3y + 4z - 6t &= 15.
\end{aligned}
$$

Ignoring the first equation and pivoting on y in the second equation to eliminate the y term in the third equation, we obtain

$$
\begin{aligned}
x + 2y - z + t &= -3 \\
y + t &= -1 \\
4z - 3t &= 12.
\end{aligned}
$$

Finally, dividing the third equation by 4, we obtain

$$
\begin{aligned}
x + 2y - z + t &= -3 \\
y + t &= -1 \\
z - \frac{3}{4}t &= 3.
\end{aligned}
$$

Now if t is assigned any value and the resulting numbers arising from the terms in t are transposed to the right side of these equations, we obtain a set of three equations in the three unknowns x, y, and z that are easily seen to possess a unique solution. This will be true for each assigned value of t. Since there are an infinite number of such values that may be assigned to t, there must be an infinite number of solutions to the original set of equations. Suppose that we assign t the value k, where k is a particular real number, then our equations assume the form:

$$
\begin{aligned}
x + 2y - z &= -3 - k \\
y &= -1 - k \\
z &= 3 + \frac{3}{4}k
\end{aligned}
$$

Since the values of z and y are already displayed, it suffices to substitute those values into the first equation to find the value of x. Thus

$$
\begin{aligned}
x &= -3 - k - 2(-1 - k) + \left(3 + \frac{3}{4}k\right) \\
&= 2 + \frac{7}{4}k
\end{aligned}
$$

The solution of the original set of equations is therefore given by

$$x = 2 + \frac{7}{4}\,k, \qquad y = -1 - k, \qquad z = 3 + \frac{3}{4}\,k, \qquad \text{and} \qquad t = k$$

Since k may be assigned any real number value, these formulas yield the desired infinite set of solutions.

As a final illustration, consider the following set of equations in which there are more equations than unknowns.

(6)
$$\begin{aligned} x - y + 2z &= 2 \\ 2x + y - z &= -3 \\ 3x - y + z &= -2 \\ -x + 2y - 3z &= -3 \end{aligned}$$

Proceeding in the usual manner, we obtain

$$\begin{aligned} x - y + 2z &= 2 \\ 3y - 5z &= -7 \\ 2y - 5z &= -8 \\ y - z &= -1 \end{aligned}$$

Or

$$\begin{aligned} x - y + 2z &= 2 \\ y - \frac{5}{3}\,z &= -\frac{7}{3} \\ 2y - 5z &= -8 \\ y - z &= -1 \end{aligned}$$

Ignoring the first equation and pivoting on y in the second equation, we obtain

$$\begin{aligned} x - y + 2z &= 2 \\ y - \frac{5}{3}\,z &= -\frac{7}{3} \\ -\frac{5}{3}\,z &= -\frac{10}{3} \\ \frac{2}{3}\,z &= \frac{4}{3} \end{aligned}$$

Multiplying the third equation by $-\frac{3}{5}$ and the fourth equation by $\frac{3}{2}$, we obtain the reduced set

$$x - y + 2z = 2$$
$$y - \frac{5}{3}z = -\frac{7}{3}$$
$$z = 2$$
$$z = 2$$

Since the fourth equation gives no new information, it could have been omitted from the original set. The solution of these equations is now easily obtained by substituting the value of z into the second equation to obtain the value of y and then substituting the values of y and z just obtained into the first equation to get the value of x. It will be found that the solution of these equations, and hence the solution of the original set, is given by $x = -1$, $y = 1$, and $z = 2$.

If the last equation in the set (6) had been modified slightly, say, by replacing -3 on the right side by $+3$, and the same calculations carried out, the last equation in the pivotal reduction would have reduced to $z = -4$. Since this value is inconsistent with the value $z = 2$ obtained from the third equation, it follows that the modified set of equations will be inconsistent. Thus there would not exist a solution in that case.

The preceding problems illustrate one of the nice features of the pivotal reduction method referred to earlier: its ability to arrive at solutions when they exist and to demonstrate the lack of a solution when the equations are inconsistent.

It may occur in carrying out the pivotal reduction technique that the coefficient of y in the second equation is zero. If so, we select another equation that has a nonzero y term in it as our second equation, and then proceed as before. Or, we may pivot on some other variable in the second equation. Thus the equations may be rearranged to suit our convenience in performing the pivotal reduction.

As an illustration of how the preceding situation may arise and how it is handled, consider the solution of the following set of equations.

$$x + 3y + 2z + t = 7$$
$$2x + 6y + 3z - 3t = 1$$
$$3x + 7y + z - 2t = -2$$
$$x + 2y - z + 2t = 1$$

Proceeding as usual, we obtain

$$x + 3y + 2z + t = 7$$
$$-z - 5t = -13$$
$$-2y - 5z - 5t = -23$$
$$-y - 3z + t = -6$$

Since the coefficient of y in the second equation is 0, we shift to the fourth equation and pivot on its y term. This gives

$$x + 3y + 2z + t = 7$$
$$-z - 5t = -13$$
$$z - 7t = -11$$
$$y + 3z - t = 6$$

Finally, we pivot on the z term in the third equation to obtain

$$x + 3y + 2z + t = 7$$
$$-12t = -24$$
$$z - 7t = -11$$
$$y + 3z - t = 6$$

From the second equation we obtain $t = 2$. The third equation then gives $z = -11 + 7(2) = 3$. These values when substituted into the fourth equation give $y = -1$. The first equation then gives $x = 2$; therefore, the complete solution is $x = 2$, $y = -1$, $z = 3$, and $t = 2$.

COMMENTS

This chapter studied the problem of how to solve sets of linear equations in an efficient systematic manner. It concentrated on explaining the pivotal reduction method. This method is capable of solving sets of equations for which the number of equations is not equal to the number of unknowns, as well as when those numbers are equal. Several simple illustrations were given of social science problems that require the solution of linear equations to give some indication of the importance of linear equations in applied problems. Later chapters will consider more sophisticated problems of this type.

The material of this chapter is a necessary preliminary to the next chapter in which methods are developed for simplifying the pivotal reduction method further so that it can be applied to solving large systems of equations. Such systems arise in many important problems.

EXERCISES

Section 1

1. Solve each of the following pairs of equations, provided that a solution exists.

(a) $3x + 2y = 1$
    $5x + 3y = 4$

(b) $2x + 5y = 4$
    $3x - 5 = 0$

(c) $x + 2y = -4$
    $3x + y = 3$

(d) $4x - 3y = 4$
    $8x - 6y = 5$

(e) $4x - 3y = 4$
    $12x - 9y = 12$

(f) $\dfrac{x}{2} = 2 - \dfrac{5}{4}y$
    $\dfrac{x}{6} = \dfrac{3}{2} - \dfrac{5}{3}y$

2. Illustrate the solution of each part of Problem 1 by graphical methods.

3. Graph each of the following pairs of equations and from your graphs estimate the solution, or solutions, if any.

(a) $y - x = 4$
    $4y - 2x = 4$

(b) $2x - 3y = 7$
    $2x - y = 3$

(c) $2y - 3x = 8$
    $6x - 4y = 5$

(d) $3x - y = -4$
    $2y - 6x = 8$

4. In the following problems $C(x)$ and $R(x)$ denote the cost function and revenue function, respectively, when x units of a product are produced. In each case, find the break-even point.

(a) $C(x) = 3x + 20$, $R(x) = 6x - 40$

(b) $C(x) = 2x + 8000$, $R(x) = 3x + 1000$

(c) $C(x) = \dfrac{3}{10}x + 20$, $R(x) = \dfrac{4}{10}x + 2$

5. In each of the following problems the first equation is an implicit form of the price-demand function and the second is an implicit form of the price-supply function. For each problem determine the price and quantity that will produce market equilibrium. Assume that the defini-

tion of market equilibrium given for linear functions also applies to nonlinear problems.

(a) $x + 2p - 15 = 0$, $x - 3p + 5 = 0$

(b) $3x + p - 21 = 0$, $3x - 4p + 9 = 0$

(c) $8p + x = 16$, $16p - x = 8$

(d) $x + 3p = 9$, $6p - x = 9$

(e) $px = 18$, $p - x + 3 = 0$

(f) $x^2 p = 2$, $x^2 + 1 - p = 0$

6. Solve the following pairs of equations. The letters a and b represent certain constants. They are introduced to illustrate how equations may have no solution, or an infinite number of solutions.

(a) $ax - 2y = 2 + b$
     $ax + 4y = 2 - 2b$

(b) $ax + by = a^2 + b^2$
     $bx - ay = a^2 + b^2$

(c) $ax + y = b$
     $8x + by = a$

7. Determine a pair of values of a and b in equations (c) of Problem 6 that will make those equations inconsistent.

8. Determine a pair of values of a and b in equations (c) of Problem 6 that will make those equations possess an infinite number of solutions.

9. A store wishes to make a mixture of two types of nuts that will sell for 80 cents a pound. The first type normally sells for 60 cents a pound and the second type for 90 cents a pound. It wishes to have 50 pounds of the mixture. How many pounds of each type should it use?

10. A company has two types of fertilizer, A and B. Fertilizer A contains 18 percent nitrogen, whereas fertilizer B contains 8 percent nitrogen. A customer wants 100 pounds of fertilizer that contains 12 percent nitrogen. If fertilizers A and B come in five-pound sacks, can the order be filled exactly by mixing A and B without using parts of a sack? If so, how many sacks of each should be used?

Section 3

1. Solve each of the following sets of equations by using the pivotal reduction technique.

(a) $x + y - z = 0$ (b) $x - 2y + 3z = 5$
$\quad 2x + y + z = 7$ $\quad 2x - 4y + 5z = 8$
$\quad 3x - y + 2z = 7$ $\quad 3x + y - 4z = 3$

(c) $x - y + 2z = 4$ (d) $3x - y + 2z = 4$
$\quad 3x + y - z = 2$ $\quad x + 4y - z = 3$
$\quad 5x - y + 3z = 10$ $\quad x - 9y + 4z = 3$

(e) $2x + 4y - z = -3$ (f) $3x - y + z = -5$
$\quad 3x - y + z = 10$ $\quad 4x + y + 2z = 3$
$\quad 4x + y - 2z = 1$ $\quad 5x + 4y - z = -2$

2. For what value, or values, of the constant c will the following set of equations (a) possess a unique solution, (b) possess an infinite number of solutions? Use the pivotal reduction technique.

$$x + 2y - z = 2$$
$$2x + y + z = 4$$
$$x - 4y + cz = 2$$

3. An individual wishes to construct a diet based on three foods A, B, and C. Packages of these foods contain the following units of protein, carbohydrates, and fat, respectively.

	P	C	F
A	5	3	4
B	2	2	1
C	3	1	4

If he wishes to consume 60 units of protein, 36 units of carbohydrates, and 50 units of fat over a certain period of time, how many packages of these three foods will satisfy his requirements?

4. A fertilizer company markets three brands of fertilizer that contain the following percentages of nitrogen and phosphates.

	N	P
A	10	3
B	5	1
C	7	5

If these three brands are marketed in 50-pound sacks and if an order is received for 5000 pounds of a fertilizer that will contain 8 percent nitrogen and 4 percent phosphates, how many sacks of each brand should be mixed to satisfy this order?

5. A store has three mixtures of nuts in one-pound boxes that consist of different amounts of cashew nuts, pecans, and almonds. These mixtures, which are denoted by A, B, and C, possess the following amounts (in ounces) of the three types of nuts.

	Cashews	Pecans	Almonds
A	9	4	3
B	6	5	5
C	4	7	5

Suppose that an order is received for 20 pounds of nuts that is to contain 134 ounces of cashews, 102 ounces of pecans, and 84 ounces of almonds. How many boxes each of A, B, and C will be needed to fill this order?

6. In Problem 5 would it be possible to fill an order of 20 pounds of nuts if it is to contain 154 ounces of cashews, 86 ounces of pecans, and 80 ounces of almonds? Explain.

Section 4

1. Solve each of the following sets of equations by using the pivotal reduction technique. Determine all the solutions, if any, when a unique solution does not exist.

(a) $x + 2y = -1$
$2x - 9y = -4$
$5x - 3y = -7$

(b) $x + 2y = 1$
$2x + 3y = 1$
$3x - 2y = 2$

(c) $x + y - z + t = -2$
$2x + 3y - z - t = 4$
$3x - y + z + 2t = 5$

(d) $x - 2y - 6z + 3t = 3$
$2x + 2y + 8z - 10t = 0$
$5x - 3y - 7z + t = 10$

2. Do as in Problem 1 for each of the following sets of equations.

(a)
$$x - y + 5z = 2$$
$$2x + 2y + 4z = 10$$
$$x + 4y + 3z = -2$$
$$2x + 7y + 2z = 6$$

(b)
$$2x - y + 3z = 8$$
$$2x + 2y + 5z = 9$$
$$5x + 3y + 2z = 6$$
$$3x + 7y + z = 2$$

(c)
$$2x + 3y + z - t = -4$$
$$3x + 2z + 3t = 13$$
$$4x - 2y + 3z - 2t = 2$$
$$5x + y - z + t = 0$$

3. A manufacturer produces two products, A and B. Each unit of products A and B requires the following amounts of time in hours in each of the indicated departments.

	Machining	Assembly	Painting
A	.2	.3	.5
B	.4	.2	.3

Suppose that the total time available per week in each department is 300 hours, 200 hours, and 300 hours, respectively. Let x and y denote the number of units of each product to be produced in a week. Determine whether there exist values of x and y that will use up all the available time. If there are, find them; if not, what compromise would you suggest?

4. A diet for a set of experimental animals is to be constructed from four foods A, B, C, and D. These foods contain the following units of protein, carbohydrates, and fat.

	P	C	F
A	5	2	4
B	2	2	2
C	3	1	5
D	2	3	2

If the diet is to contain 43 units of protein, 26 units of carbohydrates, and 42 units of fat, what amounts of these four foods will satisfy the requirements?

5. Work Problem 4 if the diet is to contain 20 units of protein, 30 units of carbohydrates, and 30 units of fat. What does your solution imply?

3 | Matrices

It is highly desirable to simplify the notation and methods of the preceding chapter in order to solve larger systems of linear equations. Some problems in the social sciences involve 50 or more variables and equations. For problems of this kind simpler methods are certainly needed.

A simplification of the desired type can be obtained by using matrix notation and methods. A *matrix* is merely a rectangular array of numbers. These numbers are usually inclosed in parentheses or brackets. We will use brackets. A particular matrix is designated by assigning it a letter name. The following are examples of matrices:

$$A = \begin{bmatrix} 3 & 4 \\ 6 & -1 \end{bmatrix}, \quad B = \begin{bmatrix} 2 & 1 & -3 \\ 3 & 0 & 5 \end{bmatrix}, \quad C = \begin{bmatrix} 1 \\ 2 \\ 2 \\ 1 \end{bmatrix}, \quad D = \begin{bmatrix} 4 & 2 & 3 \\ 6 & 1 & 0 \\ -4 & 0 & -2 \end{bmatrix}$$

A general matrix of any size may be denoted by

$$A = \begin{bmatrix} a_{11} & a_{12} & \cdots & a_{1n} \\ a_{21} & a_{22} & \cdots & a_{2n} \\ \cdot & \cdot & & \cdot \\ \cdot & \cdot & & \cdot \\ \cdot & \cdot & & \cdot \\ a_{m1} & a_{m2} & \cdots & a_{mn} \end{bmatrix}$$

A matrix with m rows and n columns, like the one illustrated here, is called an $m \times n$ (read "m by n") matrix. The numbers in a matrix are called the *elements* of the matrix. In particular, the number a_{ij} occurring in the ith row and jth column, where i and j are integers, is called the i,j element. It designates a typical, or general, element of the matrix.

After studying some of the properties of matrices we shall discover that sets of linear equations can be expressed very neatly in compact form by means of matrices and that they are very useful in carrying out the pivotal reduction method for solving those equations.

1. PROPERTIES OF MATRICES

There are three special matrices that need to be defined before we are ready to discuss properties of matrices. A matrix all of whose elements are zero is called a *zero matrix* and is denoted by a large zero, 0. A matrix that possesses the same number of rows as columns is called a *square matrix*. A square matrix of the following special form is called the *identity matrix* and is denoted by the letter I.

$$I = \begin{bmatrix} 1 & 0 & \cdots & 0 \\ 0 & 1 & \cdots & 0 \\ \cdot & \cdot & & \cdot \\ \cdot & \cdot & & \cdot \\ \cdot & \cdot & & \cdot \\ 0 & 0 & \cdots & 1 \end{bmatrix}$$

All the elements in I have the value zero except those down the main diagonal which have the value 1. With these special matrices disposed of, we are prepared to define the operations of addition and multiplication for matrices.

Two matrices can be added only if they possess the same number of rows and columns, in which case we say that they are of the same *size*. The addition of two such matrices consists of adding corresponding elements. More precisely, if we are given the $m \times n$ matrix defined in the preceding section and if B is the following $m \times n$ matrix

$$B = \begin{bmatrix} b_{11} & b_{12} & \cdots & b_{1n} \\ b_{21} & b_{22} & \cdots & b_{2n} \\ \cdot & \cdot & & \cdot \\ \cdot & \cdot & & \cdot \\ \cdot & \cdot & & \cdot \\ b_{m1} & b_{m2} & \cdots & b_{mn} \end{bmatrix}$$

then the matrix $A + B$ is defined by

$$A + B = \begin{bmatrix} a_{11} + b_{11} & a_{12} + b_{12} & \cdots & a_{1n} + b_{1n} \\ a_{21} + b_{21} & a_{22} + b_{22} & \cdots & a_{2n} + b_{2n} \\ \cdot & \cdot & & \cdot \\ \cdot & \cdot & & \cdot \\ \cdot & \cdot & & \cdot \\ a_{m1} + b_{m1} & a_{m2} + b_{m2} & \cdots & a_{mn} + b_{mn} \end{bmatrix}$$

It is clear from this definition that $A + B$ is the same as $B + A$; therefore, we may perform the addition in either order. It should also be clear that if we have a third $m \times n$ matrix, call it C, and wish to add it to $A + B$ to obtain $A + B + C$, we can add these three matrices in any desired order.

Two matrices of the same size are defined to be *equal* if, and only if, their corresponding elements are equal. Thus, $A = B$ if, and only if, $a_{ij} = b_{ij}$ for all i and j.

Let c be any real number. Then the product of c and the matrix A, written cA or Ac, is defined to be the matrix obtained by multiplying each element of A by c. Thus

$$cA = \begin{bmatrix} ca_{11} & ca_{12} & \cdots & ca_{1n} \\ ca_{21} & ca_{22} & \cdots & ca_{2n} \\ \cdot & \cdot & & \cdot \\ \cdot & \cdot & & \cdot \\ \cdot & \cdot & & \cdot \\ ca_{m1} & ca_{m2} & \cdots & ca_{mn} \end{bmatrix}$$

We may use this product property to define the difference of two matrices, written $A - B$, by defining $A - B = A + (-1)B$.

As illustrations of the preceding definitions, let A and B be the following 2×3 matrices:

$$A = \begin{bmatrix} 1 & 2 & -1 \\ 3 & 0 & 2 \end{bmatrix}, \qquad B = \begin{bmatrix} -1 & 0 & 2 \\ 1 & 1 & 0 \end{bmatrix}$$

Let us compute $A + B$, $A - B$, and $3A - 2B$. Application of the preceding definitions gives

$$A + B = \begin{bmatrix} 1 + (-1) & 2 + 0 & -1 + 2 \\ 3 + 1 & 0 + 1 & 2 + 0 \end{bmatrix} = \begin{bmatrix} 0 & 2 & 1 \\ 4 & 1 & 2 \end{bmatrix}$$

Similarly,

$$A - B = \begin{bmatrix} 1 - (-1) & 2 - 0 & -1 - 2 \\ 3 - 1 & 0 - 1 & 2 - 0 \end{bmatrix} = \begin{bmatrix} 2 & 2 & -3 \\ 2 & -1 & 2 \end{bmatrix}$$

To calculate $3A - 2B$, we first calculate $3A$ and $2B$, obtaining

$$3A = \begin{bmatrix} 3 & 6 & -3 \\ 9 & 0 & 6 \end{bmatrix} \quad \text{and} \quad 2B = \begin{bmatrix} -2 & 0 & 4 \\ 2 & 2 & 0 \end{bmatrix}$$

Then

$$3A - 2B = \begin{bmatrix} 3 - (-2) & 6 - 0 & -3 - 4 \\ 9 - 2 & 0 - 2 & 6 - 0 \end{bmatrix} = \begin{bmatrix} 5 & 6 & -7 \\ 7 & -2 & 6 \end{bmatrix}$$

The definition that we are about to give for multiplying two matrices may seem rather strange, but we shall see that it is a natural definition when matrices are applied to solving systems of equations. The multiplication of two matrices can occur only if the second matrix has the same number of rows as the first matrix has columns. Before defining multiplication in general, we consider a simple special case where

$$A = \begin{bmatrix} 2 & 1 & -3 & 1 \end{bmatrix} \quad \text{and} \quad B = \begin{bmatrix} 1 \\ -1 \\ 0 \\ 2 \end{bmatrix}$$

Since the number of rows of B is equal to the number of columns of A, namely 4, the preceding restriction is satisfied. The product of A and B, denoted by AB, is then defined to be the number obtained by multiplying each row element of A by the corresponding column element of B and summing all such products. Thus

$$AB = (2)(1) + (1)(-1) + (-3)(0) + (1)(2) = 3$$

A matrix that consists of a single row of elements is called a *row vector*, and a matrix consisting of a single column of elements is called a *column vector*. Here A is a row vector and B is a column vector, and AB is therefore the product of two vectors. To obtain a general formula for vector multiplication, let A and B be the following general row and column vectors, with the restriction that they possess the same number of elements.

$$A = \begin{bmatrix} a_1 & a_2 & \cdots & a_n \end{bmatrix}, \quad B = \begin{bmatrix} b_1 \\ b_2 \\ \cdot \\ \cdot \\ \cdot \\ b_n \end{bmatrix}$$

Then the product AB is defined by the formula

$$AB = a_1b_1 + a_2b_2 + \cdots + a_nb_n$$

With this definition available, we are ready to consider the definition of the product of two general matrices for which the second matrix has the same number of rows as the first matrix has columns. Let A and B denote two such matrices, where A is an $m \times n$ matrix and B is an $n \times r$ matrix. Here m, n, and r are any positive integers.

$$A = \begin{bmatrix} a_{11} & a_{12} & \cdots & a_{1n} \\ & & & \\ a_{i1} & a_{i2} & \cdots & a_{in} \\ & & & \\ a_{m1} & a_{m2} & \cdots & a_{mn} \end{bmatrix}, \quad B = \begin{bmatrix} b_{11} & \cdots & b_{1j} & \cdots & b_{1r} \\ b_{21} & \cdots & b_{2j} & \cdots & b_{2r} \\ & & & & \\ b_{n1} & \cdots & b_{nj} & \cdots & b_{nr} \end{bmatrix}$$

The product AB is now defined to be the $m \times r$ matrix which has as its i, j element the number obtained by multiplying the ith row vector of A by the jth column vector of B. Thus the element c_{ij} in the matrix $C = AB$ is defined to be the number given by

$$c_{ij} = a_{i1}b_{1j} + a_{i2}b_{2j} + \cdots + a_{in}b_{nj}, \quad \begin{cases} i = 1, \ldots, m \\ j = 1, \ldots, r \end{cases}$$

Operationally, we begin by multiplying the first row vector of A by each of the column vectors of B to obtain the elements of the first row of AB. There will be r such elements because there are r columns in B. Next, we multiply the second row vector of A by each of the r column vectors of B to obtain the elements in the second row of AB. This process is continued with each row of A. The result of such operations is, therefore, a matrix with m rows and r columns.

As an illustration, we calculate AB where

$$A = \begin{bmatrix} 2 & 1 & -3 & 1 \\ 4 & 0 & 2 & 0 \end{bmatrix} \quad \text{and} \quad B = \begin{bmatrix} 1 & 2 & 0 \\ -1 & 1 & 2 \\ 0 & -1 & 2 \\ 2 & 3 & -1 \end{bmatrix}$$

The first element in the first row of AB is 3 because that is the value we obtained previously when we illustrated how to multiply a row vector with a column vector. The second element in the first row of AB is given by the vector product of the first row of A with the second column of B, which is

$$(2)(2) + (1)(1) + (-3)(-1) + (1)(3) = 5$$

Similarly, the third element in the first row of AB is given by the vector product

$$(2)(0) + (1)(2) + (-3)(2) + (1)(-1) = -5$$

The elements in the second row of AB are obtained in a similar manner by using the second row vector of A and multiplying it with each of the column vectors of B. As a result of these computations we obtain

$$AB = \begin{bmatrix} 3 & 5 & -5 \\ 4 & 6 & 4 \end{bmatrix}$$

Unlike the situation with numbers, the order in which two matrices are multiplied does make a difference. It is not true in general that $BA = AB$. For the preceding illustration, the product BA does not even make sense because the number of rows of A, namely 2, does not equal the number of columns of B, namely 3, and therefore the necessary vector products cannot be calculated. Even if we have two matrices for which BA exists, it does not follow that $BA = AB$. As an illustration, let

$$A = \begin{bmatrix} 1 & 1 \\ 2 & 0 \end{bmatrix} \quad \text{and} \quad B = \begin{bmatrix} 1 & 0 \\ 1 & 1 \end{bmatrix}$$

Then

$$AB = \begin{bmatrix} 2 & 1 \\ 2 & 0 \end{bmatrix} \quad \text{and} \quad BA = \begin{bmatrix} 1 & 1 \\ 3 & 1 \end{bmatrix}$$

The following additional examples are given to enable a student to check his understanding of matrix multiplication.

Example 1. Calculate AB where

$$A = \begin{bmatrix} 2 & -1 & 3 \\ 0 & 2 & -1 \end{bmatrix} \quad \text{and} \quad B = \begin{bmatrix} 0 & 1 & 4 & -1 \\ 2 & -1 & 0 & 1 \\ -1 & 3 & 2 & 2 \end{bmatrix}$$

The answer is

$$AB = \begin{bmatrix} -5 & 12 & 14 & 3 \\ 5 & -5 & -2 & 0 \end{bmatrix}$$

Example 2. Calculate AB where

$$A = \begin{bmatrix} 1 & -3 \\ 2 & 0 \\ -1 & 2 \end{bmatrix} \quad \text{and} \quad B = \begin{bmatrix} 2 & -1 & 4 & -1 \\ 0 & 3 & 1 & 0 \end{bmatrix}$$

The answer is

$$AB = \begin{bmatrix} 2 & -10 & 1 & -1 \\ 4 & -2 & 8 & -2 \\ -2 & 7 & -2 & 1 \end{bmatrix}$$

Example 3. Calculate $(A + 2B)B$ where

$$A = \begin{bmatrix} 1 & 0 & -1 \\ 2 & 1 & 0 \\ -1 & 3 & -1 \end{bmatrix} \quad \text{and} \quad B = \begin{bmatrix} 0 & 1 & 2 \\ -3 & 1 & 0 \\ 0 & 2 & 2 \end{bmatrix}$$

The answer is

$$(A + 2B)B = \begin{bmatrix} 1 & 2 & 3 \\ -4 & 3 & 0 \\ -1 & 7 & 3 \end{bmatrix} \begin{bmatrix} 0 & 1 & 2 \\ -3 & 1 & 0 \\ 0 & 2 & 2 \end{bmatrix} = \begin{bmatrix} -6 & 9 & 8 \\ -9 & -1 & -8 \\ -21 & 12 & 4 \end{bmatrix}$$

2. EQUATIONS IN MATRIX FORM

Since the purpose of introducing matrices is to enable us to express and solve systems of linear equations more efficiently, let us consider next the problem of how to write a system of linear equations in matrix form. As a first illustration, consider one of the systems discussed in Section 1, Chapter 2:

$$2x + y = 1$$
$$3x + 2y = 5$$

In applying matrix notation to systems of equations with quite a few variables, it is advantageous to represent such variables by the letter x with appropriate subscripts. Thus, for the purpose of anticipating more complicated problems, we rewrite the preceding equations in the form:

$$2x_1 + x_2 = 1$$
$$3x_1 + 2x_2 = 5$$

We now introduce the following matrices:

$$A = \begin{bmatrix} 2 & 1 \\ 3 & 2 \end{bmatrix}, \qquad X = \begin{bmatrix} x_1 \\ x_2 \end{bmatrix}, \qquad B = \begin{bmatrix} 1 \\ 5 \end{bmatrix}$$

It should be noted that A is the matrix of the coefficients of the unknowns in the equations. In terms of these matrices the preceding set of equations can be expressed in compact form by the single matrix equation:

$$AX = B$$

This is easily verified by carrying out the indicated multiplication and applying the properties of matrices. Thus

$$AX = \begin{bmatrix} 2 & 1 \\ 3 & 2 \end{bmatrix} \begin{bmatrix} x_1 \\ x_2 \end{bmatrix} = \begin{bmatrix} 2x_1 + x_2 \\ 3x_1 + 2x_2 \end{bmatrix}$$

Equating AX and B, we obtain

$$\begin{bmatrix} 2x_1 + x_2 \\ 3x_1 + 2x_2 \end{bmatrix} = \begin{bmatrix} 1 \\ 5 \end{bmatrix}$$

Since two matrices are equal if, and only if, their corresponding elements are equal, it follows that

$$2x_1 + x_2 = 1 \qquad \text{and} \qquad 3x_1 + 2x_2 = 5$$

These are the original equations of our system.

Consider now a general system of equations in which the number of equations need not necessarily equal the number of unknowns. Let this system be written in the form:

$$a_{11}x_1 + a_{12}x_2 + \cdots + a_{1n}x_n = b_1$$
$$a_{21}x_1 + a_{22}x_2 + \cdots + a_{2n}x_n = b_2$$
$$\vdots$$
$$a_{m1}x_1 + a_{m2}x_2 + \cdots + a_{mn}x_n = b_m$$

To express these equations in compact form, we introduce the following matrices:

$$A = \begin{bmatrix} a_{11} & a_{12} & \cdots & a_{1n} \\ a_{21} & a_{22} & \cdots & a_{2n} \\ \vdots & \vdots & & \vdots \\ a_{m1} & a_{m2} & \cdots & a_{mn} \end{bmatrix}, \qquad X = \begin{bmatrix} x_1 \\ x_2 \\ \vdots \\ x_n \end{bmatrix}, \qquad B = \begin{bmatrix} b_1 \\ b_2 \\ \vdots \\ b_m \end{bmatrix}$$

Then the preceding system of equations can be expressed in the following single matrix equation:

(1) $AX = B$

This is readily verified by carrying out the multiplication AX and equating each element of the resulting $m \times 1$ matrix to the corresponding element of the vector B. For example, the first element in the product on the left is obtained by multiplying the vector in the first row of A by the column vector X to obtain $a_{11}x_1 + a_{12}x_2 + \cdots + a_{1n}x_n$. This value must then be equated to the first element on the right, namely, b_1. The result is the first of our n equations.

As previously, it should be noted that A is merely the matrix of the coefficients of the unknowns in the equations and that AX is always a column vector.

3. PIVOTAL REDUCTION IN MATRIX FORM

Not only can we express systems of equations in a neater form by means of matrices but we can also use them to simplify the pivotal reduction method for solving the equations. To illustrate how this is done, let us consider once more the system of equations that was solved in Section 2, Chapter 2:

$$x - 2y + 3z = 4$$
$$2x + y - 3z = 5$$
$$-x + y + 2z = 3$$

First we write down the matrix of coefficients, followed by the column vector of constants on the right. In doing this, we omit the bracket notation because it serves no useful purpose here, and we draw a vertical line to separate the coefficient matrix from the column vector of constants. Thus we write

$$
\begin{array}{rrr|r}
1 & -2 & 3 & 4 \\
2 & 1 & -3 & 5 \\
-1 & 1 & 2 & 3
\end{array}
$$

Now we perform the pivotal reduction method as before. First we attempt to obtain zeros for the coefficients of x in the second and third row equations by multiplying the first row by appropriate numbers and adding it to the second and third rows, respectively. Here we multiply the first row by

-2 and add it to the second row, and then we add the first row to the third row. These operations yield the following configuration:

$$\begin{array}{rrr|r} 1 & -2 & 3 & 4 \\ 0 & 5 & -9 & -3 \\ 0 & -1 & 5 & 7 \end{array}$$

Next, we divide the second row by 5 to obtain a coefficient of 1 for the y term in that equation. This gives

$$\begin{array}{rrr|r} 1 & -2 & 3 & 4 \\ 0 & 1 & -\dfrac{9}{5} & -\dfrac{3}{5} \\ 0 & -1 & 5 & 7 \end{array}$$

Our result here should be compared with equations (2), Chapter 2, as when written out in full it is identical with those equations. In this scheme we do not bother to display the variables, concentrating only on the coefficients. We now pivot on the y term of the second row to produce a zero coefficient for the y term in the third row, obtaining

$$\begin{array}{rrr|r} 1 & -2 & 3 & 4 \\ 0 & 1 & -\dfrac{9}{5} & -\dfrac{3}{5} \\ 0 & 0 & \dfrac{16}{5} & \dfrac{32}{5} \end{array}$$

Finally, we multiply the third row by $\dfrac{5}{16}$ to obtain a coefficient of 1 for the z term in the third equation. This results in

$$\begin{array}{rrr|r} 1 & -2 & 3 & 4 \\ 0 & 1 & -\dfrac{9}{5} & -\dfrac{3}{5} \\ 0 & 0 & 1 & 2 \end{array}$$

The preceding operations are precisely the same as those used earlier in demonstrating the pivotal reduction method. The difference is that here we suppress the variables and consider only the numbers involved in the operations. When we arrive at this final stage, we may write out the corresponding equations and quickly solve for the unknowns. Thus we would write

$$x - 2y + 3z = 4$$

$$y - \frac{9}{5}z = -\frac{3}{5}$$

$$z = 2$$

and solve backward to obtain the solution: $x = 4$, $y = 3$, $z = 2$.

If we wish, however, we may operate some more on these matrices and obtain the solution as the final outcome of the operations. To do so we start pivoting on the z element of the last equation and work for zeros above it, just as we started the original pivot on the first x element and worked for zeros below it. In other words, we perform the pivotal reduction method backward (or upward if you prefer) to obtain only zeros above the main diagonal. To carry out this reduction we first multiply the last row by $\frac{9}{5}$ and add it to the second row, and then multiply the last row by -3 and add it to the first row to obtain

$$\begin{array}{ccc|c} 1 & -2 & 0 & -2 \\ 0 & 1 & 0 & 3 \\ 0 & 0 & 1 & 2 \end{array}$$

Finally, we multiply the second row by 2 and add it to the first row to obtain

$$\begin{array}{ccc|c} 1 & 0 & 0 & 4 \\ 0 & 1 & 0 & 3 \\ 0 & 0 & 1 & 2 \end{array}$$

When this configuration is converted back to equation form, we obtain

$$x + 0y + 0z = 4$$
$$0x + y + 0z = 3$$
$$0x + 0y + z = 2$$

Hence we see that the solution is given by the elements of the column vector to the right of the vertical line, namely, $x = 4$, $y = 3$, and $z = 2$.

Since this matrix version of the pivotal reduction method is merely a compact way of performing the reduction, it is applicable to problems for which the number of equations does not necessarily equal the number of unknowns. To illustrate its use on such problems, consider the following system of equations of this type:

$$x - y + 2z = 2$$
$$2x + 2y - z = -3$$
$$3x - y + z = -2$$
$$-x + 2y - 3z = 3$$

We first write this in compact form as follows:

$$\begin{array}{ccc|c} 1 & -1 & 2 & 2 \\ 2 & 2 & -1 & -3 \\ 3 & -1 & 1 & -2 \\ -1 & 2 & -3 & 3 \end{array}$$

Pivoting on the x term of the first row, we get

$$\begin{array}{ccc|c} 1 & -1 & 2 & 2 \\ 0 & 4 & -5 & -7 \\ 0 & 2 & -5 & -8 \\ 0 & 1 & -1 & 5 \end{array}$$

Next, we divide the second row of numbers by 4 to obtain

$$\begin{array}{ccc|c} 1 & -1 & 2 & 2 \\ 0 & 1 & -\dfrac{5}{4} & -\dfrac{7}{4} \\ 0 & 2 & -5 & -8 \\ 0 & 1 & -1 & 5 \end{array}$$

Pivoting on the y term of the second row, we obtain

$$\begin{array}{ccc|c} 1 & -1 & 2 & 2 \\ 0 & 1 & -\dfrac{5}{4} & -\dfrac{7}{4} \\ 0 & 0 & -\dfrac{5}{2} & -\dfrac{9}{2} \\ 0 & 0 & -\dfrac{1}{4} & \dfrac{27}{4} \end{array}$$

We multiply the third row by $-\dfrac{2}{5}$ to obtain

$$\begin{array}{ccc|c} 1 & -1 & 2 & 2 \\[2mm] 0 & 1 & -\dfrac{5}{4} & -\dfrac{7}{4} \\[2mm] 0 & 0 & 1 & \dfrac{9}{5} \\[2mm] 0 & 0 & -\dfrac{1}{4} & \dfrac{27}{4} \end{array}$$

Finally, we multiply the third row by $\dfrac{1}{4}$ and add it to the fourth row to obtain

$$\begin{array}{ccc|c} 1 & -1 & 2 & 2 \\[2mm] 0 & 1 & -\dfrac{5}{4} & -\dfrac{7}{4} \\[2mm] 0 & 0 & 1 & \dfrac{9}{5} \\[2mm] 0 & 0 & 0 & \dfrac{36}{5} \end{array}$$

If this matrix version were transformed back to the corresponding reduced system of equations, we would have

$$x - y + 2z = 2$$

$$y - \frac{5}{4}z = -\frac{7}{4}$$

$$z = \frac{9}{5}$$

$$0z = \frac{36}{5}$$

It is obviously impossible to satisfy the fourth equation; hence, this system of equations has no solution and, therefore, the original system has no solution. If the system had possessed a solution, the fourth row in this final reduction process would have had zeros for all elements in that row.

The next illustration is one that involves four equations and four unknowns. Its purpose is to show that the pivotal reduction method can be carried out quite easily for systems of this size without a calculating machine, provided that the coefficients are small integers. The various steps in the solution are carried out without comment because the technique should be familiar by now.

$$x + 2y - z + 2t = 1$$
$$2x - y + 2z - t = -4$$
$$3x + 2y - z + 3t = 0$$
$$4x - 3y + 2z - t = -6$$

$$\left[\begin{array}{cccc|c}
1 & 2 & -1 & 2 & 1 \\
2 & -1 & 2 & -1 & -4 \\
3 & 2 & -1 & 3 & 0 \\
4 & -3 & 2 & -1 & -6
\end{array}\right]$$

$$\left[\begin{array}{cccc|c}
1 & 2 & -1 & 2 & 1 \\
0 & -5 & 4 & -5 & -6 \\
0 & -4 & 2 & -3 & -3 \\
0 & -11 & 6 & -9 & -10
\end{array}\right]$$

$$\left[\begin{array}{cccc|c}
1 & 2 & -1 & 2 & 1 \\
0 & 1 & -\frac{4}{5} & 1 & \frac{6}{5} \\
0 & 0 & -\frac{6}{5} & 1 & \frac{9}{5} \\
0 & 0 & -\frac{14}{5} & 2 & \frac{16}{5}
\end{array}\right]$$

$$\left[\begin{array}{cccc|c}
1 & 2 & -1 & 2 & 1 \\
0 & 1 & -\frac{4}{5} & 1 & \frac{6}{5} \\
0 & 0 & 1 & -\frac{5}{6} & -\frac{3}{2} \\
0 & 0 & 0 & -\frac{1}{3} & -1
\end{array}\right]$$

$$\left[\begin{array}{cccc|c}
1 & 2 & -1 & 2 & 1 \\
0 & 1 & -\frac{4}{5} & 1 & \frac{6}{5} \\
0 & 0 & 1 & -\frac{5}{6} & -\frac{3}{2} \\
0 & 0 & 0 & 1 & 3
\end{array}\right]$$

$$\left[\begin{array}{cccc|c}
1 & 2 & -1 & 0 & -5 \\
0 & 1 & -\frac{4}{5} & 0 & -\frac{9}{5} \\
0 & 0 & 1 & 0 & 1 \\
0 & 0 & 0 & 1 & 3
\end{array}\right]$$

$$
\begin{array}{cccc|c}
1 & 2 & 0 & 0 & -4 \\
0 & 1 & 0 & 0 & -1 \\
0 & 0 & 1 & 0 & 1 \\
0 & 0 & 0 & 1 & 3
\end{array}
$$

$$
\begin{array}{cccc|c}
1 & 0 & 0 & 0 & -2 \\
0 & 1 & 0 & 0 & -1 \\
0 & 0 & 1 & 0 & 1 \\
0 & 0 & 0 & 1 & 3
\end{array}
$$

The solution is, therefore, $x = -2$, $y = -1$, $z = 1$, and $t = 3$.

The preceding illustrations of the pivotal reduction method involved at most four variables. The same method is employed, however, to solve large systems of equations. It is not unusual to encounter social science problems that involve a large number of variables and equations. For example, there is a model in economics called the Brookings Model of the American Economy that employs approximately 80 variables. It is used to estimate quantities such as employment and consumption factors in the American economy. In psychology the analysis of batteries of tests requires the solution of 50 or more linear equations in the same number of unknowns. Large-scale systems also occur frequently in medical investigations. For example, there is a linear discriminant model that attempts to determine whether or not a patient has a certain disease and that necessitates the solution of 20 or more equations in the same number of unknowns. Problems in linear programming, which are discussed in the next chapter, sometimes involve more than 100 variables. The solution of such large systems of equations requires, of course, high speed computing equipment. Since this equipment is becoming increasingly common, investigators are becoming increasingly involved in large-scale investigations that were impossible to conduct a few years ago when only desk calculators were available to solve sets of equations.

4. THE INVERSE OF A SQUARE MATRIX

Many experimental and analytical problems in the various sciences give rise to a system of linear equations in which the constants on the right side change from experiment to experiment but in which the coefficient matrix on the left remains unchanged. Since such sets of equations have so much in common, it seems intuitively clear that it would be inefficient to solve each set by itself. There is a systematic way of solving such sets of equations, which we now explore.

We consider only those problems for which the number of equations equals the number of unknowns and for which there exists a unique solution. As an illustration, consider the problem of solving these sets of equations:

$$
\begin{aligned}
x - 2y + 3z &= 4,\ 1,\ 3,\ 1 \\
2x + y - 3z &= 5,\ 0,\ 3,\ 2 \\
-x + y + 2z &= 3,\ 2,\ 3,\ 1
\end{aligned}
$$

(2)

Here we have four sets of equations that differ only in the constants that occur on the right. We want to be able to solve all four sets simultaneously in an efficient manner. This can be accomplished by calculating what is known as the *inverse of a matrix.*

Let A be any square matrix. If there exists a matrix B such that

$$AB = I$$

where I is the identity matrix, then B is called the inverse of A. It can be shown that if such a B exists, it will also satisfy the relation

$$BA = I$$

The inverse of A is customarily denoted by the symbol A^{-1}. Thus we write $B = A^{-1}$ and realize that A^{-1} satisfies the two relations:

$$AA^{-1} = I \qquad \text{and} \qquad A^{-1}A = I$$

Since A is not a number, the exponent -1 on A should not be confused with the corresponding negative exponent on a number. Here it is merely a convenient symbol to represent the inverse.

Suppose A is a matrix for which an inverse exists. How should we calculate it? The answer to that question is given by considering the problem as one of solving several sets of equations with the columns of the identity matrix I as the constants on the right side of the equations. This fact will be demonstrated on a particular problem. In this connection consider the problem of finding the inverse of the matrix of the coefficients of the unknowns for the set of equations (2). Let

$$
A = \begin{bmatrix} 1 & -2 & 3 \\ 2 & 1 & -3 \\ -1 & 1 & 2 \end{bmatrix}
$$

We are concerned with solving the following three sets of equations:

(3)
$$\begin{aligned} x - 2y + 3z &= 1,\ 0,\ 0 \\ 2x + y - 3z &= 0,\ 1,\ 0 \\ -x + y + 2z &= 0,\ 0,\ 1 \end{aligned}$$

Suppose that we have solved these three sets of equations and have obtained the following solutions, written in the form of a three-column matrix.

(4)
$$B = \begin{bmatrix} x_1 & x_2 & x_3 \\ y_1 & y_2 & y_3 \\ z_1 & z_2 & z_3 \end{bmatrix}$$

In this notation the first column numbers give the solution for equation (3) when the first column of constants on the right is used. Similarly, this is true for the remaining columns. In terms of the notation that we arrived at in (1), this means that

$$A \begin{bmatrix} x_1 \\ y_1 \\ z_1 \end{bmatrix} = \begin{bmatrix} 1 \\ 0 \\ 0 \end{bmatrix}, \qquad A \begin{bmatrix} x_2 \\ y_2 \\ z_2 \end{bmatrix} = \begin{bmatrix} 0 \\ 1 \\ 0 \end{bmatrix}, \qquad A \begin{bmatrix} x_3 \\ y_3 \\ z_3 \end{bmatrix} = \begin{bmatrix} 0 \\ 0 \\ 1 \end{bmatrix}$$

These three matrix equations can obviously be combined into one compact matrix equation:

$$A \begin{bmatrix} x_1 & x_2 & x_3 \\ y_1 & y_2 & y_3 \\ z_1 & z_2 & z_3 \end{bmatrix} = \begin{bmatrix} 1 & 0 & 0 \\ 0 & 1 & 0 \\ 0 & 0 & 1 \end{bmatrix}$$

In view of the definition of B given in (4), this result is the same as

$$AB = I$$

This shows that the solution matrix B is the desired inverse matrix A^{-1}. Consequently, to find the inverse of our matrix it suffices to solve the three sets of equations (3) obtained by choosing as the constants on the right the column elements of the identity matrix I. We can obtain those solutions in a systematic manner by using the pivotal reduction method. This will be illustrated on matrix A.

First we write

$$\begin{array}{rrr|rrr} 1 & -2 & 3 & 1 & 0 & 0 \\ 2 & 1 & -3 & 0 & 1 & 0 \\ -1 & 1 & 2 & 0 & 0 & 1 \end{array}$$

Pivoting on the leading element, we obtain

$$
\left[
\begin{array}{ccc|ccc}
1 & -2 & 3 & 1 & 0 & 0 \\
0 & 5 & -9 & -2 & 1 & 0 \\
0 & -1 & 5 & 1 & 0 & 1
\end{array}
\right]
$$

We divide the second row by 5 to get ready for the next pivotal reduction. Thus

$$
\left[
\begin{array}{ccc|ccc}
1 & -2 & 3 & 1 & 0 & 0 \\
0 & 1 & -\dfrac{9}{5} & -\dfrac{2}{5} & \dfrac{1}{5} & 0 \\
0 & -1 & 5 & 1 & 0 & 1
\end{array}
\right]
$$

Pivoting on the y element in the second row yields

$$
\left[
\begin{array}{ccc|ccc}
1 & -2 & 3 & 1 & 0 & 0 \\
0 & 1 & -\dfrac{9}{5} & -\dfrac{2}{5} & \dfrac{1}{5} & 0 \\
0 & 0 & \dfrac{16}{5} & \dfrac{2}{5} & \dfrac{1}{5} & 1
\end{array}
\right]
$$

Multiplication of the last row by $\dfrac{5}{16}$ reduces this to

$$
\left[
\begin{array}{ccc|ccc}
1 & -2 & 3 & 1 & 0 & 0 \\
0 & 1 & -\dfrac{9}{5} & -\dfrac{2}{5} & \dfrac{1}{5} & 0 \\
0 & 0 & 1 & \dfrac{3}{16} & \dfrac{1}{16} & \dfrac{5}{16}
\end{array}
\right]
$$

Now we reverse our pivotal reduction method to work for zeros above the main diagonal. Thus pivoting on the z term in the last row, we obtain

$$
\left[
\begin{array}{ccc|ccc}
1 & -2 & 0 & \dfrac{7}{16} & -\dfrac{3}{16} & -\dfrac{15}{16} \\
0 & 1 & 0 & -\dfrac{1}{16} & \dfrac{5}{16} & \dfrac{9}{16} \\
0 & 0 & 1 & \dfrac{3}{16} & \dfrac{1}{16} & \dfrac{5}{16}
\end{array}
\right]
$$

Finally, pivoting on the y term in the second row, we obtain

$$
\begin{array}{ccc}
1 \quad 0 \quad 0 \\[4pt]
0 \quad 1 \quad 0 \\[4pt]
0 \quad 0 \quad 1
\end{array}
\left|
\begin{array}{ccc}
\dfrac{5}{16} & \dfrac{7}{16} & \dfrac{3}{16} \\[8pt]
-\dfrac{1}{16} & \dfrac{5}{16} & \dfrac{9}{16} \\[8pt]
\dfrac{3}{16} & \dfrac{1}{16} & \dfrac{5}{16}
\end{array}
\right.
$$

Since we can now read off the three sets of solutions, as indicated in (4), by reading off the elements in the three columns on the right, we have obtained the desired inverse:

(5)
$$
A^{-1} =
\begin{bmatrix}
\dfrac{5}{16} & \dfrac{7}{16} & \dfrac{3}{16} \\[8pt]
-\dfrac{1}{16} & \dfrac{5}{16} & \dfrac{9}{16} \\[8pt]
\dfrac{3}{16} & \dfrac{1}{16} & \dfrac{5}{16}
\end{bmatrix}
$$

The results of these calculations can be checked for accuracy by calculating the product AA^{-1} and verifying that it does produce the identity matrix I.

Now let us return to the problem that motivated the desirability of introducing the inverse of a matrix. Suppose that we wish to solve a single set of linear equations, expressed in matrix form as

$$AX = B$$

Assuming that A possesses an inverse, let us multiply both sides of this equation by A^{-1} to obtain the equation:

$$A^{-1}AX = A^{-1}B$$

But, since $A^{-1}A = I$, this equation is equivalent to

$$IX = A^{-1}B$$

However, it is clear from carrying out the indicated multiplication on the left that

(6)
$$
IX =
\begin{bmatrix}
1 & 0 & 0 \\
0 & 1 & 0 \\
0 & 0 & 1
\end{bmatrix}
\begin{bmatrix}
x_1 \\ x_2 \\ x_3
\end{bmatrix}
=
\begin{bmatrix}
x_1 \\ x_2 \\ x_3
\end{bmatrix}
= X
$$

We have shown that $IX = X$; consequently, the preceding result simplifies into

$$X = A^{-1}B$$

Since X is the column vector solution of our problem, this shows that the solution of the equations $AX = B$ can be found by multiplying the column vector B by the inverse matrix A^{-1}.

If we have several sets of equations that differ only in the constants on the right side, we can obtain all the solutions by multiplying each of the B vectors by A^{-1}. Therefore, once A^{-1} has been obtained, it is a simple matter to obtain the solution to each set of equations by merely inserting the proper B vector in $A^{-1}B$ and carrying out the multiplication. It is not necessary to carry out all the steps of the pivotal reduction each time. Situations in which this occurs are quite common. An example of one is given in the next section. Routine industrial analyses to determine the inputs needed to realize desired outputs, which differ from week to week, are typical of such situations.

As an illustration of how the inverse is used to solve a set of equations, consider the first set of equations given in (2):

$$x - 2y + 3z = 4$$
$$2x + y - 3z = 5$$
$$-x + y + 2z = 3$$

The solution is given by using the result in (5) and calculating

$$X = A^{-1}B = \begin{bmatrix} \dfrac{5}{16} & \dfrac{7}{16} & \dfrac{3}{16} \\ -\dfrac{1}{16} & \dfrac{5}{16} & \dfrac{9}{16} \\ \dfrac{3}{16} & \dfrac{1}{16} & \dfrac{5}{16} \end{bmatrix} \begin{bmatrix} 4 \\ 5 \\ 3 \end{bmatrix} = \begin{bmatrix} 4 \\ 3 \\ 2 \end{bmatrix}$$

This, of course, is the solution that was obtained earlier by the pivotal reduction method. The solutions of the remaining sets of equations in (2) can be obtained in the same way.

If a matrix does not possess an inverse, the pivotal reduction method for finding it will reveal that fact by showing that the sets of equations with the columns of I on the right as the constants are inconsistent. As an illustration of this possibility, let us consider the problem of finding the inverse of the matrix

$$C = \begin{bmatrix} 1 & -2 & 3 \\ 2 & 1 & -3 \\ 1 & 3 & -6 \end{bmatrix}$$

To find C^{-1} we start with

$$\left[\begin{array}{ccc|ccc} 1 & -2 & 3 & 1 & 0 & 0 \\ 2 & 1 & -3 & 0 & 1 & 0 \\ 1 & 3 & -6 & 0 & 0 & 1 \end{array} \right]$$

Pivoting on the leading element, we obtain

$$\left[\begin{array}{ccc|ccc} 1 & -2 & 3 & 1 & 0 & 0 \\ 0 & 5 & -9 & -2 & 1 & 0 \\ 0 & 5 & -9 & -1 & 0 & 1 \end{array} \right]$$

Multiplying the second row by -1 and adding it to the third row, we obtain

$$\left[\begin{array}{ccc|ccc} 1 & -2 & 3 & 1 & 0 & 0 \\ 0 & 5 & -9 & -2 & 1 & 0 \\ 0 & 0 & 0 & 1 & -1 & 1 \end{array} \right]$$

Since the third row represents the three equations

$$0x + 0y + 0z = 1, -1, 1$$

it follows that the original set of equations being solved to obtain the inverse does not possess a solution. This implies that A^{-1} does not exist, because if it did, we could obtain a unique solution of those equations by means of the formula $X = A^{-1}B$.

In summary, Sections 3 and 4 have shown that the matrix version of the pivotal reduction method not only yields a simple compact method for solving a set of equations, or for finding an inverse, but it also displays the lack of a solution when such is the case during the computations. As was pointed out at the end of Section 3, a further striking advantage is that this method is readily programmed for machine computation and is customarily used to solve large systems of equations. Large systems, however, are seldom solved by calculating the inverse matrix because they usually do not involve a large number of constant vectors on the right side.

5. LEONTIEF INPUT-OUTPUT MODEL

The following material is an interesting illustration of how the inverse of a matrix arises naturally in a theoretical model of economic theory.

Let us suppose that an economic system can be divided into a small number, n, of activities. This could be done quite easily, for example, with the economic system of the natives living on a small island. It can also be done with a more sophisticated economic system if the units of activity are made sufficiently large. Thus manufacturing could be taken as a single unit in the division, or it could be split into a few units, such as the manufacturing of basic materials, machinery, or finished products. Regardless of how this division is made, we treat each division as a segment of the economy that produces a single product. This is merely convenience of language and does not imply that only one product is actually produced. The manufacturing of basic materials, for example, would include things such as steel, aluminum, and lumber.

In this simplified model of an economic system, let there be n products, or goods, produced by the n divisions of the system. Furthermore, assume that these n goods suffice as basic materials and labor for producing the n products. This means that it is not necessary to go outside the economy to obtain goods for producing the foregoing n products. We let these n products, or goods, be denoted by G_i, $i = 1, \ldots, n$, and the n divisions of the economy that produce these corresponding goods by P_i, $i = 1, \ldots, n$. Thus the segment P_i produces the goods denoted by G_i.

Suppose that producer P_j wishes to produce one unit of G_j. This requires a certain number of units of the other goods produced in the system. We let a_{ij} denote the amount of G_i that is needed by P_j to produce one unit of G_j. It follows from the definition of a_{ij} that $a_{ij} \geq 0$. The production system for this economy can now be described by means of the following matrix.

$$
A = \begin{bmatrix}
a_{11} & a_{12} & \cdots & a_{1j} & \cdots & a_{1n} \\
a_{21} & a_{22} & \cdots & a_{2j} & \cdots & a_{2n} \\
\cdot & \cdot & & \cdot & & \cdot \\
\cdot & \cdot & & \cdot & & \cdot \\
\cdot & \cdot & & \cdot & & \cdot \\
a_{i1} & a_{i2} & \cdots & a_{ij} & \cdots & a_{in} \\
\cdot & \cdot & & \cdot & & \cdot \\
\cdot & \cdot & & \cdot & & \cdot \\
\cdot & \cdot & & \cdot & & \cdot \\
a_{n1} & a_{n2} & \cdots & a_{nj} & \cdots & a_{nn}
\end{bmatrix}
$$

This matrix is called the *technology input-output matrix* for the system. The *j*th column gives the various amounts of the *n* goods needed to produce one unit of G_j. The *i*th row gives the amounts of G_i that are needed by the *n* producers to produce one unit of each of their respective goods.

This may appear to be an unrealistic way of describing an economic system; however, by making *n* sufficiently large and by inserting the proper zero values for some of the entries, it does become quite realistic. Thus, if G_1 represents automobiles and G_2 represents steel, we would choose $a_{12} = 0$ or, at least, make a_{12} very small, because automobiles are needed in manufacturing steel only as a means of transportation. The value of a_{21}, however, would be very large because much steel is needed to produce one automobile. The units in which the a_{ij} are measured may be in dollars or in physical units of some kind. It is usually best to think of them as being measured in dollars.

Suppose now that we wish to produce x_j units of the good G_j. If a_{ij} units of G_i are needed to produce one unit of G_j, than x_j units of G_j will normally require x_j times as many units of G_j. Thus producer P_j will need $a_{ij}x_j$ units of G_i to produce x_j units of G_j. This applies to each of the *n* goods; therefore, if we wish to produce $x_1, x_2, \ldots, x_n$ units, respectively, of the *n* goods, the total amount of the good G_i that will be needed is given by

(7) $$u_i = a_{i1}x_1 + a_{i2}x_2 + \cdots + a_{in}x_n, \quad i = 1, \ldots, n$$

For the purpose of expressing this in matrix form and of obtaining a formula that gives all the *u* values, we introduce the column vectors

$$U = \begin{bmatrix} u_1 \\ u_2 \\ \cdot \\ \cdot \\ \cdot \\ u_n \end{bmatrix}, \quad X = \begin{bmatrix} x_1 \\ x_2 \\ \cdot \\ \cdot \\ \cdot \\ x_n \end{bmatrix}, \quad D = \begin{bmatrix} d_1 \\ d_2 \\ \cdot \\ \cdot \\ \cdot \\ d_n \end{bmatrix}$$

Then the relation (7), for all values of *i*, is easily seen to be expressible in the matrix form:

(8) $$U = AX$$

Relation (7) for a fixed *i* is obtained from (8) by choosing the *i*th row of AX. The vector X is called the *intensity vector*. It gives the amounts of the various goods that are to be produced.

The vector D that has been introduced here is called the *demand vector* and is defined by the relation

$$D = X - U$$

Thus D is a column vector whose ith component represents the difference between the amount of G_i that is produced and the amount that is consumed in the system. If the components of D are all positive, the system has produced more than it has consumed of the various goods, and there will be a surplus for exportation. If some component of D is negative, the economy cannot operate for long at this set of production levels because the corresponding good will eventually be exhausted.

Suppose we wish to gear our economy so that it will yield a certain demand vector D. For example, we might wish to make certain that we can export a certain dollar value of the various goods. The question then is: With a given input-output matrix A and a given demand vector D, does there exist an intensity vector X that will yield D? In matrix form, we wish to find an X that will satisfy the equation:

$$X - U = D$$

From (8), this is equivalent to

$$X - AX = D$$

Applying property (6) of the identity matrix I, we may write this in the form:

$$IX - AX = D$$

Since these column vectors may be added, this is equivalent to

$$(I - A)X = D$$

If the matrix $I - A$ possesses an inverse, we may multiply both sides of this equation by this inverse, $(I - A)^{-1}$, and obtain

$$X = (I - A)^{-1}D$$

We have now found the intensity vector that is needed to produce the demand vector D.

If all the elements of X turn out to be positive numbers, then it is possible to meet the demand D. A negative value for one of the components of X would mean that the producer of that good would produce a negative amount of the good, which means that he would need to buy that amount of his own goods from outside the system. This is not possible in our model because the importation of goods is not permitted.

As an illustration, suppose the input-output matrix is the following one in which it is assumed that there are only three divisions in the economy. These might consist, for example, of agriculture, manufacturing, and services.

$$A = \begin{bmatrix} .4 & .1 & .1 \\ .2 & .5 & .2 \\ .1 & .2 & .3 \end{bmatrix}$$

Then $I - A$ assumes the form:

$$I - A = \begin{bmatrix} .6 & -.1 & -.1 \\ -.2 & .5 & -.2 \\ -.1 & -.2 & .7 \end{bmatrix}$$

The inverse, calculated by the pivotal reduction method and using three-decimal accuracy in the calculations, is

(9) $$(I - A)^{-1} = \begin{bmatrix} 1.927 & .560 & .435 \\ .992 & 2.545 & .868 \\ .560 & .809 & 1.739 \end{bmatrix}$$

Suppose that the demand vector has components 20, 10, and 10. Then the intensity vector X is given by the matrix product:

$$X = \begin{bmatrix} 1.927 & .560 & .435 \\ .992 & 2.545 & .868 \\ .560 & .809 & 1.739 \end{bmatrix} \begin{bmatrix} 20 \\ 10 \\ 10 \end{bmatrix} = \begin{bmatrix} 48.49 \\ 53.97 \\ 36.68 \end{bmatrix}$$

This shows that the system will need to produce approximately 49, 54, and 37 units, respectively, of the three production divisions if it wishes to have a surplus of 20, 10, and 10 units, respectively, for export purposes.

The preceding ideas can be extended further by introducing the cost of labor, the price of each good, and the rate of profit. In the preceding discussion the units of the goods were assumed to be monetary. Now we assume that they are physical units, otherwise there would be no need to consider what price to charge for a product. In this more sophisticated model we let p_j represent the price to charge for one unit of G_j, we let l_j represent the cost of labor for producing one unit of G_j, and we let r_j represent the profit to be earned in producing one unit of G_j. The total cost of producing one unit of G_j, namely p_j, must therefore satisfy the equation

(10) $$p_j = p_1 a_{1j} + \cdots + p_n a_{nj} + l_j + r_j, \qquad j = 1, \ldots, n$$

The first n terms on the right represent the cost of producing one unit of G_j in terms of the various goods needed in its production, whereas the last two terms are the labor and profit costs.

It will be observed that (10), for all values of j, can be written in matrix form if we write the sum of the first n terms on the right in the form $A'P$, where A' denotes the *transpose* of the matrix A, that is, the matrix that is obtained from A by interchanging its rows and columns. Equations (10) then assume the following compact matrix form:

(11)
$$P = A'P + L + R$$

where P, L, and R are column vectors.

Suppose we have decided on what profit vector R we want. Is it possible to find a price vector P that will yield us this profit vector? The answer is given by solving (11) for P. First, we transpose $A'P$ to the left side and then write

$$(I - A')P = L + R$$

Assuming that $I - A'$ possesses an inverse, we may multiply both sides of this equation by $(I - A')^{-1}$ and obtain

(12)
$$P = (I - A')^{-1}(L + R)$$

If the right side yields a vector that contains only positive elements, it will be the desired price vector. Incidentally, when the matrix $(I - A)^{-1}$ that was calculated earlier has only positive elements, the matrix $(I - A')^{-1}$ will also have only positive elements because it is easily seen that one of these matrices is merely the transpose of the other.

For the purpose of illustrating the use of formula (12), we use the same matrix A as in the previous illustration, with the understanding that the elements now represent physical production units. Suppose that the labor and profit vectors are given by

$$L = \begin{bmatrix} 20 \\ 12 \\ 10 \end{bmatrix}, \qquad R = \begin{bmatrix} 10 \\ 8 \\ 5 \end{bmatrix}$$

Then formula (12) requires that the price vector be given by the product:

$$P = \begin{bmatrix} .6 & -.2 & -.1 \\ -.1 & .5 & -.2 \\ -.1 & -.2 & .7 \end{bmatrix}^{-1} \begin{bmatrix} 30 \\ 20 \\ 15 \end{bmatrix}$$

Since the inverse of $I - A'$ can be shown to be the transpose of the inverse of $I - A$, we merely need to take the transpose of our previously calculated inverse to obtain the inverse matrix in (12). Thus employing (9), we obtain

$$P = \begin{bmatrix} 1.927 & .992 & .560 \\ .560 & 2.545 & .809 \\ .435 & .868 & 1.739 \end{bmatrix} \begin{bmatrix} 30 \\ 20 \\ 15 \end{bmatrix} = \begin{bmatrix} 86.05 \\ 79.83 \\ 56.49 \end{bmatrix}$$

This shows that it will be necessary to charge the amounts $86, $80, and $57 apiece, respectively, for the three products of the economy to realize the profits of $10, $8, and $5 apiece, respectively, that are desired.

Although this input-output model of an economy may appear to be too simple to be realistic, it can be made more realistic by using a large number of activities. Models of this type have been constructed that contain 50 or more activities. The calculation of the inverse of a 50 × 50 matrix requires, of course, the use of high-speed computing facilities.

COMMENTS

This chapter has been concerned with studying matrices and their properties for the purpose of using them to simplify the problems of solving large systems of linear equations. They were shown to be very useful in condensing the pivotal reduction technique. They were also shown to be very useful for writing sets of linear equations in compact form and, by means of the inverse matrix, for solving these systems of equations both theoretically and practically.

Since modern computing equipment is now capable of solving large systems of equations and since this capability has enabled practical investigators to incorporate large numbers of variables into their investigations, it has become increasingly important for social scientists to understand matrix methods and to know how to apply them to their problems. In the next chapter we study one such field of application. Although we confine our study to small systems, real-life problems may involve 50 or more variables.

EXERCISES

Section 1

1. Let $A = [3 \quad 1 \quad -2]$, $B = [1 \quad -1 \quad 0]$,

$$C = \begin{bmatrix} -1 \\ 2 \\ 0 \end{bmatrix}, \quad D = \begin{bmatrix} 4 \\ 1 \\ -2 \end{bmatrix}$$

Calculate (a) $A + B$, (b) AC, (c) $AC - BD$, (d) $(A + B)(C + D)$, (e) $(2A - B)(3C + D)$.

2. Let $A = [x \quad 2 \quad -1 \quad 3]$ and

$$B = \begin{bmatrix} 2 \\ -1 \\ 0 \\ 3 \end{bmatrix}$$

Solve the equation $AB = 0$ for x.

3. Find two nonzero vectors, A and B, of the form $A = [a_1 a_2 a_3]$ and

$$B = \begin{bmatrix} b_1 \\ b_2 \\ b_3 \end{bmatrix}$$

such that $AB = 0$. What does this show?

4. A stockbroker sold one of his customers 100 shares of stock A, 50 shares of stock B, 75 shares of stock C, and 25 shares of stock D. These stocks cost $12, $20, $10, and $8 per share, respectively. Use vectors to represent the shares purchased and the prices, such that the product of the two vectors will give the total cost of the stocks. Calculate the total cost.

5. Three shoppers, A, B, and C, go into a store to buy fruit. Shopper A purchases 12 oranges, 5 grapefruit, 20 apples, 6 bananas, and 3 lemons. Shopper B purchases 20 oranges, 3 grapefruit, 10 apples, and 4 bananas. Shopper C purchases 10 oranges, 10 grapefruit, and 12 bananas. Suppose that oranges cost 10 cents each, grapefruit 20 cents each, apples 8 cents each, bananas 6 cents each, and lemons 5 cents each.

(a) Represent the purchases of each of these individuals by means of a row vector and the prices by means of a column vector.

(b) Use vector multiplication to calculate each individual's bill.

(c) Use vector addition to find the total amounts of the various fruits that were purchased by these three individuals. Use this result and vector multiplication to calculate the total amount of money spent by these individuals.

6. Do there exist values of the unknowns in the following matrix equations for which the equation is true? If so what are they?

(a) $\begin{bmatrix} 3 & -1 \\ x & 2 \end{bmatrix} = \begin{bmatrix} 3 & -1 \\ 4 & 2 \end{bmatrix}$ (b) $\begin{bmatrix} 4 & 0 \\ x & 3 \end{bmatrix} = \begin{bmatrix} 4 & 1 \\ 5 & 3 \end{bmatrix}$

(c) $\begin{bmatrix} 3 & -1 & 2 \\ 1 & x & 0 \\ y & 2 & 3 \end{bmatrix} = \begin{bmatrix} 3 & -1 & 2 \\ 1 & 4 & 0 \\ x & 2 & 3 \end{bmatrix}$

7. Let

$$A = \begin{bmatrix} 2 & -1 & 0 \\ 4 & 0 & 2 \end{bmatrix}, \qquad B = \begin{bmatrix} -1 & 2 & -1 \\ 0 & 2 & 3 \end{bmatrix},$$

$$C = \begin{bmatrix} 1 & -1 \\ 2 & 3 \\ 3 & 3 \end{bmatrix}, \qquad D = \begin{bmatrix} 1 & 2 \\ -1 & 3 \end{bmatrix}$$

Calculate, if possible, (a) $A + B$, (b) $3A - 2B$, (c) $A - D$, (d) DB, (e) $(A + B)C$.

8. Find a matrix E such that $A + E = B$, where A and B are given in Problem 7.

9. Let

$$A = \begin{bmatrix} 3 & -1 \\ 1 & 2 \end{bmatrix}, \qquad B = \begin{bmatrix} 2 & 0 \\ 3 & -1 \end{bmatrix}, \qquad C = \begin{bmatrix} 1 & 2 \\ -1 & 4 \end{bmatrix}$$

Given that A^2 denotes the product AA, calculate (a) AB, (b) BA, (c) $(AB)C$, (d) $A(BC)$, (e) A^2, (f) $(A + B)^2$.

10. What do the results in (a) and (b) of Problem 9 show?

11. What general property of matrices would you expect to be true on the basis of the results in (c) and (d) of Problem 9?

12. If A, B, and C are any 2×2 matrices, show that

$$A(B + C) = AB + AC$$

13. If A, B, and C are any 2×2 matrices, show that

$$A(BC) = (AB)C$$

14. When is it true that $A + B = 0$?

15. Let

$$A = \begin{bmatrix} a_{11} & a_{12} & a_{13} \\ a_{21} & a_{22} & a_{23} \end{bmatrix}, \qquad I = \begin{bmatrix} 1 & 0 & 0 \\ 0 & 1 & 0 \\ 0 & 0 & 1 \end{bmatrix}$$

Show that $AI = A$. Can you generalize this?

16. Let

$$A = \begin{bmatrix} 1 & 1 & 1 & 1 \\ 2 & 2 & 0 & 2 \\ 3 & 1 & 1 & 3 \end{bmatrix}, \qquad B = \begin{bmatrix} 1 & 3 & -1 \\ 0 & 1 & 2 \\ -1 & 0 & 2 \\ 2 & 0 & 1 \end{bmatrix}$$

Calculate, if possible, (a) AB, (b) BA, (c) A^2.

17. If A and B are two matrices of the same size and if r and s are any two numbers, show that
(a) $(r + s)A = rA + sA$
(b) $r(A + B) = rA + rB$

18. A, B, and C are three matrices of the same size. Show that if $A + B = A + C$, then $B = C$.

19. Let

$$A = \begin{bmatrix} 0 & 3 \\ 0 & 0 \end{bmatrix}, \qquad B = \begin{bmatrix} 2 & 1 \\ 3 & 0 \end{bmatrix}, \qquad C = \begin{bmatrix} 5 & 4 \\ 3 & 0 \end{bmatrix}$$

Show that $AB = AC$. What conclusion can you draw from this?

20. Matrix A gives three raw material needs for producing one unit of each of four products. Matrix (vector) B gives the number of units of each product that are to be produced in one week. Matrix (vector) C gives the cost per unit of each raw material. Use matrix multiplication to obtain (a) the raw material needs, and (b) the total raw material cost.

Raw Materials   Products

$$\text{Products}\begin{array}{c}1\\2\\3\\4\end{array}\begin{bmatrix}1&2&4\\2&0&1\\4&2&3\\3&1&2\end{bmatrix}=A,\ \text{Units}\ \begin{array}{cccc}1&2&3&4\end{array}[4\ \ 3\ \ 6\ \ 2]=B,$$

Cost

$$\text{Raw Materials}\begin{bmatrix}2\\4\\3\end{bmatrix}=C$$

21. A company operates four plants. Each plant produces three kinds of products. Each product requires five (or less) essential parts in its assembly. Matrix A gives the number of such parts needed to assemble one unit of each product. Each plant is scheduled to produce a fixed number of units of the three products during next week's production. This schedule is given by matrix B. Use matrix multiplication to determine how many parts of the five types will be needed to meet the scheduled production.

Products   Plants

$$\text{Parts}\begin{bmatrix}1&2&1\\2&0&3\\0&2&2\\1&0&1\\1&2&0\end{bmatrix}=A,\ \text{Products}\ \begin{bmatrix}1&2&4&0\\2&3&0&1\\3&1&2&4\end{bmatrix}=B$$

22. Let $P_1, \ldots, P_r$ denote r different plants that provide food for certain plant eating animals. These animal types are denoted by $A_1, \ldots, A_s$. Let $C_1, \ldots, C_t$ denote the types (carnivorous) of animals that live off plant eating animals. Let x_{ik} denote the amount of plant P_i eaten by an animal of type A_k during one year. Let y_{kj} denote the number of animals of type A_k eaten by a carnivore of type C_j during one year. Finally, let $X = [x_{ij}]$ and $Y = [y_{kj}]$ be the corresponding food consumption matrices for these two types of animals.
  (a) Find an expression for the amount of plant P_i that is used indirectly by a carnivore of type C_j.
  (b) If z_{ij} represents the answer in (a), show that the matrix $Z = [z_{ij}]$ is given by $Z = XY$.

Section 2

1. Express each of the following sets of equation in the matrix form $AX = B$.

   (a) $4x_1 - x_2 = 3$
   $x_1 + 2x_2 = 5$
   (c) $3x_1 + 4x_2 = 8$
   $x_1 - x_2 = 3$
   $-x_1 + 2x_2 = 1$
   $2x_1 + x_2 = 5$

   (b) $x_1 + x_2 - x_3 = 2$
   $2x_1 - x_2 + x_3 = 3$
   $x_1 - x_2 = -1$
   (d) $ax_1 + bx_2 + cx_3 = r$
   $bx_1 + cx_2 + dx_3 = s$
   $cx_1 + dx_2 + ex_3 = t$

2. Find the values of x and y that satisfy

$$\begin{bmatrix} 4 & 1 \\ 3 & 2 \end{bmatrix}\begin{bmatrix} x \\ y \end{bmatrix} = \begin{bmatrix} 2 \\ 1 \end{bmatrix}$$

3. Find the values of x and y that satisfy

$$[2 \quad -1]\begin{bmatrix} x & y \\ -y & -x \end{bmatrix} = [3 \quad 2]$$

4. Given the following matrices, solve the matrix equation $BX = C$.

$$B = \begin{bmatrix} 2 & 1 & -1 \\ 0 & 3 & 4 \\ 0 & 0 & 4 \end{bmatrix}, \quad X = \begin{bmatrix} x_1 \\ x_2 \\ x_3 \end{bmatrix}, \quad C = \begin{bmatrix} 1 \\ 0 \\ -1 \end{bmatrix}$$

5. Given the following matrix equation, write out the system of equations equivalent to it.

$$\begin{bmatrix} 3 & 1 & 2 \\ -1 & 0 & 1 \\ 2 & 1 & -3 \end{bmatrix}\begin{bmatrix} x \\ y \\ z \end{bmatrix} = \begin{bmatrix} 2 \\ 3 \\ 4 \end{bmatrix}$$

6. Given the following matrices, (a) write out the system of equations in simplified form that is equivalent to the matrix equation $AB = B + C$, and (b) solve this system.

$$A = \begin{bmatrix} 1 & 2 & 1 \\ -1 & 3 & 0 \\ 2 & 1 & -1 \end{bmatrix}, \quad B = \begin{bmatrix} x \\ y \\ z \end{bmatrix}, \quad C = \begin{bmatrix} 3 \\ -1 \\ 2 \end{bmatrix}$$

7. Given the $n \times n$ matrix $A = [a_{ij}]$ and the $n \times 1$ column vector $X = [x_i]$, write out the equations in simplified form that are equivalent to the matrix equation $AX = X$.

8. Given the following matrices, solve the matrix equation $XB = C$.

$$X = [x_1 \quad x_2 \quad x_3], \qquad C = [2 \quad -1 \quad 0]$$

$$B = \begin{bmatrix} 1 & 0 & -1 \\ 0 & 2 & 1 \\ -1 & 1 & 0 \end{bmatrix}$$

9. Given the equations

$$r + s + t = 2$$
$$2r - s + t = 3$$
$$r - s + 2t = -1$$

(a) Write down the three matrices A, X, and B such that these equations are equivalent to the matrix equation $AX = B$.

(b) Write down the three matrices B, Y, and C such that these equations are equivalent to the matrix equation $YB = C$.

Section 3

1. Solve the following set of equations by using the simplified pivotal reduction method.

$$x + y - z = 2$$
$$2x - y + z = 3$$
$$x - y - 2z = -1$$

2. Solve the following set of equations by using the simplified pivotal reduction method.

$$x + y + z + w = 3$$
$$x - y + 2z - w = 2$$
$$2x + y - z + 2w = 4$$

3. Use the simplified pivotal reduction method to determine whether the following set of equations possesses a solution.

$$x - 4y + z = 2$$
$$3x + y + 2z = 6$$
$$x - y - z = -1$$
$$2x + y - 2z = 2$$

4. Solve the following set of equations by using the simplified pivotal reduction method.

$$x + 2y - z + w = 4$$
$$2x - y + 3z - 2w = -1$$
$$x - 3y + 2z + 3w = 5$$
$$3x + y - z - w = 9$$

5. Use the simplified pivotal reduction method to determine the value of c that will allow the following set of equations to be consistent.

$$x - y + z = 1$$
$$2x + 2y - z = 2$$
$$3x - y + 2z = 3$$
$$4x + y + z = c$$

6. Solve the following set of equations by using the simplified pivotal reduction method. Comment about your result and equations of the type that have zero constants on the right.

$$x + y + 5z = 0$$
$$2x + 3y + 13z = 0$$
$$3x + 2y + 12z = 0$$

7. The letters a, b, c, and d represent certain constants.

$$x + z = a$$
$$y + z = b$$
$$x + t = c$$
$$y + t = d$$

(a) Use the simplified pivotal reduction method to solve these equations and thereby determine a restriction that must be placed on the constants if these equations are to possess a solution.
(b) If the restriction imposed in (a) is satisfied, how many solutions will these equations possess?

8. Let x, y, z, and t denote the number of units of four products that can be produced in one week if there are three types of raw materials needed for each and the number of units of such raw materials available are 30, 20, and 40, respectively. The number of units of these raw materials needed to produce one unit of each of the four products is given by the following matrix. Write down the three equations that x, y, z, and t must satisfy and solve them by the simplified pivotal reduction method. Comment.

Raw Material

	A	B	C
1	1	2	4
2	2	0	1
3	4	2	3
4	3	1	2

Product

9. Triglycerides contain three fatty acids A, B, and C; therefore, there are six possible isomers denoted by ABC, ACB, BAC, BCA, CAB, and CBA. Let $x_1, x_2, \ldots, x_6$ denote the proportions of each of these isomers in a certain species of triglycerides. The following table gives the total fatty acid composition in each of three positions for this species.

Fatty Acid

	A	B	C
1	.60	.30	.10
2	.10	.60	.30
3	.30	.10	.60

Position

The first row, for example, states that in the first position 60 percent of the members of this species have fatty acid A, 30 percent have B, and 10 percent have C. Similarly for the other two rows. Since the only isomers that have A in the first position are ABC and ACB, it follows that the total proportion of these two must be .60. Hence $x_1 + x_2 = .60$. In a similar manner, each of the entries in this table is equal to the sum of two isomer proportions. This gives the following set of equations.

$$x_1 + x_2 = .6, \ x_3 + x_5 = .1, \ x_4 + x_6 = .3$$
$$x_3 + x_4 = .3, \ x_1 + x_6 = .6, \ x_2 + x_5 = .1$$
$$x_5 + x_6 = .1, \ x_2 + x_4 = .3, \ x_1 + x_3 = .6$$

(a) Write these equations in matrix form.
(b) Solve these equations by the simplified pivotal reduction method.

Section 4

1. Find the inverse of each of the following matrices, provided that the inverse exists.

(a) $\begin{bmatrix} 1 & 2 \\ 3 & 1 \end{bmatrix}$,   (b) $\begin{bmatrix} 2 & 1 \\ 4 & 2 \end{bmatrix}$,   (c) $\begin{bmatrix} 1 & 2 & 3 \\ 2 & 3 & 4 \\ 3 & 4 & 5 \end{bmatrix}$

(d) $\begin{bmatrix} 1 & 0 & 0 \\ 0 & 1 & 0 \\ 0 & 0 & 1 \end{bmatrix}$,   (e) $\begin{bmatrix} 1 & 2 & 3 \\ 2 & 6 & 11 \\ -1 & 0 & 2 \end{bmatrix}$,   (f) $\begin{bmatrix} 1 & 1 & -1 \\ 1 & -1 & 1 \\ 1 & 1 & 1 \end{bmatrix}$

2. Find the inverse of each of the following matrices, provided that the inverse exists.

(a) $\begin{bmatrix} 1 & 0 & 1 & 0 \\ 0 & 1 & 2 & 1 \\ 0 & 1 & 1 & 1 \\ 2 & 0 & 3 & 0 \end{bmatrix}$,   (b) $\begin{bmatrix} 1 & 0 & 1 & -1 \\ 2 & 1 & 1 & 0 \\ 0 & 1 & 1 & 0 \\ 0 & -1 & 2 & 1 \end{bmatrix}$

3. Given the matrix A, (a) calculate A^{-1}, (b) verify that $A^{-1}A = I$, and (c) verify that $AA^{-1} = I$.

$$A = \begin{bmatrix} 2 & 6 \\ 4 & 8 \end{bmatrix}$$

4. Do as in Problem 3 for the following matrix.

$$A = \begin{bmatrix} 1 & 2 & -1 \\ 2 & -2 & 3 \\ 3 & 1 & -4 \end{bmatrix}$$

5. Solve Problem 1 of Section 3 by first finding the inverse of the coefficient matrix.

6. Solve Problem 4 of Section 3 by first finding the inverse of the coefficient matrix.

7. Solve the following sets of equations by first calculating the inverse of the coefficient matrix.

(a) $x + y + z = 6$
$3x + 2y - z = 5$
$2y + 3z = 11$

(b) $x + 4y + 3z = 4$
$x - 3y - 2z = -2$
$2x + 5y + 4z = 8$

8. Verify that the matrix A^{-1} given in (5) in the text is the correct inverse of the coefficient matrix for equations (3).

9. Solve the following set of equations by first calculating the inverse of the coefficient matrix.

$$x + 2y - 3z + t = 5$$
$$y + z - 2t = 4$$
$$z + 3t = 7$$
$$t = 2$$

10. Explain why a triangular matrix with nonzero elements down its main diagonal always has an inverse. It may be written in the following form, where none of the elements $a_{11}, \ldots, a_{nn}$ is zero.

$$\begin{bmatrix} a_{11} & a_{12} & \cdots & a_{1n} \\ 0 & a_{22} & \cdots & a_{2n} \\ \cdot & \cdot & & \cdot \\ \cdot & \cdot & & \cdot \\ \cdot & \cdot & & \cdot \\ 0 & 0 & \cdots & a_{nn} \end{bmatrix}$$

11. For the general 2×2 matrix A that follows, find a formula for its inverse.

$$A = \begin{bmatrix} a & b \\ c & d \end{bmatrix}$$

12. By using the formula obtained in Problem 11, write down a formula for the inverse of the matrix C, where

$$C = \begin{bmatrix} a & c \\ b & d \end{bmatrix}$$

Matrix C, which is called the transpose of matrix A, is obtained from A by interchanging the rows and columns of A. From your answer, what relationship exists between the inverse of a matrix and the inverse of its transpose? This relationship is true for $n \times n$ matrices also.

13. Show that if the matrices A and B satisfy the relation $AB = A$ and A possesses an inverse, then $B = I$.

14. Show that if the matrices A, B, and C satisfy the relation $AB = AC$ and A possesses an inverse, then $B = C$.

15. A diet is to be constructed for a set of experimental animals by using three basic foods labeled A, B, and C. The foods contain the following units of protein, carbohydrates, and fat for each gram of food.

	A	B	C
Protein	1	3	1
Carbohydrates	2	1	4
Fat	3	1	4

Suppose that the experimenter wishes to construct a diet that contains 110 units of protein, 132 units of carbohydrates, and 154 units of fat.
(a) Let x_1, x_2, and x_3 denote the amounts of the three foods to use in the diet and find the equations they must satisfy.
(b) Solve the equations in (a) by calculating the inverse matrix.
(c) Explain why the technique in (b) is useful if the experimenter is considering other diets also.

16. A coffee shop carries three kinds of coffee blends in one-pound sacks. Most of the customers choose one of those three blends; however, a few of them desire a different mix of the basic coffees going into the blends. Different mixes can often be obtained by choosing the proper amounts of each of the three basic blends, provided that the order is sufficiently large. Let A, B, and C denote the three packaged blends and suppose that a one-pound sack of each contains the following amounts, in ounces, of the basic coffees used.

	Moca	Java	Brazilian
A	2	6	8
B	4	11	1
C	8	6	2

(a) A customer orders 40 pounds of coffee that will contain 10 pounds of Moca, 20 pounds of Java, and 10 pounds of Brazilian coffee. Let x_1, x_2, and x_3 denote the number of sacks of each of the three blends that need to be used to fill this customer's order. Write down the equations that these unknowns must satisfy, and solve them by first calculating the inverse of the coefficient matrix.
(b) Use the inverse matrix obtained in (a) to find the amounts needed for another customer who wants 30 pounds of coffee with a blend of

6 pounds, 12 pounds, and 12 pounds of the three basic coffees.
Comment on your solution.

(c) Do as in part (b) for a blend of 4 pounds, 10 pounds, and 16 pounds
and comment on your solution.

Section 5

1. Assume that an economy can be expressed in terms of three basic components I, II, and III, and that the technology input-output matrix is

$$A = \begin{bmatrix} .1 & .1 & .1 \\ .1 & .2 & .3 \\ .2 & .6 & .4 \end{bmatrix}$$

If the demand vector for this system has components (3,4,6), find the
production vector X that will produce this demand.

2. Assume that the elements of matrix A in Problem 1 represent physical
units of production. Calculate the price vector P that will yield a profit
vector with components (1,2,3) if the labor cost vector has components
(4,5,6).

3. Suppose that an economy is divided into four units and that it has the following technology input-output matrix. If the demand vector has components (10,5,20,10), find the production vector X that will produce this
demand.

$$A = \begin{bmatrix} .8 & .2 & .1 & 0 \\ .2 & .5 & .2 & .1 \\ .2 & .1 & .3 & .3 \\ .1 & 0 & .1 & .4 \end{bmatrix}$$

4. Assume that the elements of matrix A in Problem 3 represent physical
units of production. Calculate the price vector P that will yield a profit
vector with components (3,2,1,1) if the labor cost vector has components
(4,3,2,2).

5. Suppose that the demand vector in Problem 1 has equal components.
What production vectors, if any, will yield such a demand vector?

6. What production vectors, if any, will yield a demand vector in Problem 3
that has equal components?

7. In the technology input-output matrix of Section 5 assume that the prod-

ucts have been arranged in order of complexity so that the production of good G_j requires only those goods that have a smaller index than j. This implies that $a_{ij} = 0$ for $i \geqslant j$. As a result, the matrix A of Section 5 will have zeros for the elements along the main diagonal and for all the elements below this diagonal. Given the following such matrix and a demand vector with components (0,2,4,8), calculate the production vector X that will produce this demand.

$$A = \begin{bmatrix} 0 & 6 & 4 & 3 \\ 0 & 0 & 0 & 2 \\ 0 & 0 & 0 & 6 \\ 0 & 0 & 0 & 0 \end{bmatrix}$$

8. Calculate the production vector X that will produce a demand vector with components (10,4,4,0,2) if the technology input-output matrix is of the type described in Problem 7 and is given by

$$A = \begin{bmatrix} 0 & 2 & 0 & 4 & 8 \\ 0 & 0 & 6 & 1 & 0 \\ 0 & 0 & 0 & 3 & 4 \\ 0 & 0 & 0 & 0 & 2 \\ 0 & 0 & 0 & 0 & 0 \end{bmatrix}$$

4 | Linear Programming

One of the most useful applications of linear functions in business decision-making problems is that of maximizing, or minimizing, a linear function of several variables when those variables satisfy certain linear restrictions. The technique for solving problems of this kind is called *linear programming*. The restrictions that the variables must satisfy are usually in the form of inequalities. Therefore, before we discuss linear programming as such, we must consider linear inequalities.

1. LINEAR INEQUALITIES IN TWO VARIABLES

Recall that the symbol $2 < 5$, which is read "2 is less than 5," is a brief way of stating that the number 2 is smaller than the number 5. We may reverse the order of the two numbers and write $5 > 2$, which is read "5 is greater than 2." The symbol $x \leq 5$, which is read "x is less than or equal to 5," implies that x is some number that is not larger than 5. The symbol $\geq$ is interpreted in a similar manner.

Inequality relationships between numbers are readily obtained by representing the numbers geometrically as points on the x axis and then writing $a < b$ if, and only if, the point a is to the left of the point b on the x axis.

It should also be recalled that it is permissible to add the same number to both sides of an inequality without changing the inequality relationship. For example, since $2 < 5$, it follows that

$$2 + 3 < 5 + 3$$

or that

$$2 - 4 < 5 - 4$$

These two inequalities simplify into

$$5 < 8$$

and

$$-2 < 1$$

An inequality relationship may be multiplied by a positive number without changing the nature of the inequality; however, multiplying both sides of an inequality by a negative number reverses the inequality sign. For example, since $2 < 5$, if we multiply both sides by 3, we obtain

$$2 \cdot 3 < 5 \cdot 3$$

which is equivalent to

$$6 < 15$$

However, if both sides are multiplied by -3, we reverse the inequality sign and obtain

$$2(-3) > 5(-3)$$

which is equivalent to

$$-6 > -15$$

The rules concerning inequalities of numbers also apply to inequalities of expressions that involve one or more variables. In this section we study these inequalities as they apply to linear functions of two variables. We begin by studying the meaning of the following particular inequality:

$$2x + y < 4$$

The problem is to determine what pairs of numbers (x,y) will satisfy this inequality. It is a simple matter to pick out particular pairs that will do so, but we wish to describe in some detail the totality of such pairs.

Suppose that we replace the inequality sign by an equality sign and consider the pairs of numbers that satisfy the resulting equality:

$$2x + y = 4$$

For convenience, this equation and the preceding inequality will be rewritten in the equivalent forms:

$$y = 4 - 2x$$

and

$$y < 4 - 2x$$

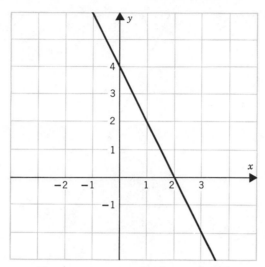

FIGURE 1   A graph of $y = 4 - 2x$.

Now consider the graphs of these two relationships. The graph of the equation is that of the straight line shown in Fig. 1. Thus any pair of numbers (x,y) that satisfies the equation corresponds to a point on the line. Since, corresponding to any fixed x value, y is the vertical distance from the x axis to the line $y = 4 - 2x$, it follows that the corresponding inequality $y < 4 - 2x$ will be satisfied if y assumes a value less than this vertical distance for this same value of x. Let a vertical line be drawn through the point on the x axis having this fixed x value. Then all points on this vertical line that lie below the line of Fig. 1 will have coordinates (x,y) that satisfy the inequality $y < 4 - 2x$. This should be obvious for a value of x for which y is positive. It will also be true if x has a value for which y is negative, because a larger negative value will correspond to a point further down the line. Therefore, allowing x to take on all possible values, the totality of pairs of numbers (x,y) that satisfy the inequality is equivalent to the totality of points that lie below the graph of the corresponding equality. Figure 2 shows part of the region where this inequality is satisfied.

Suppose, next, that we have two linear inequalities being satisfied by x and y. For example, suppose we have the following two inequalities:

$$x + 2y < 4$$

and

$$x - y < -1$$

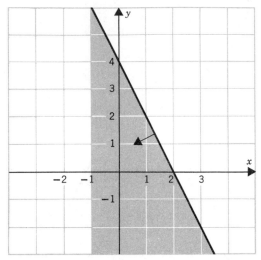

FIGURE 2   The region where $y < 4 - 2x$.

The numbers (x,y) that satisfy both inequalities can be determined geometrically by graphing the two straight lines of the corresponding equalities and then determining the region where both inequalities hold.

For ease of determining inequality regions, it is convenient to express each inequality with y on the left side and the remaining terms on the right side. Hence, we rewrite the preceding inequalities in the following equivalent forms:

$$y < 2 - \frac{1}{2} x$$

and

$$y > 1 + x$$

The graphs of the corresponding equalities are shown in Fig. 3.

Since the first inequality requires the value of y for each x to be less than the y coordinate value on the line $y = 2 - \frac{1}{2} x$, that inequality is satisfied by all points lying below this line. The second inequality requires the value of y for each x to be larger than the value $y = 1 + x$; hence, that inequality is satisfied by all points lying above this line. The points in the plane that lie in both of those regions, therefore, yield pairs of numbers (x,y) that satisfy both inequalities. The region for which this is true is shown in Fig. 4

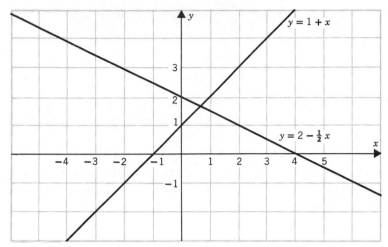

FIGURE 3   The graphs of two linear equations.

as the shaded region. It helps in determining such regions to use an arrow on each line to indicate whether the region is above or below the line.

The same procedure may be applied to any number of inequalities. It is merely necessary to graph the lines that represent the corresponding equalities and then, for each inequality, to determine whether the inequality region lies above or below the corresponding line. As a final illustration

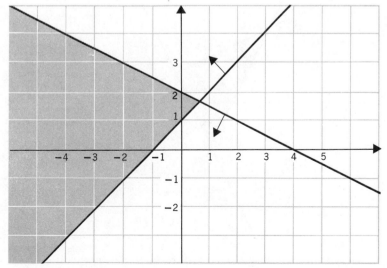

FIGURE 4   The region where two inequalities are satisfied.

of this technique, consider the following three inequalities:

$$2x - y \leqslant 1$$
$$x + 2y \leqslant 3$$
$$2x + y \geqslant -2$$

These three inequalities differ from the preceding ones in that they also include equality signs. This means that points lying on the corresponding lines are also included in the region.

The graphs of the three equalities are shown in Fig. 5.

To determine the inequality regions, we write the inequalities in a form where y is expressed as an inequality function of x; hence, we rewrite them as follows:

$$y \geqslant 2x - 1$$

$$y \leqslant -\frac{1}{2}x + \frac{3}{2}$$

$$y \geqslant -2x - 2$$

The first inequality is satisfied by any point above or on the line $y = 2x - 1$, the second inequality is satisfied by any point below or on the line $y =$

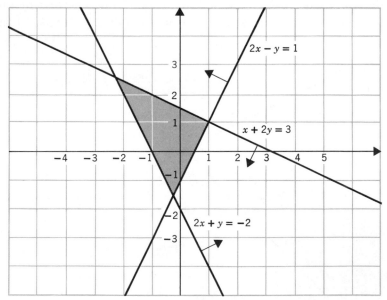

FIGURE 5   The geometry of three linear inequalities.

$-\dfrac{1}{2}x + \dfrac{3}{2}$, and the third inequality is satisfied by any point above or on the line $y = -2x - 2$. The only points that lie in all three of these regions are those lying in the shaded triangle of Fig. 5. This region includes the line segments that bound it.

2. LINEAR PROGRAMMING IN TWO DIMENSIONS

The preceding material will now be used to assist us in solving a two-dimensional linear programming problem. Although problems of this kind occur naturally in business decision making, we postpone their motivation for the present and treat the problem as a purely mathematical one.

Consider the problem of maximizing the function

$$f = 3x + 2y$$

subject to the restrictions

$$2x - y \leqslant 1$$
$$x + 2y \leqslant 3$$
$$x \geqslant 0$$
$$y \geqslant 0$$

We solve this problem by geometrical reasoning. First, we apply the methods of the preceding section to determine the region in the x, y plane where all four inequalities are satisfied. This is shown as the shaded region in Fig. 6. This region is called the *feasibility set* or the *feasibility region* because a solution to our problem is possible, or feasible, only if it corresponds to a point whose coordinates satisfy our inequalities and, hence, if it corresponds to a point lying in this region. Incidentally, in linear programming problems the restrictions are almost always of the form $\leqslant$ or $\geqslant$. Strict inequalities such as $<$ or $>$ are unrealistic and are also inconvenient mathematically; therefore, we do not use strict inequalities in the problems that follow.

If f is given a fixed value, the relation $f = 3x + 2y$ becomes an equation of a straight line. For example, let $f = 2$. Then the relation yields the line $3x + 2y = 2$ that has been labeled $f = 2$ in Fig. 6. Similarly, if $f = 4$, we obtain the line labeled $f = 4$. Both of these lines intersect the feasibility region. Since we are hunting for a point in this region that makes f as large

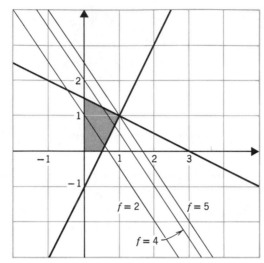

FIGURE 6   The geometry of linear programming.

as possible, we should certainly prefer any point in that region that lies on the second line over any point that lies on the first line. By continuing this type of reasoning, it is clear that we can find points in the feasibility region that will make f larger than 4 because all that we need to do is find a line that is parallel to the two preceding lines and that cuts the y axis above the $f = 4$ line and that also intersects the feasibility region. The maximum value of f will be reached when the line $3x + 2y = f$ is moved parallel to itself upward as far as possible with the restriction that it must intersect the feasibility region. This is seen to occur when it passes through the point of intersection of the lines $2x - y = 1$ and $x + 2y = 3$. Since this point of intersection is easily found to be the point (1,1), this pair of numbers is the pair that maximizes $f = 3x + 2y$, subject to our restrictions. The maximum value of f is, therefore, 5. It corresponds to the line labeled $f = 5$ in Fig. 6.

The preceding problem and its solution illustrates the general method for solving a linear programming problem in two dimensions. It consists in first determining the feasibility region that satisfies all the restrictions, and then determining that straight line, in the family of lines obtained by assigning f various values, which both intersects this region and maximizes the value of f.

If the problem is one of minimizing f we proceed in a similar manner, but then the minimizing line will occur on the opposite side of the feasibility region from that where the maximizing line is found. For example, if the problem were to minimize f in the preceding problem, it is clear from inspecting Fig. 6 that the minimum would occur at the point $(0,0)$ and that the minimum value of f would be 0.

From the nature of the geometry of the problem, it is clear that the maximizing (or minimizing) point lies on the boundary of the feasibility region. Furthermore, it will be at a corner point of this region, unless by chance the family of lines defined by the linear function f is parallel to a boundary line that contains a maximizing (or minimizing) point. If the latter situation should arise, there would not be a unique maximizing (or minimizing) point. As an illustration of this unlikely happening, suppose that we have the same inequalities as before but that our function is $f = 2x + 4y$. Assigning values to f gives a family of lines with slope $-\dfrac{1}{2}$, which is the same as the slope of the boundary line $x + 2y = 3$. As a result, the maximum value of f will be 6 and this value will be obtained at every point on the line $x + 2y = 3$ that is part of the feasibility region.

In all linear programming problems we assume that there exists a finite solution to the problem; therefore, even if the solution region is not finite we still will find a solution at a corner point.

Now consider a problem of the type for which the technique of linear programming was designed. Suppose that a small manufacturing plant produces two types of toys, which we label A and B. Because of a limited number of facilities and employees, this plant, when running at full capacity, has at most 1000 man-hours a week available in its machining department, 400 hours available in its assembly department, and 250 hours available in its painting department. The production of one toy of each of the two types requires the following amounts of man-hours in each of the three departments.

	Machining	Assembly	Painting
A	.2	.2	.1
B	.5	.1	.1

Suppose that a toy of type A sells for $10 and that a toy of type B sells for $8. Let x and y denote the number of toys of type A and B, respectively, that are to be manufactured each week. Then the problem is to determine the pair of values (x,y) that will maximize the revenue function, which is given by

$$R = 10x + 8y$$

subject to the restrictions

$$.2x + .5y \leqslant 1000$$
$$.2x + .1y \leqslant 400$$
$$.1x + .1y \leqslant 250$$
$$x \geqslant 0$$
$$y \geqslant 0$$

For convenience of graphing, the first three inequalities are multiplied by 10 and then the inequalities are rewritten in the following form:

$$y \leqslant 2,000 - \frac{2}{5}x$$
$$y \leqslant 4,000 - 2x$$
$$y \leqslant 2,500 - x$$
$$x \geqslant 0$$
$$y \geqslant 0$$

The region satisfied by these inequalities is shown as the shaded region in Fig. 7. The straight line graphs of the equalities corresponding to the first three preceding inequalities have been labeled l_1, l_2, and l_3, respectively.

When R is assigned various numerical values, the graph of the function $R = 10x + 8y$ yields a family of straight lines with slope $-\frac{5}{4}$. For example, if R is chosen equal to 10,000, we obtain the line labeled $R = 10,000$ in Fig. 7. Since we wish to maximize R and the farther up on the y axis the line cuts the larger will be the value of R, we should move this line parallel to itself up as high as possible subject to the restriction that it must intersect the feasibility region. It is clear from Fig. 7 that this highest line is the one that passes through the corner point labeled P. This maximizing line is shown in Fig. 7 by means of the broken line labeled M. Since P is the intersection of lines l_2 and l_3, its coordinates are obtained by solving the equations

$$y = 4000 - 2x$$

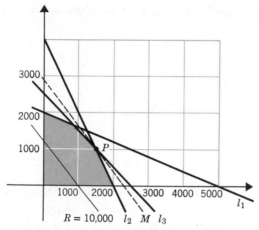

FIGURE 7 A feasibility region.

and

$$y = 2500 - x$$

The solution of these equations is $x = 1500$ and $y = 1000$. Thus, the manufacturer should produce 1500 toys of type A and 1000 toys of type B if he wishes to maximize his total revenue. His revenue per week will then be given by

$$R = 1500(10) + 1000(8) = 23{,}000$$

3. THE SIMPLEX METHOD IN TWO DIMENSIONS

Although the geometrical method for solving linear programming problems is highly satisfactory for problems that involve only two variables, it is a very difficult one to use on problems involving several variables. For those problems an algebraic method is needed. One of the most popular such algebraic devices is known as the *simplex method*. Although it was designed to solve more complicated problems, we first apply it to a two-dimensional problem in order to simplify the explanation. In particular, we apply it to the first of the two problems that were solved in the preceding section. The problem is to find the values of x and y that maximize the linear function

(1) $$f = 3x + 2y$$

subject to the restrictions

$$2x - y \leqslant 1$$
$$x + 2y \leqslant 3$$
$$x \geqslant 0$$
$$y \geqslant 0$$

The solution to this problem was obtained by inspecting the nature of the feasibility region as shown in Fig. 6 and realizing that f is maximized by that member of the family of lines $3x + 2y = f$ which passes through the corner point $(1,1)$.

It is true, in general, that a set of linear inequalities in two variables will always give rise to a feasibility region whose boundary consists of line segments. It is this property of the feasibility region which guarantees that a linear function f will take on its maximum (or minimum) value at a corner point of the boundary.

Although we are studying the problem here in two dimensions only, it can be shown by algebraic techniques that a similar property of the feasibility region is true in higher dimensions. Thus a linear function of k variables, subject to a set of linear inequalities in those variables, will assume its maximum (or minimum) value at one of the corners of the feasibility region that is determined by those inequalities. In our later applications we assume this fact without proof.

In view of the preceding comments, it should suffice to find all the corner points of the feasibility region and then calculate the value of f at each of those points to determine which of them produces the maximum (or minimum) value of f. Although such a procedure is easily carried out for a two-dimensional problem, the same is not true for higher dimensional problems because the number of corners may become very large, and much algebra is needed to locate them. The simplex method is a method that allows us to start at any given corner point and then step-by-step to proceed to a neighboring corner that yields a larger value of f until the maximizing corner is reached. For a minimizing problem, successive corners are chosen that decrease the value of f each time. We restrict ourselves to maximizing problems, because a problem in which f is to be minimized can be converted to one in which g is to be maximized by setting $g = -f$.

With these preliminaries out of the way, we are ready to solve the problem stated in (1) by the simplex method. This method begins by introducing as many new variables as are needed to convert the inequalities into equalities, except for inequalities of the type $x \geqslant 0$ and $y \geqslant 0$. An inequality such as $2x - y \leqslant 1$ can be made into an equality by introducing a new variable, say r, and writing

$$2x - y + r = 1$$

If x and y have values such that $2x - y < 1$, then r will be a positive number which, when added to $2x - y$, will make the sum equal 1. If x and y have values such that $2x - y = 1$, then r will have the value 0. Thus, r is a nonnegative number that transforms the inequality $2x - y \leqslant 1$ into the equality $2x - y + r = 1$. Similarly, we can introduce a nonnegative variable s that will transform the inequality $x + 2y \leqslant 3$ into the equality $x + 2y + s = 3$. The variables r and s that have been introduced here are called *slack variables*, because they take up the slack on the left side of an inequality of the form $\leqslant$ to make the equality sign hold. In this connection, the variables x and y are called the *basic variables*.

The linear programming problem of (1) may now be reformulated as the problem of maximizing the function

$$f = 3x + 2y$$

subject to the restrictions

$$\begin{aligned}
2x - y + r &= 1 \\
x + 2y + s &= 3
\end{aligned}$$

(2)

$$x \geqslant 0, \ y \geqslant 0, \ r \geqslant 0, \ s \geqslant 0$$

Next, we express the corner points of our feasibility region in terms of the values of the four variables x, y, r, and s. The feasibility region for this problem was obtained earlier and can be found in Fig. 6. It is duplicated in Fig. 8. Each of these corner points is the point of intersection of two lines out of the set:

$$x = 0, \ y = 0, \ 2x - y = 1, \quad \text{and} \quad x + 2y = 3$$

Since the last two lines of this set are special cases of the first two equalities in (2) and are characterized by setting $r = 0$ and $s = 0$, respectively, we may represent this set of lines by the symbols:

$$x = 0, \quad y = 0, \quad r = 0, \quad \text{and} \quad s = 0$$

The origin is a corner point and is labeled $(x = 0, y = 0)$ because it is the intersection of the y axis $(x = 0)$ and the x axis $(y = 0)$. The corner point on the x axis is labeled $(y = 0, r = 0)$ because it is the intersection of the x axis $(y = 0)$ and the line $2x - y = 1$ $(r = 0)$. The corner point labeled $(r = 0, s = 0)$ is so labeled because it is the intersection of the two lines $2x - y = 1$ $(r = 0)$ and $x + 2y = 3$ $(s = 0)$. Finally, the corner point on the y axis is labeled $(x = 0, s = 0)$ because it is the intersection of the y axis $(x = 0)$ and the line $x + 2y = 3$ $(s = 0)$. The labeling of all the corner points of the feasibility region in this manner is shown in Fig. 8.

The next step in the simplex method is to start at a corner point and proceed to a neighboring corner that will increase the value of f, assuming that the maximum value of f has not already been attained. Suppose that we start at the $(x = 0, y = 0)$ corner. Then we wish to proceed to the $(y = 0, r = 0)$ corner, or to the $(x = 0, s = 0)$ corner. It should be observed that this procedure leaves one of $x = 0$ or $y = 0$ alone and replaces the other by the zero value of a different variable. Since our function $f = 3x + 2y$ will grow faster if x is increased one unit than if y is increased one unit, and we wish to make f as large as possible, we agree to leave $y = 0$ alone and increase x as much as possible. But from Fig. 8 this means that we choose the $(y = 0, r = 0)$ corner in preference to the $(x = 0, s = 0)$ corner. Setting $y = 0$ and

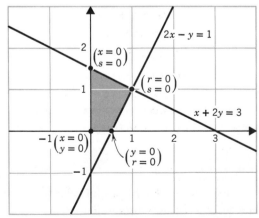

FIGURE 8 The simplex notation for corner points.

$r = 0$ in equations (2), we obtain

$$2x = 1$$
$$x + s = 3$$

Solving these equations, we find that $x = \frac{1}{2}$, $s = \frac{5}{2}$, and $f = \frac{3}{2}$. Since f had the value 0 at our starting corner ($x = 0$, $y = 0$), this shift has increased its value from 0 to $\frac{3}{2}$.

We now repeat the performance, beginning with the ($y = 0$, $r = 0$) corner and treating y and r as the basic variables and x and s as the slack variables. To do so, we must express f as a function of y and r only. This can be accomplished by expressing x as a function of y and r and substituting it into the expression for f. We therefore solve the first equation in (2) for x in terms of y and r and substitute it into the expression for f. Thus

$$f = 3\left(\frac{1 + y - r}{2}\right) + 2y = \frac{3}{2} + \frac{7}{2}y - \frac{3}{2}r$$

Now treating y and r as the basic variables, it is clear from this expression that f can be increased by increasing y from its zero value. Increasing r from its zero value, however, would decrease the value of f. This implies that we should hold r fixed at its zero value and should increase y as much as possible. Geometrically, this means that we should move from the ($y = 0$, $r = 0$) corner to the neighboring corner where $r = 0$ and y is positive, which from Fig. 8 is the ($r = 0$, $s = 0$) corner. Setting $r = 0$ and $s = 0$ in equations (2), we obtain

$$2x - y = 1$$
$$x + 2y = 3$$

Solving these equations, we get $x = 1$, $y = 1$, and $f = 5$. Since the value of f has increased from $\frac{3}{2}$ to 5, this shift has increased the value of f further.

Although we know from our earlier result that we have reached the maximizing corner, we proceed as though we were not aware of this fact. Thus, since our new basic variables are chosen to be r and s, we must express f in terms of r and s. This is accomplished by solving equations (2) for x and y in terms of r and s and substituting those values into f. The solution of equations (2) is given by

$$x = 1 - \frac{2}{5} r - \frac{1}{5} s$$

$$y = 1 + \frac{1}{5} r - \frac{2}{5} s$$

As a result, f assumes the form:

$$f = 3 \left(1 - \frac{2}{5} r - \frac{1}{5} s\right) + 2 \left(1 + \frac{1}{5} r - \frac{2}{5} s\right)$$

$$= 5 - \frac{4}{5} r - \frac{7}{5} s$$

Since r and s cannot be increased from their zero values without decreasing the value of f, it follows that $r = 0$ and $s = 0$ yield the maximum value of f, namely 5.

This technique of shifting to a neighboring corner that increases the value of f, until the maximizing corner has been reached, assumes that it is always possible to arrive at the maximizing corner in this manner. A justification for this assumption can be given that is based on the nature of the feasibility region. Since this property of our feasibility region seems obvious from the geometry of the problem for two-dimensional problems, we do not attempt a justification. The chief advantage of our method is that it eliminates the necessity of finding the coordinates of all the corner points and of checking the value of f at each of them. As we stated previously, the latter method can become a lengthy computational problem in higher dimensions.

Another striking advantage of the simplex method is that the technique can be carried out in a systematic routine manner by means of matrix methods, regardless of the number of variables involved. It is not necessary to perform any of the geometry that was used in our present problem. The geometry was introduced to explain how and why the simplex method works. In the next section we treat it from a purely algebraic point of view.

4. THE ALGEBRA OF THE SIMPLEX METHOD

Equations (2), together with the function f, are first written in this form:

(3)
$$\begin{aligned}
2x - y + r \quad\quad &= 1 \\
x + 2y \quad + s \quad &= 3 \\
-3x - 2y \quad\quad + f &= 0
\end{aligned}$$

In carrying out the following operations it is useful for us to observe that a slack variable, such as r, occurs in only one equation. Or, looking only at the coefficients of a variable in a column, a slack variable column will consist of one 1 and zeros elsewhere. Since the last row merely represents the function f in equation form, we do not treat it like the other rows and, hence, f is not treated as a slack variable here.

Recall that we started with $x = 0$, $y = 0$ and that because the coefficient of x is larger than the coefficient of y in $f = 3x + 2y$ we agreed to leave $y = 0$ alone and increase x as much as possible. Inspecting equations (3), we observe that when $y = 0$ the largest possible value of x in the first equation occurs when $r = 0$ and $x = \frac{1}{2}$, whereas in the second equation it occurs when $s = 0$ and $x = 3$. Thus we can increase x by at most $x = \frac{1}{2}$, which in turn requires that $r = 0$. Since we kept $y = 0$ and now have obtained $r = 0$, this tells us that our new corner point is the point $(y = 0, r = 0)$.

We should now repeat the process treating y and r as our new basic variables and x and s as our new slack variables. But to make x a slack variable and r a basic variable it is necessary to operate on the equations in (3) in such a way that the x column will contain two zeros and f will be a function of y and r only. This requires that we use one of the equations in (3) to eliminate the x terms in the other two equations. If r is to take the place of x as a basic variable, the equation that must be chosen for this elimination is the equation that determined how large x may grow; hence, we must choose the first equation. The technique for carrying out this elimination, that is, for making x a slack variable and r a basic variable, can be accomplished by means of the familiar pivotal reduction method. Toward this objective, we divide the first equation in (3) by 2 to obtain the set:

$$
\begin{aligned}
x - \tfrac{1}{2} y + \tfrac{1}{2} r \quad &= \tfrac{1}{2} \\
x + 2y \qquad + s \quad &= 3 \\
-3x - 2y \qquad\quad + f &= 0
\end{aligned}
$$

We then pivot on the leading x term to work for two zeros in the first column. Thus, we multiply the first equation by -1 and add it to the second equation after which we multiply the first equation by 3 and add it to the third equation to obtain

$$x - \frac{1}{2}y + \frac{1}{2}r \qquad\qquad = \frac{1}{2}$$

(4)
$$\frac{5}{2}y - \frac{1}{2}r + s \qquad = \frac{5}{2}$$

$$-\frac{7}{2}y + \frac{3}{2}r \qquad + f = \frac{3}{2}$$

This shows that x and s now behave like slack variables and that y and r are the new basic variables. If we solve the last equation for f we obtain

$$f = \frac{3}{2} + \frac{7}{2}y - \frac{3}{2}r$$

We are now ready to repeat the preceding operation with ($y = 0$, $r = 0$) as our starting corner point. Since f can be increased by increasing the value of y, but not by increasing the value of r, we inspect the first two equations in (4) to see how large an increase in y is possible when $r = 0$. It follows from the second equation that y cannot exceed the value 1 when $r = 0$ and that this occurs only if $s = 0$. The first equation places no restriction on how large y can become. Thus we hold $r = 0$ and increase y by the amount 1, thereby making $s = 0$. This implies that we shift to the ($r = 0$, $s = 0$) corner. This, in turn, means that r and s are to be the new basic variables and that x and y are to be the new slack variables. To make y a slack variable, it is necessary to eliminate y from all the equations except one. Just as before, the equation that enables us to accomplish this and at the same time make s a new basic variable is the equation that determines how large our variable can grow. This means that we must use the second equation in (4) and pivot on its y term to work for zeros in the y column.

First we multiply the second equation by $\frac{2}{5}$ and obtain the set:

$$x - \frac{1}{2}y + \frac{1}{2}r \qquad\qquad = \frac{1}{2}$$

$$y - \frac{1}{5}r + \frac{2}{5}s \qquad = 1$$

$$-\frac{7}{2}y + \frac{3}{2}r \qquad + f = \frac{3}{2}$$

Multiplying the second equation by $\frac{1}{2}$ and adding it to the first equation, and then multiplying the second equation by $\frac{7}{2}$ and adding it to the third equation, we obtain

$$x \quad +\frac{2}{5}r \qquad\qquad = 1$$

(5)
$$y - \frac{1}{5}r + \frac{2}{3}s \qquad = 1$$

$$\frac{4}{5}r + \frac{7}{5}s + f = 5$$

Now x and y behave like slack variables and r and s behave like basic variables. Solving for f in the last equation, we find

$$f = 5 - \frac{4}{5}r - \frac{7}{5}s,$$

From this equation, it is clear that f cannot be increased by increasing r or s from their zero values. Thus the maximum value of f is 5, and it occurs at the corner $(r = 0, s = 0)$. These values, when substituted into equations (5), show that the maximum value of f occurs when $x = 1$ and $y = 1$.

In carrying out these algebraic manipulations we can save much computational time by using the same compact matrix type notation that was introduced in the pivotal reduction method for solving a set of equations. In the present problem we rewrite equations (3) by merely listing the coefficients of the variables and using vertical and horizontal lines to separate the constants, and the last equation, respectively, from the rest. This gives

②	−1	1	0	0	1
1	2	0	1	0	3
−3	−2	0	0	1	0

It is well to remember that in this notation the numbers in the last row corresponding to the basic variables are the negatives of the coefficients of those variables when f is expressed in terms of them and, therefore, in determining which coefficient is the largest, we must ignore the minus signs.

Since -3 is the largest, in absolute value, of the negative terms in the last row we choose the first column. Then, ignoring the last row, we calculate the two ratios $\frac{1}{2}$ and $\frac{3}{1}$, based on the corresponding numbers in the last column and the first column. These represent the maximum amounts that the first column variable, x, can increase in the two equations if the other variables are to remain nonnegative. Since $\frac{1}{2} < \frac{3}{1}$, we select the first row and pivot on the element in that row that lies in the first column. This element has been encircled. Then we divide this row by 2 and rewrite our equations to obtain

$$
\begin{array}{ccccc|c}
① & -\dfrac{1}{2} & \dfrac{1}{2} & 0 & 0 & \dfrac{1}{2} \\
1 & 2 & 0 & 1 & 0 & 3 \\
\hline
-3 & -2 & 0 & 0 & 1 & 0
\end{array}
$$

Performing the usual pivotal reduction with ① as the pivotal element, we obtain

$$
\begin{array}{ccccc|c}
1 & -\dfrac{1}{2} & \dfrac{1}{2} & 0 & 0 & \dfrac{1}{2} \\
0 & \dfrac{5}{2} & -\dfrac{1}{2} & 1 & 0 & \dfrac{5}{2} \\
\hline
0 & -\dfrac{7}{2} & \dfrac{3}{2} & 0 & 1 & \dfrac{3}{2}
\end{array}
$$

It should be observed that this is merely a compact way of writing equations (4).

Since $-\dfrac{7}{2}$ is the only negative term in the third row, we must select the second column. We need to calculate only positive ratios to determine how large a variable may grow, which means we need to look only at the second row where we have the positive ratio $\dfrac{5}{2} \div \dfrac{5}{2} = 1$. We, therefore, pivot on the element in the second row and second column. This element has been encircled. We multiply the second row by $\dfrac{2}{5}$ and rewrite our equations to obtain

$$
\begin{array}{ccccc|c}
1 & -\dfrac{1}{2} & \dfrac{1}{2} & 0 & 0 & \dfrac{1}{2} \\
0 & ① & -\dfrac{1}{5} & \dfrac{2}{5} & 0 & 1 \\
\hline
0 & -\dfrac{7}{2} & \dfrac{3}{2} & 0 & 1 & \dfrac{3}{2}
\end{array}
$$

Pivoting on the encircled element (1) gives

$$
\begin{array}{ccccc|c}
1 & 0 & \dfrac{2}{5} & 0 & 0 & 1 \\[2ex]
0 & 1 & -\dfrac{1}{5} & \dfrac{2}{5} & 0 & 1 \\[2ex]
\hline
0 & 0 & \dfrac{4}{5} & \dfrac{7}{5} & 1 & 5
\end{array}
$$

It will be observed that this is merely a compact way of writing equations (5). Since there are no negative terms in the last row, it is not possible to increase f further and the maximum value has been reached. This maximum value is the number occurring in the last row and last column. Furthermore, the values of x and y that produce this maximum value are also displayed in the last column.

The preceding technique will now be applied to a typical problem involving three variables. Let us modify the problem that was introduced in Section 2 by assuming that the manufacturer produces three types of toys A, B, and C, and that the plant capacities in the three departments are 80 man-hours, 100 man-hours, and 40 man-hours, respectively. Small numbers are chosen here to simplify the computations. To make them seem more realistic, they may be treated as daily capacities. Suppose the amounts of man-hours required in each department to produce one unit of each toy are as follows:

	Machining	Assembly	Painting
A	.4	.3	.1
B	.2	.4	.2
C	.3	.2	.1

Assume that these three types of toys can be sold for $10, $8, and $7 per unit, respectively. Let x, y, and z denote the number of units of types A, B, and C, respectively, that are to be manufactured. Then the problem is to maximize the revenue function

$$f = 10x + 8y + 7z$$

subject to the restrictions

$$.4x + .2y + .3z \leqslant 80$$
$$.3x + .4y + .2z \leqslant 100$$
$$.1x + .2y + .1z \leqslant 40$$
$$x \geqslant 0,\, y \geqslant 0,\, z \geqslant 0$$

Multiplying the first three inequalities through by 10 and introducing three slack variables, we replace these restrictions by the following equivalent ones:

(6)
$$4x + 2y + 3z + r = 800$$
$$3x + 4y + 2z + s = 1000$$
$$x + 2y + z + t = 400$$
$$x \geqslant 0,\, y \geqslant 0,\, z \geqslant 0,\, r \geqslant 0,\, s \geqslant 0,\, t \geqslant 0$$

The simplex technique then starts with the form

④	2	3	1	0	0	800
3	4	2	0	1	0	1000
1	2	1	0	0	1	400
−10	−8	−7	0	0	0	0

It should be observed that the first three columns correspond to basic variables, that the next three columns represent slack variables, and that the column corresponding to f has been omitted because it serves no useful purpose in the calculations. Since -10 is the largest, in absolute value, of the negative terms in the last row, we choose the first column and calculate the ratios of the corresponding elements of the last column and the first column. Thus

$$\frac{800}{4} = 200, \qquad \frac{1000}{3} = 333\,\tfrac{1}{3}, \qquad \frac{400}{1} = 400$$

The first ratio is the smallest; therefore, we choose the element in the first row and first column on which to pivot. It has been encircled. Dividing the first row by 4, we obtain

①	$\frac{1}{2}$	$\frac{3}{4}$	$\frac{1}{4}$	0	0	200
3	4	2	0	1	0	1000
1	2	1	0	0	1	400
−10	−8	−7	0	0	0	0

Pivoting on ① yields

1	$\frac{1}{2}$	$\frac{3}{4}$	$\frac{1}{4}$	0	0	200
0	$\frac{5}{2}$	$-\frac{1}{4}$	$-\frac{3}{4}$	1	0	400
0	$\left(\frac{3}{2}\right)$	$\frac{1}{4}$	$-\frac{1}{4}$	0	1	200
0	-3	$\frac{1}{2}$	$\frac{5}{2}$	0	0	2000

An inspection of the last row shows that f has increased from 0 to 2000. This inspection also shows that -3 is the only negative term in the last row; therefore, we must select the second column and calculate the ratios

$$\frac{200}{\frac{1}{2}} = 400, \qquad \frac{400}{\frac{5}{2}} = 160, \qquad \frac{200}{\frac{3}{2}} = 133\frac{1}{3}$$

Since the last ratio is the smallest, we choose the element in the third row and second column on which to pivot. It has been encircled. Multiplying the third row by $\frac{2}{3}$, we obtain

1	$\frac{1}{2}$	$\frac{3}{4}$	$\frac{1}{4}$	0	0	200
0	$\frac{5}{2}$	$-\frac{1}{4}$	$-\frac{3}{4}$	1	0	400
0	①	$\frac{1}{6}$	$-\frac{1}{6}$	0	$\frac{2}{3}$	$\frac{400}{3}$
0	-3	$\frac{1}{2}$	$\frac{5}{2}$	0	0	2000

Pivoting on ① gives

1	0	$\frac{2}{3}$	$\frac{1}{3}$	0	$-\frac{1}{3}$	$\frac{400}{3}$
0	0	$-\frac{2}{3}$	$-\frac{1}{3}$	1	$-\frac{5}{3}$	$\frac{200}{3}$
0	1	$\frac{1}{6}$	$-\frac{1}{6}$	0	$\frac{2}{3}$	$\frac{400}{3}$
0	0	1	2	0	2	2400

Since the last row contains no negative terms, f cannot be increased further. Its maximum value is 2400. When it is written out in algebraic form, the last equation becomes

$$f = 2400 - z - 2r - 2t$$

The maximum occurs for $z = 0$, $r = 0$, and $t = 0$. By inserting these values into equations (6) and solving for the remaining unknowns, we obtain the desired solution:

$$x = \frac{400}{3}, \, y = \frac{400}{3}, \quad \text{and} \quad z = 0$$

Thus approximately 133 units of type A and 133 units of type B should be produced, and type C should be discontinued.

It is not necessary to substitute the values $z = 0$, $r = 0$, and $t = 0$ into equations (6) and solve for the remaining unknowns because the solutions can be read off the final simplex configuration. If the equations corresponding to this configuration were written out, we would have

$$x + \quad \frac{2}{3}z + \frac{1}{3}r \quad - \frac{1}{3}t = \frac{400}{3}$$

$$-\frac{2}{3}z - \frac{1}{3}r + s - \frac{5}{3}t = \frac{200}{3}$$

$$y + \frac{1}{6}z - \frac{1}{6}r \quad + \frac{2}{3}t = \frac{400}{3}$$

Setting $z = 0$, $r = 0$, and $t = 0$, we obtain

$$x = \frac{400}{3}, \, s = \frac{200}{3}, \quad \text{and} \quad y = \frac{400}{3}$$

But we can obtain this last set of values by merely deleting the columns of the matrix corresponding to z, r, and t and then reading off the solutions, $x = \frac{400}{3}$, $s = \frac{200}{3}$, and $y = \frac{400}{3}$ that are found in the last column of our matrix configuration.

The preceding condensed algebraic technique for carrying out the simplex method must be modified slightly for minimizing problems. Although it is true that the problem of minimizing a function f can be changed to the problem of maximizing the function g, where $g = -f$, the origin will never be a corner point of the solution region in any realistic minimizing problem and, therefore, we cannot start the simplex technique at that point.

This difficulty can be overcome by first employing the pivotal reduction method to reduce one or more of the basic variables to slack variables and by changing a corresponding number of slack variables into basic variables. If zero values of the new set of basic variables are compatible with the restriction equations, we can carry out the simplex technique in the usual manner. It is usually necessary to introduce only one or two new basic variables before a beginning can be made; however, on occasion all the original basic variables must be changed to slack variables before the restriction equations can be satisfied. In reducing original basic variables to slack variables, it is important to refrain from pivoting on an element in a row that was used before for pivoting because that will usually bring back an original basic variable that had been eliminated previously. In selecting an element on which to pivot, choose one that gives rise to simple computations. The simplex rule of choosing the element with the smallest ratio applies only after a starting corner point has been attained, and then it only applies to elements in a column for which there is a negative term in the bottom row. After a valid corner point has been found, we proceed in the same manner as for a maximization problem.

As an illustration of how a minimizing problem is treated, consider the following problem. A coal company owns two mines. The first mine can produce at most 1 ton of high grade coal, 4 tons of medium grade coal, and 6 tons of low grade coal per day. The second mine can produce at most 4 tons of high grade coal, 4 tons of medium grade coal, and 2 tons of low grade coal per day. It costs the company $100 a day to operate the first mine and $150 a day to operate the second mine. The company has an order for 80 tons of high grade coal, 160 tons of medium grade coal, and 120 tons of low grade coal. The problem is to determine how many days each mine should be operated to minimize operational costs.

Let x and y denote the number of days the respective mines need to be operated to fill this order. Then the function to be minimized is

$$f = 100x + 150y$$

These variables are subject to the restrictions

$$x + 4y \geqslant 80$$
$$4x + 4y \geqslant 160$$
$$6x + 2y \geqslant 120$$
$$x \geqslant 0, \; y \geqslant 0$$

To reduce this problem to a maximizing problem, we set $g = -f$ and proceed to maximize the function

$$g = -100x - 150y$$

subject to the same set of restrictions. We introduce the nonnegative slack variables r, s, and t, which allow us to write the restrictions in the form

$$x + 4y = 80 + r$$
$$4x + 4y = 160 + s$$
$$6x + 2y = 120 + t$$
$$x \geqslant 0, \, y \geqslant 0, \, r \geqslant 0, \, s \geqslant 0, \, t \geqslant 0$$

It should be noted that a nonnegative slack variable must be added to the smaller side of an inequality. These relations, when written in standard form, become

(7)
$$\begin{aligned} x + 4y - r \qquad\qquad &= 80 \\ 4x + 4y \qquad - s \qquad &= 160 \\ 6x + 2y \qquad\qquad - t &= 120 \end{aligned}$$

$$x \geqslant 0, \, y \geqslant 0, \, r \geqslant 0, \, s \geqslant 0, \, t \geqslant 0$$

The corresponding matrix formulation is

1	4	−1	0	0	80
4	4	0	−1	0	160
6	2	0	0	−1	120
100	150	0	0	0	0

Since there are no negative elements in the last row, it might appear that we have obtained our maximum value at $x = 0$, $y = 0$; however, $x = 0$, $y = 0$ is not a corner point of the feasibility region because the restriction equations (7) cannot be satisfied for those values of x and y. To obtain a legitimate starting corner point, we reduce x to being a slack variable and bring in r as a basic variable. That can be done by using the pivotal reduction technique on the element in the first row and first column of our matrix formulation. This gives

1	4	−1	0	0	80
0	−12	4	−1	0	−160
0	−22	6	0	−1	−360
0	−250	100	0	0	−8,000

Our new basic variables are y and r. Observe that the last row now contains a negative element and, therefore, that g has not attained its maximum value for $y = 0$ and $r = 0$. Writing the last row in equation form, we get

$$g = -8000 + 250y - 100r$$

Thus g can be increased by increasing the value of y from its zero value. An inspection of the restriction equations shows that $y = 0$, $r = 0$ does not contradict those equations, as was true earlier with $x = 0$, $y = 0$; consequently, the point $y = 0$, $r = 0$ is a legitimate corner point of the feasibility region. Therefore, we can begin the simplex technique at that point.

The usual simplex technique now proceeds as follows. Since -250 is the only negative element in the last row, exclusive of the constant term on the right, we calculate the positive ratios

$$\frac{80}{4} = 20, \qquad \frac{-160}{-12} = 13\frac{1}{3}, \qquad \frac{-360}{-22} = 16\frac{4}{11}$$

The second ratio is the smallest; therefore, we pivot on the element in the second row and second column. To do so, we first divide the second row by -12 to obtain

1	4	-1	0	0	80
0	①	$-\frac{1}{3}$	$\frac{1}{12}$	0	$\frac{40}{3}$
0	-22	6	0	-1	-360
0	-250	100	0	0	$-8{,}000$

We then pivot on the element ① and obtain

1	0	$\frac{1}{3}$	$-\frac{1}{3}$	0	$\frac{80}{3}$
0	1	$-\frac{1}{3}$	$\frac{1}{12}$	0	$\frac{40}{3}$
0	0	$-\frac{4}{3}$	$\frac{11}{6}$	-1	$\frac{-200}{3}$
0	0	$\frac{50}{3}$	$\frac{125}{6}$	0	$\frac{-14{,}000}{3}$

Since the last row contains no negative elements, exclusive of the constant term on the right, the maximum value of g has been attained. The last row, when written in equation form, becomes

$$g = -\frac{14,000}{3} - \frac{50}{3} r - \frac{125}{6} s$$

Its maximum value is $-\frac{14,000}{3}$ and is attained when $r = 0$ and $s = 0$. The negative of this value is therefore the minimum value of f. Inspection of the last column of our matrix configuration shows that the solution is given by $x = \frac{80}{3}$ and $y = \frac{40}{3}$. The cost of filling the order will, therefore, be minimized if the company runs the first mine for $26\frac{2}{3}$ days and the second mine for $13\frac{1}{3}$ days. The total production costs will then be $4666 $\frac{2}{3}$.

It may happen that the set of linear inequalities in a linear programming problem contain one or more inequalities that are already satisfied if the remaining inequalities are satisfied. These may be deleted in determining the feasibility region; however, they are not easily found when the problem involves many variables. There is usually no harm done if such redundancies are ignored. It is theoretically possible under these and other circumstances for the simplex method to cycle around without leading to the maximizing corner point. This possibility, however, seldom occurs in practice. Furthermore, there are simple devices to get around the difficulty of cycling if it should occur.

COMMENTS

This chapter has been devoted to studying how matrix methods can be used to help solve an important linear systems problem of business. That problem, which is one of maximizing or minimizing a linear function of several variables when those variables are subject to a number of linear constraints, was solved by geometrical methods for two variable systems and by the simplex technique for larger systems. The simplex technique is based on interchanging basic and slack variables until a solution is reached. This is accomplished by using the pivotal reduction method of Chapter 3.

The illustrations and the exercises clearly show that linear programming is a powerful method for increasing the efficiency of many business operations. The few simple problems that can be solved in the classroom are merely an indication of the significant possibilities in this direction. Among the important applications are the transportation problem, the optimal assignment problem, and the resource allocation problem. One of the exercises dealing with the shipment of goods from different warehouses is an illustration of a transportation problem. The optimal assignment problem is concerned with how to assign a set of m workers to a set of n jobs in the most efficient manner possible. The resource allocation problem is concerned with how to distribute the various resources of an economy so as to maximize the total gross return to the economy. All such problems can be formulated as linear programming problems.

EXERCISES

Section 1

1. Determine by graphical methods the region in the x, y plane where each of the following pairs of inequalities is satisfied.
   (a) $x + y < 3$
       $2x - y < 2$
   (c) $x + 2y < 4$
       $x - 4 > -1$
   (b) $2x + y > 3$
       $y - 3x < 1$
   (d) $2x + 3y \leqslant 6$
       $3x - 2y \geqslant 4$

2. Determine by graphical methods the region in the x, y plane where each of the following sets of inequalities is satisfied.
   (a) $2x - y < 4$, $2x + y < 6$, $x - 2y < 5$, $x > 0$.
   (b) $x + y < 3$, $2x - y < 2$, $y - 3x < 1$, $x > 0$, $y > 0$.

3. Determine by graphical methods the region in the x, y plane where each of the following sets of inequalities is satisfied.
   (a) $x + 2y \geqslant 4$, $2x + y \geqslant 3$, $x - y \leqslant 3$, $x \geqslant 0$, $y \geqslant 0$.
   (b) $x + y \geqslant 2$, $3x + y \geqslant 3$, $x + 3y \geqslant 3$, $x - 4y \leqslant 1$, $x \geqslant 0$, $y \geqslant 0$.

4. A food mixture for chickens is to be made using two basic foods, A and B. Each unit of food A weighs 4 ounces and contains 2 ounces of protein. Each unit of food B weighs 5 ounces and contains 1 ounce of protein. The mixture is to weigh at most 40 ounces and must contain at least 10 ounces of protein. Let x and y denote the units of foods A and B to use

and write down the inequalities that x and y must satisfy. Determine graphically the values of x and y that will satisfy those inequalities.

5. A nut company has purchased 2000 pounds of peanuts and 4000 pounds of mixed nuts. It sells three qualities of nuts. The best quality contains four times as many mixed nuts as peanuts, the middle quality contains equal portions of peanuts and mixed nuts, and the third quality contains twice as many peanuts as mixed nuts. Let x, y, and z denote the number of pounds of each of the three mixtures that can be made from the supply on hand and write down the restrictions that those variables must satisfy.

Section 2

1. By graphical methods find the maximum value of the function $f = 4x + 3y$ subject to the restrictions

$$2x + y \leqslant 10, \quad x + y \leqslant 6, \quad x \geqslant 0, \quad y \geqslant 0$$

2. By graphical methods find the maximum value of the function $f = 3x + 4y$ subject to the restrictions

$$2x + y \leqslant 7, \quad 3x + y \leqslant 9, \quad x \geqslant 0, \quad y \geqslant 0$$

3. By graphical methods find the minimum value of the function $f = 2x + 3y$ subject to the restrictions

$$3x + 2y \geqslant 6, \quad x + 3y \geqslant 6, \quad x \geqslant 0, \quad y \geqslant 0$$

4. By graphical methods find the minimum value of the function $f = 2x + y$ subject to the restrictions

$$x + y \geqslant 4, \quad -2x + 3y \geqslant 6, \quad x \geqslant 0, \quad y \geqslant 0$$

5. By graphical methods find the maximum value of the function $f = 2x + y$ subject to the restrictions

$$2x + 2y \leqslant 3, \quad 4x + y \leqslant 4, \quad x + 4y \leqslant 4, \quad x \geqslant 0, \quad y \geqslant 0$$

6. Add the restriction $12x \leqslant 11$ to the set of restrictions in Problem 5 and then solve it.

7. Assume that food A of Problem 4, Section 1, costs 30 cents a unit and that food B costs 10 cents a unit. Use the results of that problem to determine by graphical methods the values of x and y that will minimize the cost of the mixture to use.

Section 3

1. For Problem 1, Section 2, introduce the two slack variables r and s and label the corner points of the feasibility region in terms of zero values of the four variables x, y, r, and s.

2. For Problem 2, Section 2, introduce the two slack variables r and s and label the corner points of the feasibility region in terms of zero values of the four variables x, y, r, and s.

3. For Problem 5, Section 2, introduce the three slack variables r, s, and t and label the corner points of the feasibility region in terms of zero values of the five variables x, y, r, s, and t.

4. For Problem 6, Section 2, introduce the four slack variables r, s, t, and w and label the corner points of the feasibility region in terms of zero values of the six variables x, y, r, s, t, and w.

5. Solve the linear programming problem associated with Problem 1 by the simplex method starting at the corner $(x = 0, y = 0)$.

6. Solve the linear programming problem associated with Problem 2 by the simplex method starting at the corner $(x = 0, y = 0)$.

7. Solve the linear programming problem associated with Problem 3 by the simplex method starting at the corner $(x = 0, y = 0)$.

8. Solve the linear programming problem associated with Problem 4 by the simplex method starting at the corner $(x = 0, y = 0)$.

9. In Problem 5 suppose that you were to start at the corner where the line $x + y = 6$ cuts the y axis. Explain without going through any calculations why the next corner could not be $(x = 0, y = 0)$.

Section 4

1. Work Problem 5, Section 3, by using the condensed matrix technique for the simplex method.

2. Work Problem 6, Section 3, by using the condensed matrix technique for the simplex method.

3. Work Problem 7, Section 3, by using the condensed matrix technique for the simplex method.

4. Work Problem 8, Section 3, by using the condensed matrix technique for the simplex method.

5. Use the condensed matrix technique for the simplex method to maximize the function $f = 3x + y + 2z$ subject to the restrictions

$$x + 2y - z \leqslant 8$$
$$2x - y + 3z \leqslant 2$$
$$-x + 3y + 4z \leqslant 4$$
$$x \geqslant 0, \ y \geqslant 0, \ z \geqslant 0$$

6. Use the condensed matrix technique for the simplex method to minimize the function $f = x + 4y + 5z$ subject to the restrictions

$$x + y + z \geqslant 3$$
$$-x + 2y + 2z \geqslant 2$$
$$y + 2z \geqslant 4$$
$$x \geqslant 0, \ y \geqslant 0, \ z \geqslant 0$$

Since this is a minimization problem, $(x = 0, y = 0, z = 0)$ cannot be used as a starting point. To obtain a start, make z into a slack variable by pivoting on the z term in the first row.

7. A diet for a type of domestic animal is to be designed by using two kinds of foods. Food I contains 3 units of protein, 4 units of carbohydrates, and 2 units of fat. Food II contains 2 units of protein, 2 units of carbohydrates, and 4 units of fat. The two foods cost 50 cents and 30 cents per pound, respectively. Suppose an animal requires at least 10 units of protein, 12 units of carbohydrates, and 8 units of fat per week.
(a) Let x and y denote the amounts of the two foods needed and write down the total cost function, f, in terms of them. Write down the restrictions that must be satisfied.
(b) Use the condensed matrix simplex method to find the values of x and y that minimize f. Since this is a minimizing problem, $(x = 0, y = 0)$ cannot be used as a starting point. Try making y a slack variable.

8. A coal company owns three mines. The daily production capacity of each mine in terms of the number of tons of the three grades of coal obtained by running the mine in the normal fashion for one day is given in the following table:

Quality	Mine I	Mine II	Mine III
High grade	2	6	4
Medium grade	3	2	6
Low grade	8	3	2

The daily operating costs for these three mines are \$100, \$120, and \$140, respectively. An order comes in for 20 tons of high grade coal, 20 tons of medium grade coal, and 30 tons of low grade coal. Use the condensed simplex technique to determine how many days each mine should be operated to minimize the production costs for this order.

9. Suppose a company has two factories F_1 and F_2 and two warehouses W_1 and W_2, and that factory F_1 has 100 tons of a product on hand and factory F_2 has 120 tons on hand. Suppose further that warehouse W_1 needs at least 80 tons and warehouse W_2 needs at least 60 tons of the product to keep up with its orders. The shipping costs per ton of product are

$$F_1 \text{ to } W_1 : \$3, \qquad F_2 \text{ to } W_1 : \$4$$
$$F_1 \text{ to } W_2 : \$5, \qquad F_2 \text{ to } W_2 : \$7$$

Let x, y, z, and w denote the respective amounts in tons to be shipped from F_1 to W_1, F_1 to W_2, F_2 to W_1, and F_2 to W_2, and let f denote the total shipping costs.

(a) Express f as a function of the four variables and write down the restrictions those variables must satisfy.

(b) Use the condensed simplex technique to determine the values of those variables that will minimize f. It will be necessary to change several of the basic variables to slack variables before the feasibility region is reached. Check to see whether the restrictions can be satisfied before proceeding with the simplex technique. *Hint:* try changing y and z to slack variables before checking.

5 | Probability

Many experiments in the various sciences are of the type in which essentially the same outcome occurs each time the experiment is run. Most experiments, however, yield results that differ in an unpredictable way when repeated runs of the experiment are made. These experiments are often called random experiments. Games of chance, such as tossing a coin, rolling a pair of dice, or spinning a roulette wheel are examples of experiments for which the outcome is unpredictable and which, therefore, would be classified as random. An experiment that consists of a newspaper reporter's asking individuals passing him at a street corner whether they favor a bill being discussed by the city council is also a random type experiment.

Although it is not possible to predict the outcome of a random type experiment, experience has shown that these experiments often have a certain amount of long-run regularity. For example, although it is not possible to predict whether a coin will come up heads or tails when the coin is tossed once, experience ususally shows that about half of the tosses will produce heads when a number of tosses are made. Thus it is possible to make certain kinds of predictions when the experiment is run many times even though it is not possible to predict the outcome of any one of the individual experiments.

Probability is a branch of mathematics that deals with random experiments of the preceding type. Its objective is to make predictions about the long-run outcomes of such experiments. The preceding illustrations of random experiments are quite academic; however, the social sciences abound in illustrations of important random experiments to which probability can make useful contributions. For the present we limit ourselves to illustrations from games of chance because they are much simpler to discuss than the more sophisticated practical experiments.

In developing the theory of probability it is convenient to discuss only experiments that can be repeated or that can be conceived of as being repeatable. Tossing a coin, reading the daily temperature on a thermometer, or counting the number of bad eggs in a carton are examples of a simple repetitive experiment. An experiment in which several rabbits are fed different rations in an attempt to determine the relative growth properties of rations may be performed only once with those same animals; nevertheless, we may conceive of this experiment as the first in an unlimited number of similar experiments and, therefore, it may be considered as being repeatable.

Consider a simple repetitive experiment such as tossing a coin twice or, what is equivalent, tossing two distinct coins simultaneously. In this experiment there are four possible outcomes of interest, which we denote by

$$HH, HT, TH, TT$$

The symbol HT, for example, means that a head is obtained on the first toss and a tail on the second toss. If the experiment had consisted of tossing the coin three times, there would have been eight possible outcomes of the experiment, which would be represented by

$$HHH, HHT, HTH, THH, HTT, THT, TTH, TTT$$

The three letters in a group express the outcomes of the three tosses in the given order. An experiment such as reading the temperature on a thermometer, however, has conceivably an infinite number of possible outcomes. Since it is not possible to read a thermometer beyond a certain degree of accuracy, even this experiment is realistically one in which there are only a finite number of possibilities. In this discussion of probability we consider only experiments with a finite number of possible outcomes.

For any experiment to which probability is to be applied, it is first necessary to decide what possible outcomes of the experiment are of interest, and to make a list of all such outcomes. This list must be such that when the experiment is performed, exactly one of the outcomes will occur. In the experiment of tossing a coin three times, interest was centered on whether the coin showed a head or a tail on each of the tosses; therefore, all the possible outcomes are those that were listed previously. A game of chance experiment that will be used frequently for illustrative purposes is the experiment of drawing a ball from a box of balls of different colors. Thus, suppose a box contains three red, two black, and one green ball. Then we are interested in knowing what color a drawn ball is and not in knowing which particular ball is obtained. Here, there are three possible outcomes of the experiment corresponding to the three colors. Another interesting game of chance experiment is the experiment of rolling two dice. If it is assumed that we can distinguish between the two dice and that interest centers on what number of points shows on each of the dice, then there are 36 possible outcomes, because each die has six possible outcomes and these outcomes can be paired in all possible ways. Table 1 gives a list of the possible outcomes.

TABLE 1

11	21	31	41	51	61
12	22	32	42	52	62
13	23	33	43	53	63
14	24	34	44	54	64
15	25	35	45	55	65
16	26	36	46	56	66

The first number of each pair denotes the number that came up on one of the dice, and the second number denotes the number that came up on the other. If the two dice are not distinguishable, it is necessary to roll them in order instead of rolling them simultaneously.

1. THE GEOMETRY OF PROBABILITY

It is convenient in developing the theory of probability to visualize things geometrically and to represent each of the possible outcomes of an experiment by means of a point. Thus in the experiment of tossing a coin three times we would use eight points. It makes no difference what points are chosen as long as we know which point corresponds to which possible outcome. Each point is labeled with a letter or synbol to indicate the outcome that it represents. Since possible outcomes are often called simple events, the letter e with a subscript corresponding to the number of the outcome in the list of possible outcomes is customarily used in labeling a point. In the coin-tossing experiment, for example, we could label the eight points by means of $e_1, e_2, \ldots, e_8$. Thus e_1 would represent the event of obtaining HHH and e_2 that of obtaining HHT. Since a label such as HHT is self-explanatory, there seems little point in introducing additional labels here. However, it is easier to write e_2 than HHT; consequently, the e symbol does possess an advantage. It is convenient in developing the rules of probability to give a name to this basic set of points. This is done formally as follows:

Definition. The set of points representing the possible outcomes of an experiment is called the sample space for the experiment.

A sample space for the coin-tossing experiment is shown in Fig. 1, in which the numbering of the points is indicated by the symbols directly above the points.

$$\begin{array}{cccccccc}
\text{HHH} & \text{HHT} & \text{HTH} & \text{THH} & \text{HTT} & \text{THT} & \text{TTH} & \text{TTT} \\
\bullet & \bullet & \bullet & \bullet & \bullet & \bullet & \bullet & \bullet \\
e_1 & e_2 & e_3 & e_4 & e_5 & e_6 & e_7 & e_8
\end{array}$$

FIGURE 1   A sample space for a coin-tossing experiment.

For the experiment of drawing a colored ball from the box of balls described earlier, the sample space consists of three points, which have been labeled in the order of the colors listed and which are shown in Fig. 2.

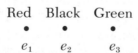

$$\begin{array}{ccc}
\text{Red} & \text{Black} & \text{Green} \\
\bullet & \bullet & \bullet \\
e_1 & e_2 & e_3
\end{array}$$

FIGURE 2   A sample space for a colored ball experiment.

A convenient sample space for the experiment of rolling two dice is a double array of 36 points, with six rows and six columns, attached to the elements of Table 1. This sample space is not illustrated here because Table 1 and your imagination should suffice. The symbols of Table 1 are highly descriptive of the experimental outcomes and are also very simple; there is no point in introducing a new set of symbols.

Thus far there has been no apparent reason for introducing a geometrical representation of the outcomes of an experiment; however, later in the development of the theory the advantage of this approach will be evident.

The next step in the construction of a mathematical model for an experiment is to attach numbers to the points in the sample space that represent the relative frequencies with which those outcomes are expected to occur. If the experiment of tossing a coin three times were repeated a large number of times and a cumulative record kept of the proportion of those experiments that produced, say, three heads, we would expect that proportion to approach $\frac{1}{8}$ because each of the eight possible outcomes would be expected to occur about equally often. Actual experiments of this kind usually show that such expectations are justified, provided the coin is well balanced and is tossed vigorously. In view of these considerations, we should attach the number $\frac{1}{8}$ to each of the points in the sample space shown in Fig. 1. The number assigned to the point labeled e_i in a sample space is called the probability of the event e_i and is denoted by the symbol $P\{e_i\}$. Thus in the coin-tossing experiment each of the events $e_1, e_2, \ldots, e_8$ would possess the probability $\frac{1}{8}$.

The situation for the experiment corresponding to the sample space shown in Fig. 2 is somewhat different from the preceding one. It is no longer true that each of the possible outcomes would be expected to occur with the same relative frequency. If the balls are well mixed in the box before each drawing and the drawn ball is always returned to the box so that the composition of the box is unchanged, we should expect to obtain a black ball twice as often as a green ball and a red ball three times as often as a green ball. This implies that in repeated sampling experiments we should expect the relative frequencies for the three colors red, black, and green to be close to $\frac{3}{6}, \frac{2}{6}$, and $\frac{1}{6}$, respectively.

The experiment of rolling two dice is treated in much the same manner as the coin-tossing experiment. Symmetry and experience suggest that each point in the sample space corresponding to Table 1 should be assigned the probability $\frac{1}{36}$.

The preceding experiments illustrate how one proceeds in general to assign probabilities to the points of a sample space. If the experiment is one for which symmetry and similar considerations suggest what relative frequencies are to be expected for the various outcomes, then those expected relative frequencies are chosen as the probabilities for the corresponding points. This was the basis for the assignment of probabilities in the coin-tossing experiment, the colored ball experiment, and the die-rolling experiment. If no such symmetry considerations are available but experience with the given experiment is available, then the relative frequencies obtained from this experience can be used for the probabilities to be assigned.

Since the probabilities assigned to the points of a sample space are either the expected relative frequencies based on symmetry considerations or the long-run experimental relative frequencies, probabilities must be numbers between 0 and 1 and their sum must be 1, because the sum of a complete set of relative frequencies is always 1. In the experiments related to coin tossing, colored balls, and dice rolling, the probabilities obviously sum to 1 because they were constructed that way.

Now, in any given experimental situation, whether academic or real, it is the privilege of the experimenter to assign any probabilities he desires to the possible outcomes of the experiment, provided that they are numbers between 0 and 1 and provided that they sum to 1. Of course, if he is sensible he tries to assign numbers that represent what he believes or knows to be the long-run relative frequencies for those outcomes, otherwise his mathematical model is not likely to represent the actual experiment satisfactorily from a frequency point of view and, therefore, his conclusions derived on the basis of the model are likely to be erroneous.

It is usually quite easy to assign satisfactory probabilities to the possible outcomes of games of chance; however, this is not true for most real-life experiments. For example, if the experiment consists in selecting an individual at random from the population of a city and interest is centered on whether the individual will die during the ensuing year, then there is no satisfactory way of assigning a probability to this possibility other than by using the experience of insurance companies. If we were interested in determining proper insurance premiums, it would be necessary for us to assign probabilities of death at the various ages. These are usually chosen to be the values obtained from extensive experience of insurance companies over the years. Since mortality rates have been decreasing over the years for most age groups, any mortality table based on past experience is likely to be out of date for predicting the future. Thus the probabilities assigned on the basis of past experience may not be very close to the actual relative frequencies existing today and, therefore, the premiums calculated from them will not be very accurate. Fortunately for the insurance companies, premiums calculated on the basis of past experience are larger than they would be if they had been based on more up-to-date experience.

In view of the foregoing discussion, it follows that the probability of a simple event is to be interpreted as a theoretical, or idealized, relative frequency of the event. This does not mean that the observed relative frequency of the event will necessarily approach the probability of the event for an increasingly large number of experiments because we may have chosen an incorrect model; however, we hope that it will. Thus, if we have a supposedly honest die, we would hope that the observed relative frequency of, say, a four showing will approach the probability $\frac{1}{6}$ as an increasingly large number of rolls is made; but we should not be too upset if it did not approach $\frac{1}{6}$ because of the imperfections in any manufactured article and because of the difficulty of simulating an ideal experiment. Here, it should be observed that the operators of gambling houses have done well financially by assuming that dice do behave as expected.

Constructing theoretical models to explain nature is the chief function of scientists. If the models are realistic, the conclusions derived from them are likely to be realistic. A probability model is such a model designed to enable us to draw conclusions about relative frequencies of experimental outcomes.

Although probability is being interpreted here in terms of expected relative frequency, it is also used by some individuals as a measure of their degree of belief in the future occurrence of an event, even when there is no conceptually repeatable experiment involved. Thus they might be willing to assign a probability to the event that there will be no major war in the world during the next 10 years. Such a probability should correspond to the betting odds that the individual will give that a war will not occur. Probabilities of this type, which are called personal, or subjective, are frequently employed in business decision-making problems. It is perfectly proper for a businessman to assign probabilities that correspond to his betting odds in a given problem and treat them as probabilities of the same type as those that we assign in games of chance. However, in doing so, he must realize that the conclusions arrived at, which are based on those probabilities, are only as reliable as his probability assignments are realistic. Thus the essential difference between probability models applied to games of chance and business situations is in the reliability of the assigned probabilities and the validity of the model. Games of chance models have had a lot of successful experience to justify their use.

2. PROBABILITY OF AN EVENT

Now that a geometrical model has been constructed for an experiment, consisting of a set of points with labels $e_1, e_2, \ldots$ to represent all the possible outcomes and a corresponding associated set of probabilities $P\{e_1\}$, $P\{e_2\}, \ldots$, the time has come to discuss the probability of composite events. The possible outcomes $e_1, e_2, \ldots$ .of a sample space are called *simple events*. A *composite event* is defined as a collection of simple events. For example, the event of obtaining exactly two heads in the coin-tossing experiment of Fig. 1 is a composite event that consists of the three simple events e_2, e_3, and e_4. Composite events are usually denoted by capital letters such as A, B, or C.

Now since the probabilities assigned to the simple events of Fig. 1, namely $\frac{1}{8}$, represent expected relative frequencies for their occurrence, we should expect the composite event consisting of the simple events e_2, e_3, and e_4 to occur in about three eights of such experiments in the long run of such experiments. Thus, we should expect to sum the probabilities of the simple events. In view of these expectations, and because probability has been introduced as an idealization of relative frequency, the following definition of probability for a composite event should seem very reasonable.

(1) **Definition.** *The probability that a composite event A will occur is the sum of the probabilities of the simple events of which it is composed.*

As an illustration, if A is the event of obtaining two heads in tossing a coin three times, it follows from this definition and the sample space in Fig. 1 that

$$P\{A\} = P\{e_2\} + P\{e_3\} + P\{e_4\} = \frac{3}{8}$$

As another illustration, let B be the event of getting a red or a green ball in the experiment for which Fig. 2 is the sample space. Since B is composed of the events e_1 and e_3, it follows that

$$P\{B\} = P\{e_1\} + P\{e_3\} = \frac{4}{6}$$

As a final illustration for which the composite events are not quite so obvious, consider once more the experiment of rolling two dice. Table 1, with points associated with each outcome and with the probability $\frac{1}{36}$ attached to each point, can serve as the sample space here. First, let C be the event of getting a total of seven points on the two dice. The simple events that yield a total of seven points are the following: 16, 25, 34, 43, 52, 61. The sum of the corresponding six probabilities of $\frac{1}{36}$ each therefore gives $P\{C\} = \frac{6}{36} = \frac{1}{6}$. Next, let E be the event of getting a total number of points that is an even number. Simple events such as 11, 13, 22, and so on, satisfy the requirement of yielding an even-numbered total. The sum of two even digits, or the sum of two odd digits, will yield an even number. From Table 1 we observe that there are 18 points of this type; hence, it follows that $P\{E\} = \frac{18}{36} = \frac{1}{2}$. Finally, let G be the event that both dice will show at least four points. Here the simple events such as 44, 45, 56, and so on will satisfy. Table 1 shows that there are nine such events; therefore, $P\{G\} = \frac{9}{36} = \frac{1}{4}$.

In many games-of-chance experiments the various possible outcomes are expected to occur with the same relative frequency; therefore, all the points of the sample space for these experiments are assigned the same probability, namely $\frac{1}{n}$, where n denotes the total number of points in the sample space. This was true, for example, in the experiments of coin tossing and dice rolling. It was not true, however, for the colored ball experiment. When the experiment is of this simple type, that is, when all the simple event probabilities are equal, the calculation of the probability of a composite event is very easy. It consists of merely adding the probability $\frac{1}{n}$ as many times as there are simple events comprising the composite event. Thus, if the composite event A consists of a total of $n(A)$ simple events, the value of $P\{A\}$ can be expressed by the formula

$$(2) \qquad\qquad P\{A\} = \frac{n(A)}{n}$$

In the experiment of rolling two dice, for example, the probability of obtaining a total of seven points is obtained by counting the number of points in the sample space given in Table 1 that produce a seven total, of which there are 6, and dividing this number by the total number of points, namely 36.

Although it is not often possible in real-life problems to use formula (2), it is easier to work with than is the general definition (1) involving the addition of probabilities; therefore, it alone is used in the next few sections to derive basic formulas. The formulas obtained in this manner can be shown to hold true equally well for the general definition and, therefore, are applicable to all types of problems. Since only the formulas are needed in applied problems, there will be no appreciable loss in the understanding of how to solve practical problems by following this procedure.

3. ADDITION RULE

Applications of probability are often concerned with a number of related events rather than with just one event. For simplicity, consider two such events, A_1 and A_2, associated with an experiment. We may be interested in knowing whether both A_1 and A_2 will occur when the experiment is performed. This joint event is denoted by the symbol $(A_1$ and $A_2)$ and its probability by $P\{A_1$ and $A_2\}$. On the other hand, we may be interested in knowing whether at least one of the events A_1 and A_2 will occur when the experiment is performed. This event is denoted by the symbol $(A_1$ or $A_2)$ and its probability by $P\{A_1$ or $A_2\}$. At least one of the two events will occur if A_1 occurs but A_2 does not, if A_2 occurs but A_1 does not, or if both A_1 and A_2 occur. Thus the word "or" here means or in the sense of either one, the other, or both. The purpose of this section is to obtain a formula for $P\{A_1$ or $A_2\}$.

If two events A_1 and A_2 possess the property that the occurrence of one prevents the occurrence of the other, they are called *mutually exclusive* events. For example, let A_1 be the event of getting a total of seven in rolling two dice, and A_2 the event of getting a total of eleven: then A_1 and A_2 are mutually exclusive events. For mutually exclusive events there are no outcomes that correspond to the occurrence of both A_1 and A_2; therefore, the

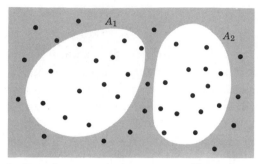

FIGURE 3 Sample space with two mutually exclusive events.

two events do not possess any points in common in the sample space. This is shown schematically in Fig. 3. In this diagram the points lying inside the two light regions labeled A_1 and A_2 represent the simple events that yield the composite events A_1 and A_2, respectively.

If $n(A_1)$ denotes the number of points lying inside the region labeled A_1 and $n(A_2)$ the number lying inside the region labeled A_2, then the total number of points associated with the occurrence of either A_1 or A_2 is the sum of those two numbers; consequently, if n denotes the total number of sample points, it follows from formula (2) that

$$P\{A_1 \text{ or } A_2\} = \frac{n(A_1) + n(A_2)}{n}$$

$$= \frac{n(A_1)}{n} + \frac{n(A_2)}{n}$$

Since the last two fractions are precisely those defining $P\{A_1\}$ and $P\{A_2\}$, this result yields the desired addition formula, which we can express as:

(3) **Addition Rule.** *When A_1 and A_2 are mutually exclusive events*

$$P\{A_1 \text{ or } A_2\} = P\{A_1\} + P\{A_2\}$$

For more than two mutually exclusive events, it is necessary only to apply this formula as many times as required. A slightly more complicated formula can be derived for events that are not mutually exclusive; however, there is no occasion to use such a formula in later sections and thus it is omitted here. It can be found in one of the exercises at the end of this chapter.

In the preceding illustration of rolling two dice, in which A_1 and A_2

denoted the events of getting a total of seven and eleven points, respectively, the probability of getting either a total of seven or a total of eleven can be obtained by means of formula (3). From Table 1 it is clear that A_1 and A_2 contain no points in common and, from counting points, that $P\{A_1\} = \frac{6}{36}$ and $P\{A_2\} = \frac{2}{36}$; therefore

$$P\{A_1 \text{ or } A_2\} = \frac{6}{36} + \frac{2}{36} = \frac{8}{36}$$

This result is, of course, the same as that obtained by counting the total number of points, namely eight, that yield the composite event $(A_1 \text{ or } A_2)$ and by applying formula (2) directly.

As another illustration, what is the probability of getting a total of at least 10 points in rolling two dice? Let A_1, A_2, and A_3 be the events of getting a total of exactly 10 points, 11 points, and 12 points, respectively. From Table 1 it is clear that these events have no points in common and that their probabilities are given by $P\{A_1\} = \frac{3}{36}$, $P\{A_2\} = \frac{2}{36}$, and $P\{A_3\} = \frac{1}{36}$. Therefore, by formula (3), the probability that at least one of those mutually exclusive events will occur is given by

$$P\{A_1 \text{ or } A_2 \text{ or } A_3\} = \frac{3}{36} + \frac{2}{36} + \frac{1}{36} = \frac{1}{6}$$

This result also could have been obtained directly from Table 1 by counting favorable and total outcomes. Although the formula does not seem to possess any advantage here over direct counting for problems related to Table 1, it is a very useful formula for problems in which probabilities of events are available but for which tables of possible outcomes are not.

4. MULTIPLICATION RULE

The purpose of this section is to obtain a formula for $P\{A_1 \text{ and } A_2\}$ in terms of probabilities of single events. To do so, it is necessary to introduce the notion of conditional probability. Suppose that we are interested in knowing whether A_2 will occur subject to the condition that A_1 is known to have occurred, or else is certain to occur. The geometry of this problem is shown in Fig. 4. It is assumed here that A_1 and A_2 are not mutually exclusive events.

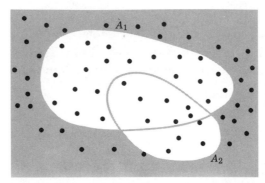

FIGURE 4   A sample space for conditional probability.

Since A_1 must occur, the only experimental outcomes that need be considered are those corresponding to the occurrence of A_1. The sample space for this problem is therefore reduced to the simple events that comprise A_1. They are represented in Fig. 4 by the points lying inside the region labeled A_1. Among those points, the ones that also lie inside the region labeled A_2 correspond to the occurrence of both A_1 and A_2. They are the points that lie in the overlapping parts of A_1 and A_2. Let $n(A_1)$ denote the number of points lying inside A_1 and let $n(A_1 \text{ and } A_2)$ denote the number that lie inside both A_1 and A_2. Then, from formula (2), the probability that A_2 will occur if the sample space is restricted to be the set of points inside A_1 is given by the ratio $n(A_1 \text{ and } A_2)/n(A_1)$. But this probability is what is meant by the probability that A_2 will occur subject to the restriction that A_1 must occur. If this latter probability is denoted by the new symbol $P\{A_2 \mid A_1\}$, then

(4)
$$P\{A_2 \mid A_1\} = \frac{n(A_1 \text{ and } A_2)}{n(A_1)}$$

As an illustration of the application of this formula, a calculation will be made of the probability that the sum of the points obtained in rolling two dice is seven, if it is known that the dice showed at least three points each. Let A_1 denote the event that two dice will show at least three points each and let A_2 denote the event that two dice will show a total of seven points. The sample space for this problem is shown in Fig. 5; it is obtained directly from Table 1.

The points that comprise the event A_1 are all the points except those in the first two rows and the first two columns of Fig. 5. They are shown inside the light rectangle of Fig. 5. Here $n(A_1) = 16$. The points that comprise the

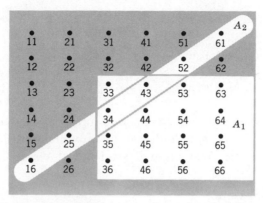

FIGURE 5 Sample space for a conditional probability problem.

event A_2 are the diagonal points shown in Fig. 5. The number of points that lie inside A_2 which also lie inside A_1 is seen to be $n(A_1 \text{ and } A_2) = 2$. As a result, formula (4) gives the result:

$$P\{A_2 \mid A_1\} = \frac{2}{16} = \frac{1}{8}$$

What this means in an experimental sense is that in the repeated rolling of two dice, we discard all those experimental outcomes in which either die showed a number of points less than three. Then among the experimental outcomes that are retained we calculate the proportion of outcomes that yielded a total of seven points. This proportion in the long run of experiments should approach $\frac{1}{8}$. Interestingly, the chances of getting a total of seven points is less when we know that both dice show at least three points than under ordinary rolls.

Now consider formula (4) in general terms once more. From Fig. 4 and formula (2) it is clear that

$$P\{A_1\} = \frac{n(A_1)}{n}$$

and that

$$P\{A_1 \text{ and } A_2\} = \frac{n(A_1 \text{ and } A_2)}{n}$$

By dividing the second of these two expressions by the first and canceling n, we obtain

$$\frac{P\{A_1 \text{ and } A_2\}}{P\{A_1\}} = \frac{n(A_1 \text{ and } A_2)}{n(A_1)}$$

This result in conjunction with (4) yields the formula

(5) $$P\{A_2 \mid A_1\} = \frac{P\{A_1 \text{ and } A_2\}}{P\{A_1\}}$$

This formula defines the *conditional probability* of A_2 given A_1 and, when it is written in product form, yields the fundamental multiplication formula for probabilities, which may be expressed as follows:

(6) Multiplication Rule

$$P\{A_1 \text{ and } A_2\} = P\{A_1\}P\{A_2 \mid A_1\}$$

In words, this formula states that the probability that both of two events will occur is equal to the probability that the first event will occur, multiplied by the conditional probability that the second event will occur when it is known that the first event is certain to occur. Either one of the two events may be called the first event because this is merely convenient language for discussing them, and no time order is implied in the way they occur. Even though no time order is implied for the two events A_1 and A_2 in the symbol $P\{A_2 \mid A_1\}$, it is customary to call this conditional probability "the probability that A_2 will occur when it is known that A_1 has occurred." Thus, if you are being dealt a five-card poker hand, someone might ask: What is the probability that your hand will contain the ace of spades if it is known that you have received the ace of hearts? This is merely convenient language for discussing probabilities of a poker hand that must contain the ace of hearts, and there is no implication that if the ace of spades is in the hand it was obtained after obtaining the ace of hearts. For many pairs of events, however, there is a definite time-order relationship. For example, if A_1 is the event that a high school graduate will go to college and A_2 is the event that he will graduate from college, then A_1 must precede A_2 in time.

As an illustration of the multiplication rule, let us calculate the probability of getting two red balls in drawing two balls from a box containing three red, two black, and one green ball. We assume here that the first ball drawn is not returned to the box before the second drawing is made. This experiment differs from the one that is considered in Fig. 2, since there repetitions of the experiment involve returning the drawn ball each time.

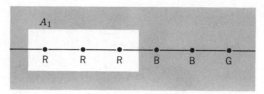

FIGURE 6 Sample space for first drawing.

Let A_1 denote the event of getting a red ball on the first drawing and A_2 that of getting a red ball on the second drawing. To be able to continue using formula (2) rather than the more general definition (1), it is necessary to give each ball a number and to use six points in the sample space as shown in Fig. 6. The first three will represent red balls, the next two the black balls, and the last one the green ball. Then, by formula (2),

$$P\{A_1\} = \frac{3}{6}$$

For the purpose of calculating $P\{A_2 | A_1\}$ it is sufficient to consider only those experimental outcomes for which A_1 has occurred. Since the first ball drawn is not returned to the box, this means considering only those experiments in which the first ball drawn was one of the three red balls. Thus, the second part of the experiment can be treated as a new single experiment in which one ball is to be drawn from a box containing two red, two black, and one green ball. As a result, $P\{A_2 | A_1\}$ represents the probability of getting a red ball in drawing one ball from this new box of balls. Here there are five points in the sample space as shown in Fig. 7, the first two representing red balls, the next two black balls, and the last one a green ball. Hence, by formula (2),

$$P\{A_2 | A_1\} = \frac{2}{5}$$

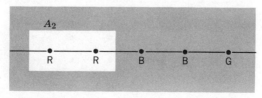

FIGURE 7 Sample space for second drawing.

Application of formula (6) to these two results then gives

$$P\{A_1 \text{ and } A_2\} = \frac{3}{6} \cdot \frac{2}{5} = \frac{1}{5}$$

The advantage of using formula (6) on this problem becomes apparent if we try to solve this problem by applying formula (2) directly to the sample space that corresponds to this two-stage experiment. That sample space consists of 30 points and resembles the sample space shown in Fig. 5, except that the main diagonal points are missing because they correspond to getting the same numbered colored ball on both drawings. This two-dimensional sample space is shown in Fig. 8. Using Fig. 8, the solution is obtained by counting the points common to A_1 and A_2 and dividing by the total number of points. This gives $\frac{6}{30} = \frac{1}{5}$. The advantage of formula (6) is most pronounced in two-stage and multiple-stage experiments, which usually possess complicated sample spaces with a large number of points, because this formula reduces the calculation of probabilities to calculations for one-stage experiments only. The sample spaces for one-stage experiments are usually quite simple and much easier to visualize than those for multiple-stage experiments. Hereafter, in calculating probabilities, the techniques based on formula (6) and single-stage experiments will be used almost exclusively to avoid the time-consuming method based on applying formula (2) directly to the sample space of the entire experiment. It should be understood, however, that it is always possible to calculate any type of probability that may arise by working exclusively with the original sample space. A student should occasionally work a problem in this manner to test

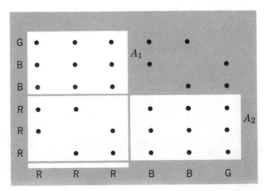

FIGURE 8   Sample space for a two-stage experiment.

his understanding of basic concepts. It would be well for him, for example, to solve the preceding illustrative problem by constructing the sample space for that two-stage experiment.

As another illustration of the multiplication rule, calculate the probability of getting two prizes in taking two punches on a punchboard which contains 5 prizes and 20 blanks. If A_1 denotes the event of getting a prize on the first punch and A_2 the event of getting a prize on the second punch, then formula (6) gives

$$P\{A_1 \text{ and } A_2\} = \frac{5}{25} \cdot \frac{4}{24} = \frac{1}{30}$$

The value of $P\{A_2 | A_1\} = \frac{4}{24}$ arises from the fact that, since the first punch yielded a prize, there are only 4 prizes left and only 24 punches left.

As a final illustration, consider the problem of calculating the probability of getting two aces in drawing two cards from a bridge deck. If A_1 denotes the event of getting an ace on the first drawing and A_2 the event of getting an ace on the second drawing, then formula (6) gives

$$P\{A_1 \text{ and } A_2\} = \frac{4}{52} \cdot \frac{3}{51} = \frac{1}{221}$$

The value of $P\{A_2 | A_1\}$ is $\frac{3}{51}$ because there are only three aces left after the first drawing and only 51 cards remaining.

If the events A_1 and A_2 are such that the probability that A_2 will occur does not depend on whether or not A_1 occurs, then A_2 is said to be independent of A_1 and we can write

$$P\{A_2 | A_1\} = P\{A_2\}$$

For this case, the multiplication rule reduces to

$$P\{A_1 \text{ and } A_2\} = P\{A_1\}P\{A_2\}$$

Since the event $(A_1 \text{ and } A_2)$ is the same as the event $(A_2 \text{ and } A_1)$, A_1 and A_2 may be interchanged in (6) to give

$$P\{A_1 \text{ and } A_2\} = P\{A_2\}P\{A_1 | A_2\}$$

Comparing the right sides of these two formulas shows that $P\{A_1 | A_2\} = P\{A_1\}$. This demonstrates that A_1 is independent of A_2 when A_2 is independent of A_1. Because of this mutual independence, it is proper to say that A_1 and A_2 are independent, without specifying which is independent of the other. As a result,

When A_1 and A_2 are independent

(7) $$P\{A_1 \text{ and } A_2\} = P\{A_1\}P\{A_2\}$$

In view of this result, we can state that two events are independent if, and only if, the probability of their joint occurrence is equal to the product of their individual probabilities. This formula generalizes in the obvious manner to more than two independent events. Although it is easy to state the condition that must be satisfied if two events are to be independent, it is not always so easy in real life to decide whether two events are independent. As a rather far-fetched example, suppose that the two events are A_1: the stock market will rise next week, and A_2: a stockholder will catch a cold next week. It would seem obvious that the probability that one of these events will occur would be the same whether or not the other event occurred, hence, that these are independent events. However, if it should happen that stocks rose considerably in value, the chances are that many stockholders might go out to celebrate their good fortune and thereby might increase their chances of catching a cold during that week, in which case these events would not be independent in a probability sense.

In games of chance, such as roulette, it is always assumed that consecutive plays are independent events. If we were not willing to accept this assumption, then we would be forced to assume that the roulette wheel possesses a memory or that the operator of the wheel is secretly manipulating it.

As illustrations of the application of the preceding rules of probability, consider a few simple card problems.

Two cards are drawn from an ordinary deck of 52 cards, the first card drawn being replaced before the second card is drawn.

Example 1. What is the probability that both cards will be spades? Let A_1 denote the event of getting a spade on the first draw, and A_2 the event of getting a spade on the second draw. Since the first card drawn is replaced, the probability of getting a spade on the second draw should not depend on whether or not a spade was obtained on the first draw; hence, A_2 may be assumed to be independent of A_1. Formula (7) then gives

$$P\{A_1 \text{ and } A_2\} = \frac{13}{52} \cdot \frac{13}{52} = \frac{1}{16}$$

Example 2. What is the probability that the cards will be either two spades or two hearts? Let B_1 be the event of getting two spades, and B_2 the event of getting two hearts. Then, from the preceding result, it follows that

$$P\{B_1\} = P\{B_2\} = \frac{1}{16}$$

Since the events B_1 and B_2 are mutually exclusive and the problem is to calculate the probability that either B_1 or B_2 will occur, formula (3) applies; hence,

$$P\{B_1 \text{ or } B_2\} = \frac{1}{16} + \frac{1}{16} = \frac{1}{8}$$

As before, let two cards be drawn from a deck, but this time the first card drawn will not be replaced.

Example 3. What is the probability that both cards will be spades? Here A_2 is not independent of A_1 because if a spade is obtained on the first draw the chances of getting a spade on the second draw will be smaller than if a nonspade had been obtained on the first draw. For this problem formula (6) must be used. Then

$$P\{A_1 \text{ and } A_2\} = \frac{13}{52} \cdot \frac{12}{51} = \frac{1}{17}$$

The second factor is $\frac{12}{51}$ because there are only 51 cards after the first drawing, all of which are assumed to possess the same chance of being drawn, and there are only 12 spades left.

Example 4. As an illustration that does not involve games of chance and that involves more than two independent events, consider the following problem. Assuming that the ratio of male children is $\frac{1}{2}$ (which is only approximately true), find the probability that in a family of six children (a) all the children will be of the same sex, and (b) five of the children will be boys and one will be a girl.

(a) Let A_1 be the event that all the children will be boys and A_2 the event that they will all be girls. Because A_1 and A_2 are mutually exclusive events

$$P\{A_1 \text{ or } A_2\} = P\{A_1\} + P\{A_2\}$$

Since the six individual births may be assumed to be six independent events with respect to the sex of the child, it follows by using the more general version of the multiplication formula (7) that

$$P\{A_1\} = P\{A_2\} = \left(\frac{1}{2}\right)^6$$

Hence

$$P\{A_1 \text{ or } A_2\} = \left(\frac{1}{2}\right)^6 + \left(\frac{1}{2}\right)^6 = \frac{1}{32}$$

(b) Let A_1 be the event that the oldest child is a girl and the others are boys, A_2 the event that the second oldest is a girl and the others are boys, and similarly for events A_3, A_4, A_5, and A_6. Since the event of having five boys and one girl will occur if, and only if, one of the six mutually exclusive events $A_1, \ldots, A_6$ occurs, it follows from (3) that

$$P\{5 \text{ boys and } 1 \text{ girl}\} = P\{A_1\} + \cdots + P\{A_6\}$$

But

$$P\{A_1\} = \cdots = P\{A_6\} = \left(\frac{1}{2}\right)^6$$

Hence

$$P\{5 \text{ boys and } 1 \text{ girl}\} = 6 \left(\frac{1}{2}\right)^6 = \frac{3}{32}$$

Example 5. As a final illustration we work several problems related to the drawing of two cards from a box containing the following five cards: the ace of spades, the ace of clubs, the two of hearts, the two of diamonds, and the three of spades. An ace is considered as a one. Spades and clubs are black cards, whereas hearts and diamonds are red cards. The two cards are to be drawn from this box without replacing the first card drawn before the second drawing. By using the addition and multiplication formulas, calculate the probability that (a) both cards will be red, (b) the first card will be an ace and the second card will be a two, (c) both cards will be the same color, (d) one card will be a spade and the other will be a club, (e) a total of four points on the two cards will be obtained, and (f) exactly one ace will be obtained, if it is known that both cards are black.

(a) Applying formula (6) and considering the experiment in two stages

$$P\{RR\} = \frac{2}{5} \cdot \frac{1}{4} = \frac{1}{10}$$

(b) Applying formula (6)

$$P\{A2\} = \frac{2}{5} \cdot \frac{2}{4} = \frac{1}{5}$$

(c) The two events RR and BB constitute the two mutually exclusive ways in which the desired event can occur; hence, by applying formula (3), we obtain

$$P\{RR \text{ or } BB\} = P\{RR\} + P\{BB\} = \frac{2}{5} \cdot \frac{1}{4} + \frac{3}{5} \cdot \frac{2}{4} = \frac{2}{5}$$

(d) The two events SC and CS will satisfy; hence

$$P\{SC \text{ or } CS\} = P\{SC\} + P\{CS\} = \frac{2}{5} \cdot \frac{1}{4} + \frac{1}{5} \cdot \frac{2}{4} = \frac{1}{5}$$

(e) A total of four will be obtained if both cards are two's, or if one card is a three and the other card is a one. If a subscript is used to denote the number on a card, the events that satisfy are the following ones: H_2D_2, D_2H_2, S_3S_1, S_1S_3, S_3C_1, C_1S_3. Since these constitute the mutually exclusive ways in which the desired event can occur and since each of these possesses the same probability, namely, $\frac{1}{5} \cdot \frac{1}{4} = \frac{1}{20}$, it follows that

$$P\{4 \text{ total}\} = \frac{6}{20} = \frac{3}{10}$$

(f) Let A_1 denote the event that both cards will be black and A_2 the event that exactly one ace will be obtained. Then $P\{A_2 | A_1\}$ is the probability needed to solve the problem. From formula (5), this requires the computation of $P\{A_1\}$ and $P\{A_1 \text{ and } A_2\}$. First, $P\{A_1\} = \frac{3}{5} \cdot \frac{2}{4} = \frac{3}{10}$. Next, both A_1 and A_2 will occur if one of the following mutually exclusive events occurs: S_1S_3, S_3S_1, C_1S_3, S_3C_1. Since each of these events has the probability $\frac{1}{5} \cdot \frac{1}{4} = \frac{1}{20}$ and there are four of them, it follows that $P\{A_1 \text{ and } A_2\} = \frac{4}{20} = \frac{1}{5}$. Hence

$$P\{A_2 | A_1\} \frac{\frac{1}{5}}{\frac{3}{10}} = \frac{2}{3}$$

This last problem could have been solved more quickly by looking at the sample space, or arguing that since both drawn cards must be black we are justified in treating the problem as one in which two cards are to be drawn from a box containing only the three black cards. The solution given, however, illustrates how to use the formula for conditional probabilities on the original sample space.

Although the preceding rules of probability were derived on the assumption that all the possible outcomes of the experiment in question were expected to occur with the same relative frequency, the rules hold for more general experiments. The derivation becomes more complicated then, but it does not lead to a better understanding of the rules and, therefore, is omitted. These rules can even be applied to events related to experiments involving an infinite number of possible outcomes. These more general experiments are considered in the next chapter.

5. BAYES' FORMULA

There is a certain class of important problems based on the application of formula (5) that lead to rather involved computations; therefore, it is convenient to have a formula for solving problems of this kind in a systematic manner. These problems may be illustrated by the following academic one. Suppose that a box contains two red balls and one white ball and a second box contains two red balls and two white balls. One of the boxes is selected by chance and a ball is drawn from it. If the drawn ball is red, what is the probability that it came from the first box? Let A_1 denote the event of choosing the first box and let A_2 denote the event of drawing a red ball. Then the problem is to calculate the conditional probability $P\{A_1 \mid A_2\}$. This will be done by the use of formula (5) with A_1 and A_2 interchanged in it. Since the phrase by chance is understood to mean that each box has the same probability of being chosen, it follows that the probability of drawing the first box is $\frac{1}{2}$, and that of drawing the second box is the same. The calculation of the numerator term in (5) can be accomplished by using formula (6) in the order of events listed there. Thus

$$P\{A_2 \text{ and } A_1\} = P\{A_1 \text{ and } A_2\} = \frac{1}{2} \cdot \frac{2}{3} = \frac{1}{3}$$

The denominator, $P\{A_2\}$, can be calculated by considering the two mutually exclusive ways in which A_2 can occur, namely, getting the first box and then a red ball or getting the second box and then a red ball. By formula (3), $P\{A_2\}$ will be given by the sum of the probabilities of those two mutually exclusive possibilities; hence

$$P\{A_2\} = \frac{1}{2} \cdot \frac{2}{3} + \frac{1}{2} \cdot \frac{2}{4} = \frac{7}{12}$$

Application of the modified version of formula (5) then yields the desired result:

$$P\{A_1 \mid A_2\} = \frac{\dfrac{1}{3}}{\dfrac{7}{12}} = \frac{4}{7}$$

This problem could be worked very easily by looking at the sample space for the experiment; however, the objective here is to work with formula (5) and attempt to obtain a formula for treating more complicated problems of the same type as the present one.

The foregoing problem is a special case of problems of the following type. We are given a two-stage experiment. The first stage can be described by stating that exactly one of, say, k possible outcomes must occur when the complete experiment is performed. Those possible outcomes are denoted by $e_1, e_2, \ldots, e_k$. In the second stage there are, say, m possible outcomes, exactly one of which must occur. These are denoted by $o_1, o_2, \ldots, o_m$. The values of the probabilities for each of the possible outcomes $e_1, e_2, \ldots, e_k$ are given. As before, they are denoted by $P\{e_1\}, P\{e_2\}, \ldots, P\{e_k\}$. The values of all the conditional probabilities of the type $P\{o_j \mid e_i\}$, which represents the probability that the second-stage event o_j will occur when it is known that the first-stage event e_i occurred, are also given. The problem is to calculate the probability that the first-stage event e_i occurred when it is known that the second-stage event o_j occurred. This conditional probability would be written $P\{e_i \mid o_j\}$. For simplicity, the calculations will be carried out for $P\{e_1 \mid o_1\}$; the calculations for any other pair would be the same.

In terms of the present notation, formula (5) assumes the form

(8)
$$P\{e_1 \mid o_1\} = \frac{P\{e_1 \text{ and } o_1\}}{P\{o_1\}}$$

Formula (6) gives

(9)
$$P\{e_1 \text{ and } o_1\} = P\{e_1\}P\{o_1 \mid e_1\}$$

From the information given in this problem it will be observed that the two probabilities on the right side of (9) are known; therefore the numerator in (8) can be obtained from (9). The value of $P\{o_1\}$ in (8) can be computed by considering all the mutually exclusive ways in which o_1 can occur in conjunction with the first stage of the experiment. The second-stage event o_1 will occur if the first-stage event e_1 occurs and then o_1 occurs, or if the first-stage event e_2 occurs and then o_1 occurs, . . . , or if the first-stage event e_k occurs and then o_1 occurs. Hence, applying formula (3) to these k mutually exclusive events, we may write

$$P\{o_1\} = P\{e_1 \text{ and } o_1\} + P\{e_2 \text{ and } o_1\} + \cdot \cdot \cdot + P\{e_k \text{ and } o_1\}$$

If e_1 is replaced by e with the appropriate subscript in (9) that formula can be used to calculate each of the probabilities on the right side of this equality. As a result, we obtain the formula

$$P\{o_1\} = P\{e_1\}P\{o_1 \mid e_1\} + P\{e_2\}P\{o_1 \mid e_2\} + \cdot \cdot \cdot + P\{e_k\}P\{o_1 \mid e_k\}$$

This can be written in condensed form by using the summation symbol Σ as follows:

$$P\{o_1\} = \sum_{i=1}^{k} P\{e_i\}P\{o_1 \mid e_i\}$$

Here e_i denotes event number i in the set $e_1, e_2, \dots, e_k$ and the summation symbol tells us to add all the probabilities of the type displayed beginning with e_1 and ending with e_k. This result together with (9) when applied to (8) will give the desired result, which is known as Bayes' formula.

(10) **Bayes' Formula**

$$P\{e_1 \mid o_1\} = \frac{P\{e_1\}P\{o_1 \mid e_1\}}{\displaystyle\sum_{i=1}^{k} P\{e_i\}P\{o_1 \mid e_i\}}$$

Returning to the problem that was solved earlier without this formula, we observe that there were two events e_1 and e_2 in the first stage corresponding to choosing the first or the second box and that $P\{e_1\} = P\{e_2\} = \frac{1}{2}$. The second stage also consisted of two events o_1 and o_2 corresponding to obtaining a red or a white ball. The conditional probabilities of obtaining a red ball based on what transpired at the first stage were given by $P\{o_1 | e_1\} = \frac{2}{3}$ and $P\{o_1 | e_2\} = \frac{2}{4}$. We observe that the substitution of these values in (10) yields the result that was obtained previously.

Consider now a more practical application of this formula. Suppose a test for detecting a certain rare disease has been perfected that is capable of discovering the disease in 97 percent of all afflicted individuals. Suppose further that when it is tried on healthy individuals, 5 percent of them are incorrectly diagnosed as having the disease. Finally, suppose that, when it is tried on individuals who have certain other milder diseases, 10 percent of them are incorrectly diagnosed. It is known that the percentages of individuals of the three types being considered here in the population at large are 1 percent, 96 percent, and 3 percent, respectively. The problem is to calculate the probability that an individual, selected at random from the population at large and tested for the rare disease, actually has the disease if the test indicates he is so afflicted.

Here there are three events e_1, e_2, and e_3 in the first stage corresponding to the three types of individuals in the population. Their corresponding probabilities are $P\{e_1\} = .01$, $P\{e_2\} = .96$, and $P\{e_3\} = .03$. There are two events o_1 and o_2 in the second stage corresponding to whether the test claims that the individual has the disease or not. The conditional probabilities are given by $P\{o_1 | e_1\} = .97$, $P\{o_1 | e_2\} = .05$, and $P\{o_1 | e_3\} = .10$. In terms of the present notation, the problem is to calculate $P\{e_1 | o_1\}$. A direct application of formula (10) based on the preceding probabilities will supply the answer:

$$P\{e_1 | o_1\} = \frac{(.01)(.97)}{(.01)(.97) + (.96)(.05) + (.03)(.10)} = .16$$

This result may seem rather surprising because it shows that only 16 percent of the individuals whom the test would indicate have the disease actually do have it when the test is applied to the population at large. The 84 percent who were falsely diagnosed might resent the temporary mental anguish caused by their belief that they had the disease before further tests revealed the falsity of the diagnosis. They might also resent the necessity of having been required to undergo further tests when it turned out that those tests were really unnecessary. A calculation like the preceding one might therefore cause authorities to ponder a bit before advocating mass testing.

6. COUNTING TECHNIQUES — TREES

It is convenient in solving some of the more difficult problems involving two-or-more stage experiments to have systematic methods for calculating the compound event probabilities that arise. A pictorial method that has proved particularly useful is based on what is known as a *probability tree*.

For the purpose of illustrating how such a tree is constructed, let us again consider the first example to which formula (6) was applied. A box contains three red, two black, and one green ball, and two balls are to be drawn. This is a two-stage experiment for which the various possibilities that can occur may be represented by a horizontal tree like that shown in Fig. 9. Each stage of a multiple-stage experiment has as many branches as there are possibilities at that stage. Here there are three main branches at the first stage and three branches at each of the second stages, except for the last one

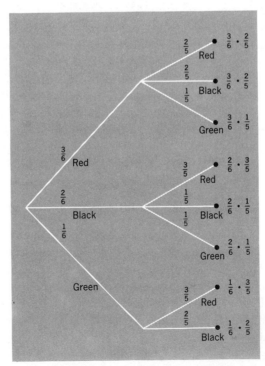

FIGURE 9   A probability tree.

where there are only two branches because it is impossible to obtain a green ball at the second drawing if a green ball is obtained on the first drawing. The total number of terminating branches in such a tree gives the total number of possible outcomes in the compound experiment, and therefore the end points of those branches may be treated as the sample points of a sample space.

The probability that is attached to any branch of the tree is the conditional probability that the event listed under that branch will occur, subject to the condition that the preceding branch events all occurred. Thus, the $\frac{2}{5}$ listed above the top terminal branch is the conditional probability that a red ball will be obtained on the second drawing if a red ball was obtained on the first drawing. The probability listed at the end of a terminal branch is the probability of obtaining the sequence of events that are required to arrive at the terminal point and is obtained by multiplying the probabilities associated with the branches leading to that terminal point. Thus, the first terminal probability $\frac{3}{6} \cdot \frac{2}{5}$ is the probability of obtaining a red ball at the first drawing times the conditional probability of doing so at the second drawing. By means of this tree and its probabilities, it is relatively easy to answer various probability questions.

Probability trees yield a simple pictorial method for calculating probabilities for which Bayes' formula would normally be employed. As an example, consider the previous problem that was used to motivate Bayes' formula, namely, the problem of calculating $P(A_1 | A_2)$ where A_1 is the event of selecting Box 1 and A_2 is the event of getting a red ball. The tree corresponding to this two-stage experiment is shown in Fig. 10 with the proper probabilities attached to each branch and to each of the four sample points.

Now the topmost branch corresponds to the compound event A_1 and A_2; therefore, $P(A_1 \text{ and } A_2) = \frac{1}{2} \cdot \frac{2}{3}$. Furthermore, it follows from definition (1) that $P(A_2) = \frac{1}{2} \cdot \frac{2}{3} + \frac{1}{2} \cdot \frac{2}{4}$, because the event A_2 consists of the two sample points associated with the word "Red." Hence, by formula (5)

$$P\{A_1 | A_2\} = \frac{P\{A_1 \text{ and } A_2\}}{P\{A_2\}} = \frac{\frac{1}{2} \cdot \frac{2}{3}}{\frac{1}{2} \cdot \frac{2}{3} + \frac{1}{2} \cdot \frac{2}{4}} = \frac{4}{7}$$

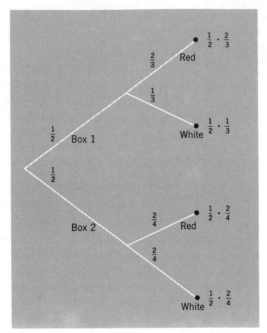

FIGURE 10   A tree for a Bayesian problem.

The technique is now seen to be the following one. After constructing the probability tree, select the terminal branch that corresponds to the occurrence of both A_1 and A_2. Then divide the probability associated with that terminal branch by the sum of the probabilities of all the terminal branches that are associated with the event A_2.

This technique will also be applied to the second illustration associated with Bayes' formula. First, it is necessary to construct a tree such as that shown in Fig. 11. To obtain $P\{e_1 \mid o_1\}$ it now suffices to divide the probability $(.01)(.97)$ associated with the top terminal branch by the sum of the probabilities of the terminal branches associated with the letter o_1. This gives

$$P\{e_1 \mid o_1\} = \frac{(.01)(.97)}{(.01)(.97) + (.96)(.05) + (.03)(.10)} = .16$$

In addition to their use as a device to help calculate probabilities, trees are often used to assist in visualizing problems that involve several stages. As an illustration of this type of use, consider the following problem. A business firm must decide whether to build a small plant or a large

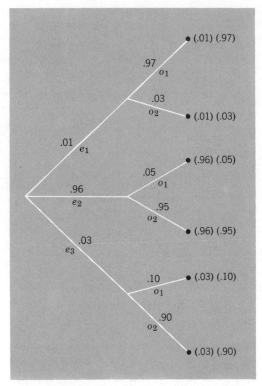

FIGURE 11   A tree for a Bayesian problem.

plant to manufacture a new product, which it expects to market during the next 15 years. It anticipates that the demand for the new product will be high during the first two years. If the product turns out to be satisfactory, the demand will likely continue high during the following years; however, if it is unsatisfactory, the demand will drop to a low level. The initial demand may, however, be low in spite of the firm's expectations, in which case the demand during subsequent years is almost certain to be low. If the product is satisfactory, the firm will have to expand its plant after two years, provided that it started with a small plant; otherwise competition will enter the field and take away the excess available market.

We treat this problem as a several-stage experiment of which the first stage is concerned with deciding whether to build a large or a small plant. The second stage corresponds to whether the initial demand is high or low. The third stage is concerned with deciding whether to expand the plant,

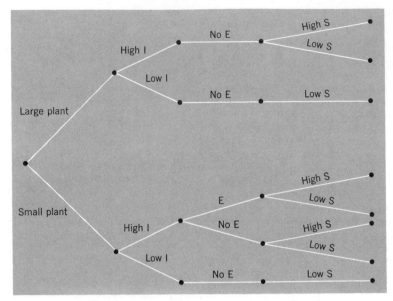

FIGURE 12   A decision-experience tree.

and the fourth stage indicates how the product is expected to sell in subsequent years. A tree to represent these possibilities is a combination of a decision and experience tree. It can be constructed as shown in Fig. 12.

This tree makes the analysis of the firm's problem easy to visualize. If the firm were able to assign probabilities to the various experience branches of the tree, it would be able to calculate the probabilities of the various possibilities and use them to help in deciding whether to construct a large, or a small, plant. Methods based on such probabilities, which are discussed briefly in the next chapter, are particularly helpful in the solution of this type of problem.

7. COUNTING TECHNIQUES–COMBINATIONS

If there are many stages to an experiment and several possibilities at each stage, the probability tree associated with the experiment would become too large to be manageable. For problems of this kind the counting of sample points is simplified by means of algebraic formulas. Toward this objective, consider a two-stage experiment for which there are r possibilities at the first stage and s possibilities at the second stage. A tree representing this experiment is shown in Fig. 13. Since each of the r main branches has s terminal branches attached to it, the total number of possibilities here is rs, and therefore this would be the number of sample points in the sample space for this two-stage experiment. If a third stage with t possibilities were added, the total number of sample points would become rst. This can be extended in an obvious manner to any number of stages.

Now consider the application of this counting method to the problem of determining in how many ways it is possible to select r objects from n distinct objects. Toward this objective, let us first consider the particular

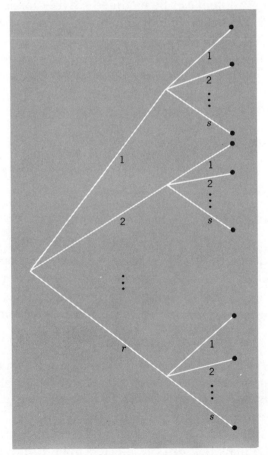

FIGURE 13   A tree for a two-stage experiment.

problem of determining how many three-letter words can be formed from the five letters a, b, c, d, e if a letter may be used only once in a given word and if any set of three letters is called a three-letter word.

Forming a three-letter word may be thought of as a three-stage experiment in which the first stage is that of choosing the first letter of the word. In this problem there are five possibilities for the first stage, but only four possibilities for the second stage because the letter chosen at the first stage is not available at the second stage. There are only three possibilities left at the third stage; therefore, according to the preceding counting method, the total number of three-letter words that can be formed is given by $5 \cdot 4 \cdot 3 = 60$.

The preceding problem can be modified slightly by asking: How many committees consisting of three individuals can be selected from a group of five individuals? If the letters a, b, c, d, e are associated with the five individuals, then a three-letter word will correspond to a committee of three; however, two words using the same three letters but in a different order will correspond to the same committee. For example, *ace* and *cea* are distinct words but do not represent different committees. Since the three letters a, c, and e will produce only one committee but will give rise to $3 \cdot 2 \cdot 1 = 6$ distinct three-letter words, and this will be true for every selection of three letters, it follows that there will be only $\frac{1}{6}$ as many committees of three as there are three-letter words; hence, there will be $\frac{60}{6} = 10$ such committees.

Now suppose we are given n distinct objects, say n distinct letters a, b, c, . . . , and that r of those objects are to be chosen to form r-letter words. This can be treated as an r-stage experiment in which there are n possibilities at the first stage, $n - 1$ possibilities at the second stage, and so on. Then, the total number of words that can be formed is given by the formula

$$(11) \qquad {}_nP_r = n(n-1)(n-2) \; \cdots \; (n-r+1)$$

The symbol ${}_nP_r$ is called the number of permutations of n objects taken r at a time. An arrangement along a line of a set of objects is called a *permutation* of those objects; therefore, an r-letter word is a permutation of the r letters used to construct the word.

The corresponding problem of counting the number of committees consisting of r individuals that can be selected from n individuals is readily solved by means of the preceding formula. Just as in the preceding special problem, it suffices to realize that a committee is concerned only with which r letters are selected and not in how they are arranged along a line. Since r distinct letters can be arranged along a line to produce $r(r-1)(r-2) \cdots 1$ distinct words but only one committee, the number of words must be divided by $r(r-1)(r-2) \cdots 1$ to give the number of distinct committees. Thus the total number of committees that can be formed is given by the formula:

$$(12) \qquad \binom{n}{r} = \frac{n(n-1)(n-2) \cdots (n-r+1)}{r(r-1)(r-2) \cdots 1}$$

The symbol $\binom{n}{r}$ is called the number of combinations of n things taken r at a time. A *combination* of r objects is merely a selection of r objects without regard to the order in which they are selected or arranged after selection. A permutation of r objects may be thought of as the outcome of a two-stage experiment in which the first stage consists in choosing a combination of r objects and the second stage consists in arranging that combination along a line.

A convenient symbol to use in connection with formula (12) is the factorial symbol, which is an exclamation mark after an integer, indicating that the number concerned should be multiplied by all the positive integers smaller than it. Thus $4! = 4 \cdot 3 \cdot 2 \cdot 1$ and $r! = r(r-1)(r-2) \cdots 1$. To allow r to assume the value 0 in the following formulas, we define $0! = 1$. Formula (12) can therefore be written in the form

$$\binom{n}{r} = \frac{n(n-1) \cdots (n-r+1)}{r!}$$

The numerator of this expression can also be expressed with factorial notation by observing that

$$\frac{n!}{(n-r)!} = \frac{n(n-1) \cdots (n-r+1)(n-r)(n-r-1) \cdots 1}{(n-r)(n-r-1) \cdots 1}$$

$$= n(n-1) \cdots (n-r+1)$$

By using factorial symbols in this manner, we may write formula (12) in the compact form

$$(13) \qquad \binom{n}{r} = \frac{n!}{r!(n-r)!}$$

The usefulness of formula (12) or (13) for counting purposes is illustrated in the following problems.

Example 1. What is the probability of getting five spades in a hand of five cards drawn from a deck of 52 playing cards. Here the total number of possible outcomes corresponds to the total number of five-card hands that can be formed from a deck of 52 distinct cards. Since arrangement is of no interest here, this is a combination counting problem. The total number of possible hands is, according to formula (12), given by

$$\binom{52}{5} = \frac{52 \cdot 51 \cdot 50 \cdot 49 \cdot 48}{5 \cdot 4 \cdot 3 \cdot 2 \cdot 1}$$

The number of outcomes that correspond to the occurrence of the desired event is equal to the number of ways of selecting five spades from 13 spades. This is given by

$$\binom{13}{5} = \frac{13 \cdot 12 \cdot 11 \cdot 10 \cdot 9}{5 \cdot 4 \cdot 3 \cdot 2 \cdot 1}$$

The desired probability is given by the ratio of these two numbers; hence, it is equal to

$$\frac{13 \cdot 12 \cdot 11 \cdot 10 \cdot 9}{52 \cdot 51 \cdot 50 \cdot 49 \cdot 48} = \frac{33}{66640} = .0005$$

The moral seems to be that we should not expect to obtain a five-card poker hand containing only spades. Even if we settle for a hand containing five cards of the same suit, the probability is only four times as large, namely .002, which is still hopelessly small.

Example 2. If a list of 20 individuals who have volunteered to supply blood when it is needed for a transfusion includes 15 with type B blood, and if three are selected at random from the list, what is the probability that (a) all three will be of type B, (b) two will be of type B and one will not, (c) at least one will be of type B?

(a) The total number of possible outcomes is the total number of ways of choosing three individuals from a group of 20 which is given by $\binom{20}{3}$. The number of outcomes that correspond to the occurrence of the desired event is the number of ways of choosing three individuals from a group of 15, which is given by $\binom{15}{3}$. The probability of this favorable event occurring is therefore given by the ratio:

$$\frac{\binom{15}{3}}{\binom{20}{3}} = \frac{15 \cdot 14 \cdot 13}{20 \cdot 19 \cdot 18} = \frac{91}{228} = .40$$

(b) Here the number of favorable outcomes is given by the number of ways of choosing two individuals from a group of 15, multiplied by the number of ways of choosing one individual from a group of five. That product is $\binom{15}{2}\binom{5}{1}$. The probability of this event occurring is therefore given by the ratio:

$$\frac{\binom{15}{2}\binom{5}{1}}{\binom{20}{3}} = \frac{15 \cdot 14 \cdot 5 \cdot 3}{20 \cdot 19 \cdot 18} = \frac{35}{76} = .46$$

(c) This event will occur if the event of getting 0 such individuals does not occur. The latter probability is given by $\dfrac{\binom{5}{3}}{\binom{20}{3}}$; hence, the desired probability is given by

$$1 - \frac{\binom{5}{3}}{\binom{20}{3}} = 1 - \frac{5 \cdot 4 \cdot 3}{20 \cdot 19 \cdot 18} = 1 - \frac{1}{114} = \frac{113}{114} = .99$$

The following examples are additional illustrations of various types of probability problems whose solutions require the use of the preceding counting techniques. They are solved with very little comment and without evaluating the combination symbols as it is assumed that a student who has understood the earlier material can follow the steps given here and can carry out the computations.

Example 3. What is the probability that a poker hand of five cards will contain three aces and two kings?

There are $\binom{4}{3}$ ways of choosing the aces and $\binom{4}{2}$ ways of choosing the kings; hence, the desired probability is given by

$$\frac{\binom{4}{3}\binom{4}{2}}{\binom{52}{5}}$$

Example 4. A committee of six students is to be selected from a group of 20 students. If seven of the 20 are graduate students and the others are undergraduates, what is the probability that if the committee is chosen by chance it will contain exactly three undergraduate and three graduate students?

The answer is given by

$$\frac{\binom{13}{3}\binom{7}{3}}{\binom{20}{6}}$$

Example 5. A box of 30 radio tubes contains 25 good tubes and five defective tubes. If three tubes are selected by chance from this box, what is the probability that at least two of them will be good?

The answer is obtained by adding the probability of getting exactly two good tubes and of getting exactly three good tubes; hence, it is given by

$$\frac{\binom{25}{2}\binom{5}{1}}{\binom{30}{3}} + \frac{\binom{25}{3}}{\binom{30}{3}} = \frac{\binom{25}{2}\binom{5}{1} + \binom{25}{3}}{\binom{30}{3}}$$

Example 6. In a bridge game you know that one of your opponents has at least three aces. If you have no aces, what is the probability that this opponent has all four aces?

Since a bridge hand contains 13 cards and your hand is known and may be discarded from consideration, the deck may be treated as one containing four aces in a total of 39 cards. This is a conditional probability to which formula (5) is applicable; hence

$$P\{4\text{ aces} \mid \geqslant 3\text{ aces}\} = \frac{P\{4\text{ aces and} \geqslant 3\text{ aces}\}}{P\{\geqslant 3\text{ aces}\}} = \frac{P\{4\text{ aces}\}}{P\{\geqslant 3\text{ aces}\}}$$

The last result follows from the fact that if four aces are obtained, it automatically follows that at least three aces must be obtained. The answer is therefore given by

$$\frac{P\{4 \text{ aces}\}}{P\{\geqslant 3 \text{ aces}\}} = \frac{\binom{4}{4}\binom{35}{9}}{\binom{4}{3}\binom{35}{10} + \binom{4}{4}\binom{35}{9}}$$

COMMENTS

This chapter has been concerned with studying the concept of probability and with explaining its basic principles. The basic rules of probability were derived and applied to the solution of simple problems, mostly of the academic type but some of a more practical nature. These same rules are used in the next chapter to derive models for solving a few of the more sophisticated probability problems of the social sciences. Thus, this is essentially a preliminary chapter, needed to construct such models.

Probability is the fundamental tool in making decisions when those decisions must be based on incomplete data. Since many problems arising in the social sciences are of this type, it follows that a knowledge of probability is vitally important to their study.

EXERCISES

Section 1

1. List all the possible outcomes if a coin is tossed four times.

2. A box contains one red, one black, and one green ball. Two balls are to be drawn in sequence from this box without replacing the first ball drawn before the second drawing. Construct a sample space for this experiment similar to Table 1 in the text.

3. What probabilities should be assigned to the points of the sample space corresponding to the experiment of Problem 1?

4. What probabilities would you assign to the points of the sample space of Problem 2?

5. What sample space would you have constructed and what probabilities would you have assigned in Problem 2 if the first ball had been returned to the box before the second drawing?

6. A box contains two black and one white ball. Two balls are to be drawn from this box. Construct a sample space for this experiment by (a) using six points, (b) using three points.

7. What probabilities would you assign to the points of the two sample spaces constructed in Problem 6?

8. What expected relative frequencies would you guess should be assigned to the possible outcomes of an experiment consisting of selecting a female student at random and noting whether she would be rated as a blonde, redhead, or brunette? How would you go about improving on your guess?

9. If you were interested in studying the fluctuations of the stock market over a period of time, would you expect the relative frequency of rises to be about the same as the relative frequency of declines? Would it be possible for this to be true and yet have the stock market rise in value regularly over a period of a year?

Section 2

1. A box contains four red, three black, two green, and one white ball. A ball is drawn from the box. What is the probability that the ball will be (a) red, (b) red or black?

2. An honest die is rolled twice. Using Table 1 in the text, calculate the probability of getting (a) a total of five, (b) a total of less than five, (c) a total that is an even number.

3. An honest coin is tossed four times. Using the model of Problems 1 and 3, Section 1, calculate the probability of getting (a) four heads, (b) three heads and one tail, (c) at least two heads.

4. For the experiment of rolling two honest dice, calculate the probability that (a) the sum of the numbers will not be 10, (b) neither 1, 2, nor 3 will appear, (c) each die will show two or more points, (d) the numbers on the two dice will not be the same, (e) exactly one die will show fewer than three points.

5. A box contains two black, two white, and one green ball. Two balls are to be drawn from the box. Construct a two-dimensional sample space for this experiment by using 20 sample points and assigning equal probabilities to those points. Using this sample space and definition (1), or formula (2), calculate the probability of getting (a) two black balls, (b) one black and one white ball, (c) one black and one green ball.

6. Work Problem 5 if the first ball drawn is returned to the box before the second ball is drawn.

7. A department fills orders which vary in number from day to day between 0 and 10 according to the following probabilities: .02, .07, .15, .20, .19, .16, .10, .06, .03, .01, and .01. During a day selected at random, what is the probability of obtaining:
(a) Four orders or less?
(b) At least one order?
(c) Less than six orders but more than two orders?

Section 3

1. Let A_1 and A_2 be the events associated with Fig. 5. Use that sample space to calculate $P\{A_1 \text{ or } A_2\}$. Since A_1 and A_2 are not mutually exclusive, formula (3) cannot be applied here.

2. With the aid of Fig. 4, show that the addition rule for two events A_1 and A_2 that are not mutually exclusive is given by the formula:

$$P\{A_1 \text{ or } A_2\} = P\{A_1\} + P\{A_2\} - P\{A_1 \text{ and } A_2\}$$

Section 4

1. Two balls are to be drawn from a bag containing two white and four black balls. (a) What is the probability that the first ball will be white and the second black? (b) What is this probability if the first ball is replaced before the second drawing?

2. Two balls are to be drawn from a bag containing two white, three black, and four green balls. (a) What is the probability that both balls will be green? (b) What is this probability if the first ball is replaced before the second drawing? (c) What is the probability that both balls will be the same color, assuming that the first ball is not replaced?

3. A box contains five coins, four of which are honest coins but the fifth of which has heads on both sides. If a coin is selected from the box and then is tossed two times, what is the probability that two heads will be obtained?

4. Assume that there are an equal number of male and female students in a graduating high school class and that the probability is $\frac{1}{3}$ that a graduating male student and $\frac{1}{4}$ that a graduating female student will have taken three years of mathematics. What is the probability that a graduating senior selected at random:
   (a) Will be a male and will have had three years of mathematics?
   (b) Will have had three years of mathematics?

5. The following numbers were obtained from a mortality table based on 100,000 individuals chosen at age 10.

Age	Number Alive	Deaths per 1000 During That Year
17	94,818	7.688
18	94,089	7.727
19	93,362	7.765
20	92,637	7.805
21	91,914	7.855

If these numbers are used to define probabilities of death for the corresponding groups and if A, B, and C denote individuals of ages 17, 19, and 21, respectively, calculate the probability that during this year: (a) A will die and B will live; (b) A and B will both die; (c) A and B will both live; (d) at least one of A and B will die; (e) at least one of A, B, and C will die.

6. A testing organization wishes to rate a particular brand of table radios. Six radios are selected at random from the stock of radios and the brand is judged to be satisfactory if nothing is found wrong with any of the six radios.
   (a) What is the probability that the brand will be rated as satisfactory if 10 percent of the radios actually are defective?
   (b) What is this probability if 20 percent are defective?

7. Let A_1 and A_2 be the events associated with Fig. 5. Use the general formula of Problem 2, Section 3, to calculate $P\{A_1 \text{ or } A_2\}$ and compare your answer with that of Problem 1, Section 3.

8. A committee consisting of seven male and five female students decides to select its chairman and secretary by chance. This is done by choosing two cards from a set of 12 cards that are associated with the 12 names of the committee members and then designating the first name drawn as chairman. What is the probability that
   (a) both officers will be male?
   (b) the chairman will be male and the secretary female?
   (c) the chairman will be female and the secretary male?
   (d) both officers will be female?

9. If three cards are drawn from a deck, what is the probability that
   (a) the third card will be a spade but the first two cards will not be spades?
   (b) the third card will be a spade but the first card will not be a spade?

10. Records show that 90 percent of the parts turned out by a machine pass inspection. What is the probability that if the next three parts that come off the assembly line are inspected
   (a) all of them will pass inspection?
   (b) two of them will pass inspection and one will fail?
   (c) at least one part will pass inspection?

11. Experience at a college has shown that 60 percent of the students coming from large high schools eventually graduate, that 50 percent of those from medium size high schools graduate, and that 40 percent of those from small high schools graduate. If the percentages of students from the three-sized high schools are 50, 30 and 20, respectively, what is the probability that
   (a) a student selected at random from an entering class will eventually graduate?
   (b) a student selected at random from a graduating class came from a large high school?

12. A coin is tossed four times. What is the probability of getting an equal number of heads and tails?

Section 5

1. Suppose that 10 percent of car owners who have an accident will have had at least one other accident during the preceding year and suppose that a driving simulator test is failed by 80 percent of these drivers but by only 30 percent of the drivers who have had only the one accident. If an individual car owner selected at random from those having had an accident takes the test and fails, what is the probability that he is the type who will have had additional accidents the preceding year?

2. Assume that there are an equal number of male and female students in a high school and that the probability is $\frac{1}{5}$ that a male student and $\frac{1}{25}$ that a female student will be a physical science major. What is the probability that
   (a) a student selected at random will be a male science student?
   (b) a student selected at random will be a science student?
   (c) a science student selected at random will be a male student?

3. Suppose that a college aptitude test designed to separate high school students into promising and not promising groups for college entrance has had the following experience. Among the students who made satisfactory grades in their first year at college, 80 percent passed the aptitude test. Among the students who did unsatisfactory work their first year, 30 percent passed the test. It is assumed that the test was not used for admission to college. If it is known that only 70 percent of first-year college students do satisfactory work, what is the probability that a student who passed the test will be a satisfactory student?

4. The probability that a pool motor will fail within one year is $\frac{1}{20}$ for a new motor and $\frac{1}{10}$ for a rebuilt motor. A repairman selects a motor from a supply of 40 motors and installs it. Unknown to him, 10 of those motors are rebuilt motors. If the installed motor fails within one year, what is the probability that it was one of the rebuilt motors?

5. Records show that 5 percent of the parts turned out by a new machine are defective. A second older machine producing the same kind of parts produces 10 percent defectives. A box containing 100 parts is received by a purchaser. He selects two parts at random and tests them, finding

that one is good and one is defective. If this box was selected at random from the warehouse and the warehouse contains twice as many boxes of parts from the old machine as from the new machine, what is the probability that the purchaser received the inferior quality product?

Section 6

1. Draw a tree to represent the possible outcomes of a game of tennis between players A and B if play continues until one of the players has won two sets. How many possibilities are there?

2. Work Problem 1 if the match continues until three sets have been won.

3. If the probability is $\frac{2}{3}$ that player A will defeat player B in a single set and successive sets are treated as independent events, what is the probability in Problem 1 that player A will win?

4. Players A and B match pennies; hence, their chances of winning a single match are equal. They play until one of them has won two matches in a row or until one of them has won three matches.
   (a) Draw a tree to show the various possibilities.
   (b) What is the probability that at most three matches will be required to determine a winner?

5. A record is kept of individuals who have been interviewed by a polling organization. An individual is classified according to sex, age, and political party. There are three categories of age, young, middle aged, and old, and three categories of political party, Democrat, Republican, and Independent. Draw a tree to represent the various possibilities here. How many folders would be needed to file all individuals in their proper class?

6. A "chain-letter" telephone campaign to sponsor a candidate is begun by a committee of five individuals. Each member agrees to call 10 friends and to urge each such friend to call 10 of his friends, who in turn are each requested to call 10 of their friends.
   (a) If there are no duplications of individuals called and all those who are called cooperate, how many individuals will have been made aware of the campaign in this manner?
   (b) If the committee wishes to contact 30,000 individuals in this

manner, how many links in the chain will be required, assuming no duplications and cooperation?

7. Suppose the probability is p that the weather (sunshine or cloudy) will be the same on any given day as it was on the preceding day. If it is sunny today, what is the probability that it will be sunny the day after tomorrow? Use a tree.

8. Employees in a certain firm are given an aptitude test when first employed. Experience has shown that of the 60 percent who pass the test, 80 percent of them are good workers, whereas of the 40 percent who fail, only 30 percent are rated as good workers. What is the probability that an employee selected at random will be a good worker? Use a tree.

9. Work Problem 2(b), Section 5, by means of a tree.

10. Work Problem 3, Section 5, by means of a tree.

11. Use the tree constructed in Problem 8 to calculate the probability that a worker selected at random from the satisfactory group failed the aptitude test.

12. Suppose that 90 percent of the individuals who buy a new car are good credit risks. Suppose that 80 percent of the good credit risks carry credit cards but that only 20 percent of the bad credit risks do so.
    (a) Draw a probability tree to show the various possibilities.
    (b) Use the tree to calculate the probability that an individual with a credit card is a poor credit risk.

Section 7

1. Ten cities have applied for federal funds to conduct an experiment in preschool child care. There are sufficient funds for only three of these pilot programs. In how many ways can the cities be chosen?

2. A true-false examination has 20 statements each of which is to be marked as being true, T, or false, F. If a student has no knowledge of the material and, therefore, guesses at each question, how many different ways can he mark those 20 statements?

3. If a box contains 40 good and 10 defective fuses and 10 fuses are selected, what is the probability that they will all be good?

4. If a poker hand of five cards is drawn from a deck, what is the probability that it will contain exactly one ace?

5. A buyer will accept a lot of 80 articles if a sample of four selected at random contains no defectives. What is the probability that he will accept the lot if it contains eight defectives?

6. An investor selected four stocks at random for investment from a list of 12 recommended stocks. During the following year, eight of the 12 stocks rose in value, whereas the other four declined in value. What is the probability that all four of the investor's stocks rose in value?

7. An office staff consists of 50 individuals, of whom 10 would be classified as executives. At an anniversary party for the company, three door prizes are to be given. What is the probability that
(a) all three prizes will be won by executives?
(b) at most one prize will be won by executives?

8. A box of 20 spare parts for certain machines contains 16 good parts and four defective parts. If three parts are picked at random from the box, what is the probability that
(a) they will all be good?
(b) they will all be defective?
(c) at least two will be good?

9. Four cards are to be drawn from an ordinary deck of 52 cards. What is the probability that
(a) all four cards will be spades?
(b) all four cards will be of the same suit?
(c) none of the four cards will be a spade?

10. A bridge hand of 13 cards is drawn from a deck of 52 cards. What is the probability that the hand
(a) will contain exactly one ace?
(b) will contain at least one ace?
(c) will contain at least six spades?
(d) will contain 13 spades?

11. An employer wishes to select two of his employees to be on a public relations committee. They are to be selected from a group of 40 male and 20 female employees. If they are chosen by chance, what is the probability that at least one female will be chosen?

12. If a bridge player and his partner have nine spades between them, what is the probability that the four spades held by their opponents will be split three and one?

13. Find the probability that a poker hand of five cards will contain only black cards if it is known to contain at least four black cards.

14. Your bridge hand has no aces. If you know that your partner has at least two aces, what is the probability that his hand has exactly three aces?

6 | Additional Topics in Probability

The preceding chapter was concerned with the definition and the basic rules of probability. In this chapter we observe how those rules can be used to solve certain types of problems of a probabilistic nature. One type of problem is of particular interest to business students, another is of interest to life science students, and a third is of interest to students who are curious about gambling odds at casinos.

Probability is a very powerful tool and has important applications in many diverse fields. It is, for example, the foundation of statistical methods which have become indispensible in designing and analyzing social science surveys and experiments. In this chapter we study only a few isolated types of problems that have been selected for their intrinsic interest and because they are essentially nonstatistical in nature. Statistical type problems are better discussed in a course on statistical methods.

1. RANDOM VARIABLES

In experiments of the repetitive type for which a probability model is to be constructed, interest usually centers on a particular property of the outcome of the experiment. For example, in rolling two dice, interest usually centers on the total number of points showing because that is all that really matters in the game of craps. Similarly, in taking samples of college students, interest might center on how many hours per week a student studies, or on a student's grade point average.

We use the letter x to represent a quantity chosen for study in such an experiment. Thus, in the preceding illustrations, x might represent the sum of the points showing on the two dice, or the number of hours per week a student studies, or a student's grade point average. In connection with such experiments, which involve probabilities of uncertain outcomes, the variable x is called a *random variable*.

For the purpose of seeing how a random variable is introduced in a simple experiment, return to the sample space for the coin-tossing experiment that is given in Fig. 1 of Chapter 5. If x is used to denote the total number of heads obtained, then each point of that sample space will possess the value of x shown directly above the corresponding point in Fig. 1 below. It will be observed that the random variable x can assume any one of the values 0, 1, 2, or 3, but no other values.

3	2	2	2	1	1	1	0
•	•	•	•	•	•	•	•
HHH	HHT	HTH	THH	HTT	THT	TTH	TTT

FIGURE 1   The values of a random variable for a coin experiment.

As another illustration, let x denote the total number of points obtained in rolling two honest dice. The sample space for this experiment is given by Table 1 of Chapter 5. That sample space has been duplicated here in Fig. 2 but with the omission of the labels attached to the points. The numbers attached to the points are the values of the random variable x for this experiment. It will be observed that this random variable x can assume any one of the values 2,3, . . . ,12.

•	•	•	•	•	•
2	3	4	5	6	7
•	•	•	•	•	•
3	4	5	6	7	8
•	•	•	•	•	•
4	5	6	7	8	9
•	•	•	•	•	•
5	6	7	8	9	10
•	•	•	•	•	•
6	7	8	9	10	11
•	•	•	•	•	•
7	8	9	10	11	12

FIGURE 2   The values of a random variable for a dice experiment.

In each of the preceding illustrations it will be observed that the value of the random variable x depends only on the particular sample point chosen. This means that x is a function of the sample points of the sample space. A formal definition is as follows:

Definition. A random variable is a numerical valued function defined on a sample space.

The word random, or chance, is used to designate variables of this type to point out that the value such a variable assumes in an experiment depends on the outcome of the experiment, and this in turn depends on chance.

Now that interest is being centered on the values of a random variable for an experiment rather than on all the possible outcomes, a new simpler sample space can be constructed for the experiment, which can be substituted for the original sample space. Thus in the coin-tossing experiment the only events of interest are the composite events given by $x = 0,1,2$, and 3. But it follows from Fig. 1, and definition (1) of Chapter 5, that the probabilities for these composite events are given by $P\{0\} = \frac{1}{8}, P\{1\} = \frac{3}{8}, P\{2\} = \frac{3}{8}$ and $P\{3\} = \frac{1}{8}$, respectively. These composite events can now be treated as simple events in a new sample space of four points with each point associated with a value of the random variable x. The probabilities just calculated for those composite events are then assigned to the four points of the new sample space. Figure 3 shows this new sample space with its associated probabilities.

In a similar manner, a new sample space for the random variable representing the total number of points showing on two dice can be constructed by means of Fig. 2, and definition (1) of Chapter 5. Here the composite events are those corresponding to the random variable x assuming the values $2,3, \ldots ,12$. The probabilities for those composite events are readily calculated by means of definition (1) of Chapter 5 to be $\frac{1}{36}$, $\frac{2}{36}, \frac{3}{36}, \frac{4}{36}, \frac{5}{36}, \frac{6}{36}, \frac{5}{36}, \frac{4}{36}, \frac{3}{36}, \frac{2}{36}$, and $\frac{1}{36}$, respectively. As a result a new

FIGURE 3   Sample space for a coin-tossing random variable.

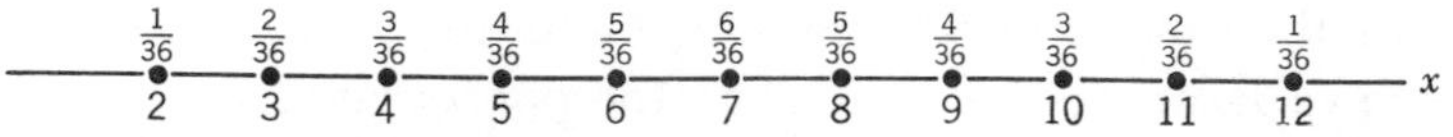

$$\begin{array}{ccccccccccc} \frac{1}{36} & \frac{2}{36} & \frac{3}{36} & \frac{4}{36} & \frac{5}{36} & \frac{6}{36} & \frac{5}{36} & \frac{4}{36} & \frac{3}{36} & \frac{2}{36} & \frac{1}{36} \\ \bullet & \bullet & \bullet & \bullet & \bullet & \bullet & \bullet & \bullet & \bullet & \bullet & \bullet \\ 2 & 3 & 4 & 5 & 6 & 7 & 8 & 9 & 10 & 11 & 12 \end{array} \longrightarrow x$$

FIGURE 4   Sample space for a dice-rolling random variable.

sample space consisting of eleven points based on the values of the random variable x can be constructed as shown in Fig. 4.

After a random variable and its corresponding sample space have been introduced and the probabilities to be assigned to those sample points calculated, we may dispense with the original sample space and work only with the new one. In this connection, if we let $x_1, x_2, \ldots, x_k$ represent the possible values that the random variable x can assume, then we let $P\{x_i\}$ denote the probability that is to be attached to the sample point corresponding to x_i. The set of values $P\{x_i\}, i = 1, \ldots, k$, is called the *probability distribution* of the random variable x. Thus, for the random variable x corresponding to Fig. 3, the probability distribution of x is given by the set of values

$$P\{0\} = \frac{1}{8}, \quad P\{1\} = \frac{3}{8}, \quad P\{2\} = \frac{3}{8}, \quad P\{3\} = \frac{1}{8}$$

In studying probability distributions, it is useful to represent them graphically by means of line charts. In these charts, the ordinate at x_i is made proportional to the probability $P\{x_i\}$. The graphs of the probability distributions defined by Figs. 3 and 4 are shown in Figs. 5 and 6, respectively.

2. EXPECTED VALUE

In this section we study one property of random variables that is particularly useful, called the *expected value* of the random variable. For the purpose of describing what the phrase "expected value" means, consider a game in which we toss three honest coins and receive one dollar for each head that shows. How much money should we expect to win if we are permitted to play this game once? From Fig. 5 the probabilities associated with getting 0, 1, 2, or 3 heads in tossing three honest coins are $\frac{1}{8}, \frac{3}{8}, \frac{3}{8}$, and $\frac{1}{8}$, respectively. Intuitively, therefore, we should expect to get

0 dollars $\frac{1}{8}$ of the time, 1 dollar $\frac{3}{8}$ of the time, 2 dollars $\frac{3}{8}$ of the time, and 3

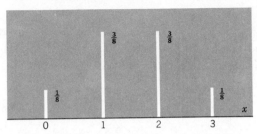

FIGURE 5   Distribution for a coin-tossing random variable.

dollars $\frac{1}{8}$ of the time if the game were played a large number of times. We should therefore expect to average the amount

$$\$0 \cdot \frac{1}{8} + \$1 \cdot \frac{3}{8} + \$2 \cdot \frac{3}{8} + \$3 \cdot \frac{1}{8} = \$1.50$$

This amount is what is commonly called the expected amount to be won if the game is played only once. This example is an illustration of the general concept of expected value which follows.

Suppose that the random variable x must assume one of the values $x_1, x_2, \ldots, x_k$ when the experiment is performed and that the probabilities associated with those values are $P\{x_1\}, P\{x_2\}, \ldots, P\{x_k\}$. Then the expected value of this random variable is defined to be the quantity

(1) $$E[x] = \sum_{i=1}^{k} x_i P\{x_i\}$$

In the preceding illustration the random variable x was the amount of money to be won in tossing three coins; hence, x had to assume one of the values 0, 1, 2, or 3, and the corresponding probabilities were $\frac{1}{8}, \frac{3}{8}, \frac{3}{8}$, and $\frac{1}{8}$.

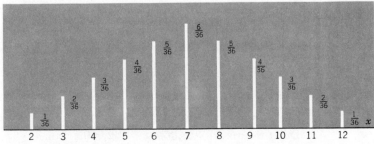

FIGURE 6   Distribution for a random variable related to dice rolling.

As another illustration, suppose that an individual is permitted to roll a pair of dice once and is paid in dollars the total number of points that he obtains. What is his expected value? Here we apply formula (1) to the probability distribution given in Fig. 6 to obtain

$$E[x] = 2 \cdot \frac{1}{36} + 3 \cdot \frac{2}{36} + 4 \cdot \frac{3}{36} + 5 \cdot \frac{4}{36} + 6 \cdot \frac{5}{36} + 7 \cdot \frac{6}{36}$$

$$+ 8 \cdot \frac{5}{36} + 9 \cdot \frac{4}{36} + 10 \cdot \frac{3}{36} + 11 \cdot \frac{2}{36} + 12 \cdot \frac{1}{36} = 7$$

Hence, he should expect to win $7. If the entrance fee to play this game once is larger than $7, a player can expect to lose money playing it. A game in which the entrance fee is the same as the expected value of the game is called a *fair game*. Casino games are not fair.

3. AN APPLICATION OF EXPECTED VALUE

In this section we encounter a problem of a probability nature that can often be solved satisfactorily by means of expected values. Although we study only this particular problem, the technique employed in solving it can be used to solve much more complicated problems of the same general type.

Suppose a sports promoter is concerned about the possibility of rain occurring at a future sporting event that he is sponsoring. He can protect himself financially against such an occurrence by buying insurance to cover his potential losses. Assume that he will net $20,000 provided that it does not rain but only $2000 if it does and that he is contemplating a $20,000 insurance policy costing $5000. This information is conveniently displayed in a two-by-two table, called a *payoff table*, which lists his net profit under the various possibilities. Table 1 gives this information.

TABLE 1

Action	Weather	
	Clear	Rain
Insure	15,000	17,000
Don't insure	20,000	2,000

If we had considered the possibility of a light rain as well as a heavy rain with correspondingly adjusted losses, we would have had three columns in our table and, hence, would have had a two-by-three payoff table. We could have added another row also by considering, for example, two kinds of insurance coverage, partial or complete. Thus the size of a payoff table depends on the nature of the problem.

The basis for preferring one of the two possible actions over the other will be that of expected profit. Hence, the decision to take out insurance will be made if, and only if, the expected profit in doing so is larger than the expected profit when no insurance is purchased. For the purpose of calculating these two expected profits, let p denote the probability that the weather will be clear on the day of the sporting event. Then, letting x denote the amount of profit, we apply formula (1) to each of the two rows of Table 1. These expected values will be denoted by $E_I[x]$ and $E_N[x]$, corresponding to insurance and no insurance, respectively. We obtain

$$E_I[x] = 15{,}000\ p + 17{,}000\ (1-p)$$

and

$$E_N[x] = 20{,}000\ p + 2000\ (1-p)$$

Assume that the value of p is not known. Then these two possible actions will be equally profitable provided that p satisfies the equation obtained by setting $E_I[x] = E_N[x]$. Hence p must satisfy the equation

$$15{,}000\ p + 17{,}000\ (1-p) = 20{,}000\ p + 2000\ (1-p)$$

The solution of this equation is easily seen to be $p = .75$. The value of p that produces equality between the two expected values in a problem such as this is called the *break-even point*.

It should be clear intuitively that if the probability of clear weather is larger than .75, the promoter should not take out insurance, because the better the weather the less need for insurance and the two possible actions are equally profitable when $p = .75$. This fact is readily verified algebraically, however, by solving the inequality $E_N[x] > E_I[x]$, which is

$$20{,}000\ p + 2000\ (1-p) > 15{,}000\ p + 17{,}000\ (1-p)$$

Dividing by 1000, we obtain

$$20\ p + 2(1-p) > 15\ p + 17\ (1-p)$$

Transposing the p terms to the left side and solving for p, we obtain $p > .75$.

In view of the preceding calculations, the sports promoter should consult the weather records for the period of time when his event is to be held to see whether the proportion of clear (no rain) days exceeds .75. If it does and he is willing to use expected profit for making decisions, then he should not buy insurance. This assumes that the event is to take place so far in the future that no reliable weather forecast is available and that the insurance rates are based on this fact also. Insurance rates are usually based on past experience, but they could be based on a forecast if the event were to occur within a sufficiently short time interval for which a valid forecast is feasible.

The preceding technique is easily extended to tables that contain any number of rows and columns. It is necessary to calculate as many expected values as there are rows in the table. If there are more than two columns, it is necessary to know the probabilities for each of the column events, or to have estimates of those probabilities based on past experience or records.

As an illustration of a problem in which there are more than two column possibilities, consider the following table of values that represents the anticipated profits to be made in two business ventures, A and B, corresponding to three economic conditions during the next five years. If the probabilities for those three economic conditions are estimated to be .2, .7, and .1, respectively, which business venture should be selected?

Economic Conditions

	Poor	Average	Excellent
A	−5000	20,000	60,000
B	5000	20,000	30,000

Calculations give

$$E_A = -5000\ (.2) + 20{,}000\ (.7) + 60{,}000\ (.1) = 19{,}000$$

and

$$E_B = 5000\ (.2) + 20{,}000\ (.7) + 30{,}000\ (.1) = 18{,}000$$

Hence, venture A should be selected.

Making a decision in problems of this type on the basis of expected value assumes that the individual concerned is willing to gamble and suffer occasional losses when his gambles fail rather than settle for security through insurance. However, a conservative individual might well prefer a considerably lower expected profit under insurance than that obtainable by not taking out insurance.

4. BERNOULLI TRIALS

In this section we consider a particular random variable that is a generalization of the coin-tossing random variable introduced in Section 2. Toward this objective consider an experiment in which the outcome can always be classified as a success or a failure. In the coin-tossing experiment, success would correspond to getting a head, and failure would correspond to getting a tail. Let p denote the probability of getting a success and let $q = 1 - p$ denote the failure probability. Further, let the experiment be repeated n times and let x denote the total number of successes that will be obtained in the n repetitions of the experiment. In the coin-tossing experiment, for example, we would have $p = q = \frac{1}{2}$ and $n = 3$. In terms of this notation the basic problem is to find the probability distribution of the random variable x.

It is assumed in problems of this type that the n experiments are independent in a probability sense and, therefore, that the multiplication rule for independent events may be applied to them. The n independent repetitions of the experiment are usually called the n *trials* of the experiment. A sequence of independent trials such as this in which the probability of success is the same for all trials is called a sequence of *Bernoulli trials*. The name is in honor of a Swiss mathematician who pioneered in the study of probability. The coin-tossing experiment is an illustration of a sequence of three Bernoulli trials for which the probability of success in any given trial is $\frac{1}{2}$.

The technique for finding the probability distribution of x is a generalization of that used in Section 1 for the coin-tossing problem. We first calculate the probabilities for all possible sequences of outcomes and then add the probabilities of those sequences that yield the same value of x. Suppose, for example, that we wish to calculate $P\{k\}$, where this symbol denotes the probability that the random variable x will assume the value k, and where k is some integer between 0 and n. One possible sequence that will make $x = k$ is the following one in which all the successes occur first, followed by all failures:

$$\overbrace{S\ S\ \cdots\ S}^{k}\quad \overbrace{F\ F\ \cdots\ F}^{n-k}$$

Another such sequence is the following one in which a failure occurs first, followed by k consecutive successes, then followed by the remaining failures. Thus

$$F\ \overbrace{S\ S\ \cdots\ S}^{k}\quad \overbrace{F\ F\ \cdots\ F}^{n-k-1}$$

Because of the independence of the trials, the probability of obtaining the first of these two sequences is given by

$$\overbrace{p\cdot p\ \cdots\ p}^{k}\quad \overbrace{q\ q\ \cdots\ q}^{n-k} = p^k q^{n-k}$$

The probability for the second sequence is given by

$$q\cdot\overbrace{p\cdot p\ \cdots\ p}^{k}\cdot\overbrace{q\cdot q\ \cdots\ q}^{n-k-1} = p^k q^{n-k}$$

The probability for the two sequences is the same, and clearly will be the same for every sequence that satisfies the condition of having k successes and $n-k$ failures.

The number of ways in which the desired event can occur is equal to the number of different sequences that can be written down of the type just displayed, those containing k letters S and $n-k$ letters F. But this number is equal to the number of ways of choosing k positions out of n positions along a line in which to place the letter S. The remaining $n-k$ positions will automatically be assigned the letter F. Since we are interested only in which of the n positions are to be selected and not in the order in which we choose them, this is a combination problem of choosing k things from n things. From formula (13), Chapter 5, the number of such sequences is given by

$$\binom{n}{k} = \frac{n!}{k!\,(n-k)!}$$

Since each of these sequences represents one of the mutually exclusive ways in which the desired event can occur, and each such sequence has the same probability of occurring, namely $p^k q^{n-k}$, it follows that the desired probability is obtained by adding this probability as many times as there are sequences. But the number of sequences was just found to be $\binom{n}{k}$; therefore, $P\{k\}$ is obtained by multiplying $p^k q^{n-k}$ by $\binom{n}{k}$. Hence

$$(2) \qquad P\{k\} = \frac{n!}{k!\,(n-k)!}\,p^k q^{n-k}, \qquad k = 0,1, \ldots ,n$$

This formula gives the probability that of $x = k$, that is, of obtaining k successes in n Bernoulli trials for which the probability of success in a single trial is p. The random variable x is commonly called the *binomial variable* and formula (2) is a formula for the *binomial distribution*.

Although the problems used to motivate this derivation have been related here to games of chance, there are many types of practical problems that can be solved by means of formula (2). We consider only a few simple problems that require little computation to illustrate its use.

Example 1. The probability that parents with a certain type of blue-brown eyes will have a child with blue eyes is $\frac{1}{4}$. If there are six children in the family, what is the probability that at least half of them will have blue eyes? To solve this problem the six children in the family will be treated as six independent trials of an experiment for which the probability of success in a single trial is $\frac{1}{4}$. Thus $n = 6$ and $p = \frac{1}{4}$ here. It is necessary to calculate $P\{3\}$, $P\{4\}$, $P\{5\}$, and $P\{6\}$ and sum because these probabilities correspond to the mutually exclusive ways in which the desired event can occur. By formula (2),

$$P\{3\} = \frac{6!}{3!3!} \left(\frac{1}{4}\right)^3 \left(\frac{3}{4}\right)^3 = \frac{540}{4096}$$

$$P\{4\} = \frac{6!}{4!2!} \left(\frac{1}{4}\right)^4 \left(\frac{3}{4}\right)^2 = \frac{135}{4096}$$

$$P\{5\} = \frac{6!}{5!1!} \left(\frac{1}{4}\right)^5 \left(\frac{3}{4}\right)^1 = \frac{18}{4096}$$

$$P\{6\} = \frac{6!}{6!0!} \left(\frac{1}{4}\right)^6 \left(\frac{3}{4}\right)^0 = \frac{1}{4096}$$

The probability of getting at least three successes is obtained by adding these probabilities; consequently, by writing $x \geqslant 3$ to represent at least three successes, we obtain

$$P\{x \geqslant 3\} = \frac{694}{4096} = .169$$

This result shows that there is a very small chance that a family such as this will have so many blue-eyed children. In only about 17 of 100 such families will at least half of the children be blue-eyed.

Example 2. A manufacturer of certain parts for automobiles guarantees that a box of his parts will contain at most two defective items. If the box holds 20 parts and experience has shown that his manufacturing process produces 2 percent defective items, what is the probability that a box of his parts will satisfy the guarantee? This problem can be considered as a binomial distribution problem for which $n = 20$ and $p = .02$. A box will satisfy the guarantee if the number of defective parts is 0, 1, or 2. By means of formula (2) the probabilities of these three events are given by

$$P\{0\} = \frac{20!}{0!\,20!}\,(.02)^0(.98)^{20} = (.98)^{20} = .668$$

$$P\{1\} = \frac{20!}{1!\,19!}\,(.02)^1(.98)^{19} = 20(.02)(.98)^{19} = .273$$

$$P\{2\} = \frac{20!}{2!\,18!}\,(.02)^2(.98)^{18} = 190(.02)^2(.98)^{18} = .053$$

The calculations here were made with the aid of logarithms. Since these are mutually exclusive events, the probability that there will be at most two defective parts, written $x \leqslant 2$, is the sum of these probabilities; hence, the desired answer is

$$P\{x \leqslant 2\} = .994$$

This result shows that the manufacturer's guarantee will almost always be satisfied.

Example 3. As a final illustration, consider the following problem concerning whether it pays to guess on an examination. Suppose an examination consists of 10 questions of the multiple-choice type, with each question having five possible answers but only one of the five being the correct answer. If a student receives 3 points for each correct answer and -1 point for each incorrect answer and if on each of the 10 questions his probability of guessing the correct answer is only $\frac{1}{3}$, what is his probability of obtaining a total positive score on those 10 questions?

If x denotes the number of questions answered correctly, then a positive score will result if $3x > 10 - x$ because the left side of this inequality gives the total number of positive points scored and the right side gives the total number of penalty points. This inequality will be satisfied if $x > \frac{10}{4}$, which implies that at least three correct answers must be obtained to realize a positive score. The desired probability is, therefore, given by

$$P\{x \geqslant 3\} = 1 - \sum_{x=0}^{2} \frac{10!}{x!\,(10-x)!} \left(\frac{1}{3}\right)^{x} \left(\frac{2}{3}\right)^{10-x}$$

$$= 1 - \left\{\left(\frac{2}{3}\right)^{10} + 10 \left(\frac{1}{3}\right) \left(\frac{2}{3}\right)^{9} + 45 \left(\frac{1}{3}\right)^{2} \left(\frac{2}{3}\right)^{8}\right\}$$

$$= .70$$

Thus he has an excellent chance of gaining points if his probability of guessing a correct answer is as high as $\frac{1}{3}$. If he knew nothing about the material and selected one of the five alternatives by chance, his probability would, of course, be only $\frac{1}{5}$ for each question. It is assumed here, however, that he knows enough about the subject to be able to discard two of the five possibilities as being obviously incorrect and to make a guess regarding the other three. If he had no such knowledge, so that his probability would be $\frac{1}{5}$, then similar calculations would show that it would not pay to guess.

5. MARKOV CHAINS

In this section we construct a probability model for a sequence of events that is a generalization of Bernoulli trials. Recall that Bernoulli trials are independent events and that the probability of success is constant from trial to trial. We generalize this model in two ways. First, we permit the number of possible outcomes to be some positive integer k where $k \geqslant 2$. Thus we no longer can speak only of success or failure at a trial. Let $A_1, A_2, \ldots, A_k$ denote those possible outcomes. Second, we drop the assumption that the trials are independent events and introduce a certain amount of dependence. A large share of the interesting random variables in the social sciences are variables that are observed at regular time intervals over a given period of time. Very often those variables are not independent. For example, the price of a given individual stock may vary from week to week in a random manner, but its price during one week will usually depend rather heavily on what its price was during the preceding week. As another illustration, if a random variable represents the number of items, out of a set of n learned items, that will be recalled after, say, t time intervals have elapsed, then the value of that variable will certainly depend on how many items were recalled after $t - 1$ time intervals.

In some problems of the preceding type the dependence is a local one in the sense that the value of the random variable at the end of t time intervals depends on its value at the end of $t - 1$ time intervals but not on any earlier time interval values. This would not usually be true, however, for the stock market because many buyers look at the past performance of a stock, and not merely at its price last week, in determining whether to buy it this week. In this section we consider this special model in which the dependence of a random variable on earlier random variables extends only to the immediately preceding one. Thus we assume that the probability of outcome A_j occurring at a given trial depends on what outcome occurred at the immediately preceding trial, but on no others.

Let p_{ij} denote the probability that outcome A_j will occur at a given trial if outcome A_i occurred at the immediately preceding trial. The possible outcomes $A_1, A_2, \ldots, A_k$ are called the states of the system. Thus p_{ij} is the probability of going from state A_i to state A_j at the next trial of the sequence. A sequence of experiments of the preceding type is called a *Markov sequence* or a *Markov chain*.

It should be noted that a Bernoulli sequence is a special case of a Markov sequence in which there are only two states A_1 and A_2 corresponding to success and failure and in which $p_{11} = p$, $p_{12} = q$, $p_{21} = p$, and $p_{22} = q$.

The probabilities p_{ij}, which are called *transition probabilities*, are usually displayed in matrix form as follows:

$$P = \begin{bmatrix} p_{11} & p_{12} & \cdots & p_{1k} \\ p_{21} & p_{22} & \cdots & p_{2k} \\ \cdot & \cdot & & \cdot \\ \cdot & \cdot & & \cdot \\ \cdot & \cdot & & \cdot \\ p_{k1} & p_{k2} & \cdots & p_{kk} \end{bmatrix}$$

These probabilities apply to the system at any point in time. They express the probability relationship that exists at two neighboring time points, regardless of the chosen point in time.

The preceding probabilities are one-step transition probabilities. There are, however, also two-step and, more generally, n-step transition probabilities. A two-step transition probability, denoted by $p_{ij}^{(2)}$, is the probability that the system will be in state A_j two time intervals later if it is now in state A_i. Formulas for expressing multiple-step transition probabilities in terms of one-step probabilities can be obtained by using matrix methods and the rules of probability. For example, to obtain a formula for $p_{ij}^{(2)}$, we proceed as follows.

To arrive at A_j from A_i in exactly two steps, we must first go from A_i to A_r, where r is one of the integers $1,2, \ldots k$, and then go from A_r to A_j at the next step. The probability of doing this is $p_{ir}p_{rj}$. Since there are k mutually exclusive ways in which the desired event can occur corresponding to $r = 1,2, \ldots ,k$, the sum of those probabilities must be the value of $p_{ij}^{(2)}$. Hence, we have the formula

(3)
$$p_{ij}^{(2)} = \sum_{r=1}^{k} p_{ir}p_{rj}$$

Now consider the evaluation of P^2, that is, of the matrix product

$$P^2 = \begin{bmatrix} p_{11} & \cdots & p_{1k} \\ \cdot & & \cdot \\ \cdot & & \cdot \\ \cdot & & \cdot \\ p_{i1} & \cdots & p_{ik} \\ \cdot & & \cdot \\ \cdot & & \cdot \\ \cdot & & \cdot \\ p_{k1} & \cdots & p_{kk} \end{bmatrix} \begin{bmatrix} p_{11} & \cdots & p_{1j} & \cdots & p_{1k} \\ \cdot & & \cdot & & \cdot \\ \cdot & & \cdot & & \cdot \\ \cdot & & \cdot & & \cdot \\ p_{k1} & \cdots & p_{kj} & \cdots & p_{kk} \end{bmatrix}$$

To obtain the element in the ith row and jth column of P^2, it is necessary to multiply the ith row vector in the first matrix by the jth column vector in the second matrix. This gives the vector product

$$p_{i1}p_{1j} + p_{i2}p_{2j} + \cdots + p_{ik}p_{kj}$$

But this is the same as the expression defining $p_{ij}^{(2)}$ in (3); hence, it follows that P^2 is the matrix whose typical element is $p_{ij}^{(2)}$. We may express this relationship by writing

(4) $$P^2 = [p_{ij}^{(2)}]$$

This shows that the two-step transition probabilities may be obtained by merely multiplying the one-step transition probability matrix by itself. In a similar manner, it follows that the n-step transition probabilities, $p_{ij}^{(n)}$, are given by the relationship

$$P^n = [p_{ij}^{(n)}]$$

As an illustration of a Markov chain, we consider a problem related to politics. Suppose that of the sons of Republicans, 60 percent vote Republican, 30 percent vote Democratic, and 10 percent vote Socialist; of the sons of Democrats, 60 percent vote Democratic, 20 percent vote Republican, and 20 percent vote Socialist; of the sons of Socialists, 50 percent vote Socialist, 40 percent vote Democratic, and 10 percent vote Republican. With this information, and assuming that the Markov chain properties hold true, we wish to perform the following:

(a) Write down the transition matrix.
(b) Calculate the two-step transition matrix.
(c) Determine the probability that the grandson of a Republican will vote Democratic.
(d) Find the answer in (c) for a great-grandson.

(a) Using the order Republican, Democrat, Socialist, the transition matrix is

$$P = \begin{bmatrix} .6 & .3 & .1 \\ .2 & .6 & .2 \\ .1 & .4 & .5 \end{bmatrix}$$

(b) By using formula (4)

$$P^2 = \begin{bmatrix} .6 & .3 & .1 \\ .2 & .6 & .2 \\ .1 & .4 & .5 \end{bmatrix} \begin{bmatrix} .6 & .3 & .1 \\ .2 & .6 & .2 \\ .1 & .4 & .5 \end{bmatrix}$$
$$= \begin{bmatrix} .43 & .40 & .17 \\ .26 & .50 & .24 \\ .19 & .47 & .34 \end{bmatrix}$$

(c) The answer here is given by $p_{12}^{(2)}$; hence, picking out the element in the first row and second column of P^2, we obtain $p_{12}^{(2)} = .40$.

(d) Here we need the element $p_{12}^{(3)}$, which can be obtained by multiplying the first row vector of P by the second column vector of P^2. This gives

$$(.6)(.40) + (.3)(.50) + (.1)(.47) = .437$$

In the preceding problem, suppose that the proportions of Republicans, Democrats, and Socialists in a community are given by the row vector $A = [a_1 \ a_2 \ a_3]$, where $a_1 + a_2 + a_3 = 1$. Then the matrix product

$$AP = [a_1 \ \ a_2 \ \ a_3] \begin{bmatrix} p_{11} & p_{12} & p_{13} \\ p_{21} & p_{22} & p_{23} \\ p_{31} & p_{32} & p_{33} \end{bmatrix}$$

is a row vector whose components give the proportions of Republicans, Democrats, and Socialists in the first generation of offspring (sons). For example, the first component in this product, namely

$$a_1 p_{11} + a_2 p_{21} + a_3 p_{31}$$

gives the probability that a first generation offspring will be a Republican, because it is the sum of the probabilities of the mutually exclusive ways in which a son will become a Republican. He may start with a Republican father and then become a Republican; or he may start with a Democratic father and then become Republican; or he may start with a Socialist father and then become a Republican. Similarly for the other two components. In a similar manner, if we wish to calculate the proportions of offspring that will be Republicans, Democrats, or Socialists at the nth generation, we merely need to calculate the matrix product AP^n.

As an illustration, suppose that the preceding initial proportions are given by the row vector $A = [.4 \ \ .5 \ \ .1]$. Then after one generation the proportions become

$$AP = [.4 \ \ .5 \ \ .1] \begin{bmatrix} .6 & .3 & .1 \\ .2 & .6 & .2 \\ .1 & .4 & .5 \end{bmatrix}$$

$$= [.35 \ \ .46 \ \ .19]$$

After two generations the proportions become

$$AP^2 = [.4 \ \ .5 \ \ .1] \begin{bmatrix} .43 & .40 & .17 \\ .26 & .50 & .24 \\ .19 & .47 & .34 \end{bmatrix}$$

$$= [.321 \ \ .457 \ \ .222]$$

It would be interesting to carry these calculations further for succeeding generations to observe whether these proportions approach some fixed values. Calculating high powers of a matrix requires high-speed computing equipment and may be very expensive. Fortunately, however, there exists a mathematical theorem that tells us what happens to the transition probabilities in a Markov chain as n becomes increasingly large, provided that P contains at least one column of positive elements. This theorem, which we do not prove, can be expressed as follows:

Theorem. *As $n \to \infty$ each row vector of P^n approaches the probability vector X that is a solution of the matrix equation $XP = X$.*

It is easy to show that if P is a transition probability matrix, which implies that its elements are nonnegative and that the sum of the elements in any row is 1, then P^n will also be a matrix of this type. Our theorem therefore states that the resulting transition probabilities are the same for all rows of the matrix.

As an illustration, we compute this transition matrix, which is often called the limiting transition matrix, for the preceding political problem. The equation that needs to be solved is the following one:

$$[x_1 \quad x_2 \quad x_3] \begin{bmatrix} .6 & .3 & .1 \\ .2 & .6 & .2 \\ .1 & .4 & .5 \end{bmatrix} = [x_1 \quad x_2 \quad x_3]$$

Multiplying the matrices on the left and equating components on both sides, we obtain

$$.6x_1 + .2x_2 + .1x_3 = x_1$$
$$.3x_1 + .6x_2 + .4x_3 = x_2$$
$$.1x_1 + .2x_2 + .5x_3 = x_3$$

These equations are equivalent to the set

$$-.4x_1 + .2x_2 + .1x_3 = 0$$
$$.3x_1 - .4x_2 + .4x_3 = 0$$
$$.1x_1 + .2x_2 - .5x_3 = 0$$

They simplify to

$$-4x_1 + 2x_2 + x_2 = 0$$
$$3x_1 - 4x_2 + 4x_3 = 0$$
$$x_1 + 2x_2 - 5x_3 = 0$$

Since the solution to our problem requires that X be a probability vector, we must have $x_1 + x_2 + x_3 = 1$. The solution of the preceding equations that also satisfies this restriction is readily seen to be given by

$$x_1 = \frac{12}{41}, \quad x_2 = \frac{19}{41}, \quad x_3 = \frac{10}{41}$$

Our theorem then states that as $n \to \infty$, the elements of P^n approach the elements of the matrix

$$\begin{bmatrix} \frac{12}{41} & \frac{19}{41} & \frac{10}{41} \\ \frac{12}{41} & \frac{19}{41} & \frac{10}{41} \\ \frac{12}{41} & \frac{19}{41} & \frac{10}{41} \end{bmatrix}$$

Now let us calculate the long-run proportions of Republicans, Democrats, and Socialists. Just as was done earlier to obtain the first and second generation proportions, this can be accomplished by calculating the product of the row vector $[.4 \quad .5 \quad .1]$ and the matrix just obtained. This gives

$$[.4 \quad .5 \quad .1] \begin{bmatrix} \frac{12}{41} & \frac{19}{41} & \frac{10}{41} \\ \frac{12}{41} & \frac{19}{41} & \frac{10}{41} \\ \frac{12}{41} & \frac{19}{41} & \frac{10}{41} \end{bmatrix} = \begin{bmatrix} \frac{12}{41} & \frac{19}{41} & \frac{10}{41} \end{bmatrix}$$

It should be observed that since the rows of the limiting transition matrix are all equal and the elements of the row vector A sum to 1, it follows that the result of this type of matrix product must be a row of the limiting transition matrix. Hence, the preceding solution

$$x_1 = \frac{12}{41}, \quad x_2 = \frac{19}{41}, \quad x_3 = \frac{10}{41}$$

gives the proportions of Republicans, Democrats, and Socialists that will be realized in the long run if the transition probabilities do not change over time. Incidentally, this also shows that the initial set of proportions, .4, .5, and .1, have no effect on the long-run set of proportions.

The preceding result, showing that the limiting proportions are the same as the limiting values of the transition probabilities, is quite general. It did not depend on the particular numbers used in the illustration. Thus it is true in general that the initial proportions in any problem have no effect on the long-run proportions. Those proportions depend only on the nature of the transition probabilities.

6. GAMBLER'S RUIN

In this section we study the chances of an individual's winning at a gambling casino when playing a simple game such as roulette. Our discussion is intended as an educational study and is of practical value only insofar as it warns of the futility of hoping to win consistently at a casino, and thus of squandering one's money.

Consider a game in which the probability of winning at a single play of the game is p. Suppose that we are playing against an opponent who has \$B to bet and that we have \$A to bet against him. At each play of the game \$1 is bet; therefore, if we lose we pay our opponent \$1, otherwise he pays us \$1. The game ceases when we have won \$B or when we have lost \$A. Let $N = A + B$ denote the total amount of money involved in the game.

Suppose that the game has progressed a while and that we now possess \$x, where x is some integer between 0 and N. Let p_x denote the probability that we shall eventually be ruined, that is, that we shall lose \$A to our opponent before we win \$B from him. This can occur in only two ways. Either we win the next play of the game in which case we have $x + 1$ dollars, and then are eventually ruined, or we lose the next play, in which case we have $x - 1$ dollars, and then are eventually ruined. Since these are the two mutually exclusive ways in which ruin can occur, the probability of eventual ruin is the sum of the probabilities of those two mutually exclusive ways. Thus, applying the rules of probability, we obtain

(5) $$p_x = p \cdot p_{x+1} + q \cdot p_{x-1}$$

We have assumed that x is some integer between 0 and N. From the meaning of p_x, however, it follows that

(6) $$p_0 = 1 \quad \text{and} \quad p_N = 0$$

The problem is to use the relationships given by equations (5) and (6) to obtain a formula for p_x. To accomplish this objective, we first rewrite equation (5) as follows:

$$(p + q)p_x = pp_{x+1} + qp_{x-1}$$

Or

$$p(p_x - p_{x+1}) = q(p_{x-1} - p_x)$$

Letting $r = p/q$, we can write this last result in the form:

$$p_{x-1} - p_x = r(p_x - p_{x+1})$$

Then, letting $x = N - 1, N - 2, \ldots, 1$, and using equation (6), we obtain

(7)
$$
\begin{aligned}
p_{N-2} - p_{N-1} &= r(p_{N-1} - p_N) &= rp_{N-1} \\
p_{N-3} - p_{N-2} &= r(p_{N-2} - p_{N-1}) &= r^2 p_{N-1} \\
&\quad\vdots &\quad\vdots \\
p_0 - p_1 &= r(p_1 - p_2) &= r^{N-1}p_{N-1}
\end{aligned}
$$

Summation of both sides gives

$$p_0 - p_{N-1} = (r + r^2 + \cdots + r^{N-1})p_{N-1}$$

By transposing the p_{N-1} term on the left to the right side and using (6), we can reduce this to

(8)
$$1 = (1 + r + r^2 + \cdots + r^{N-1})p_{N-1}$$

Next, we choose only the last A equations from (7). This gives us the set

$$p_{A-1} - p_A = r(p_A - p_{A+1}) = r^{N-A}p_{N-1}$$

$$
\begin{aligned}
&\quad\vdots &\quad\vdots &\quad\vdots \\
p_0 - p_1 &= r(p_1 - p_2) &= r^{N-1}p_{N-1}
\end{aligned}
$$

Summing these equations, we obtain

$$p_0 - p_A = (r^{N-A} + \cdots + r^{N-1})p_{N-1}$$

The elimination of p_{N-1} by means of (8) gives

$$p_0 - p_A = \frac{r^{N-A} + \cdots + r^{N-1}}{1 + r + \cdots + r^{N-1}}$$

Finally, applying (6), we obtain the formula

$$p_A = 1 - \frac{r^{N-A} + \cdots + r^{N-1}}{1 + r + \cdots + r^{N-1}}$$

(9)
$$= \frac{1 + r + \cdots + r^{N-A-1}}{1 + r + \cdots + r^{N-1}}$$

$$= \frac{1 + r + \cdots + r^{B-1}}{1 + r + \cdots + r^{N-1}}$$

If $p = \frac{1}{2}$, then $r = 1$ and the right side reduces to $\frac{B}{N}$. If $p \neq \frac{1}{2}$, we can apply the formula for the sum of a geometric progression to both the numerator and denominator of formula (9) to reduce the right side to the form $(1 - r^B)/(1 - r^N)$. We summarize the preceding derivation in the form of a theorem as follows:

Theorem. *If two individuals have $A and $B, respectively, with which to wager against each other in a game in which $1 is wagered each time and if p is the probability that the first individual will win from the second individual in a single play of the game, then the probability that the first individual will eventually be ruined is given by the formula*

(10)
$$p_A = \frac{1 - r^B}{1 - r^N}$$

when $p \neq \frac{1}{2}$ and by

$$p = \frac{B}{N}$$

when $p = \frac{1}{2}$. Here $N = A + B$ and $r = \frac{p}{1 - p}$.

Incidentally, there is no requirement in formula (10) that $p < \frac{1}{2}$. We can use it when playing against a friendly opponent who is less skilled than we are, in which case $p > \frac{1}{2}$. Furthermore, we are free to call either player the first individual and use the relationship $p_B = 1 - p_A$, which is based on the assumption that one of the players must eventually be ruined.

As an illustration of how formula (10) is applied, consider a gambling situation in which individual I has \$30 and individual II has \$20, and the probability that I will win against II in a single game is .45. Then $r = \dfrac{.45}{.55} = \dfrac{9}{11}$, A = 30, B = 20, N = 50, and formula (10) becomes

$$p_I = \frac{1 - \left(\dfrac{9}{11}\right)^{20}}{1 - \left(\dfrac{9}{11}\right)^{50}}$$

Calculations by means of logarithms, for example, will show that $p_I = .982$. Thus, it is quite certain that individual I will lose his \$30 before he will be able to capture the \$20 of individual II. The odds are too heavily against him, and his larger capital is not of much use in compensating for this. Increasing A by a very large amount would have practically no effect on the answer. Decreasing B, however, would be very beneficial to I. Thus if B were 10 instead of 20, calculations will show that $p_I = .866$.

If an individual is gambling against a casino, which implies that B is very large and that $p < \dfrac{1}{2}$, then p_A will be practically equal to 1 because when $r < 1$ the numbers r^B and r^N will both differ very little from 0. Thus, for all practical purposes, it is certain that a gambler will eventually lose when gambling against a casino.

7. A LEARNING MODEL

Experiments with animals to determine their rate of learning often involve counting the number of trials that are needed by the animal to master a learning experiment. For example, the number of trials needed by a rat to find its way through a maze is used as a comparative measure of its learning ability. Such a measure is useful in studying various influences on the learning process.

Let p_k denote the probability that an animal will make the correct decision in a learning experiment when the experiment is run the kth time. On each succeeding run of the experiment the animal learns and, therefore, should increase its probability of making the correct decision at the next trial. One model that is used in learning experiments of this kind assumes that the increase in this probability is proportional to the maximum amount that the probability can increase. Since the probability is p_k at the kth trial, the maximum amount that the probability can increase from the kth to the $k + 1$st trial is $1 - p_k$. The increase in p_k is therefore given by $c(1 - p_k)$ where c is the proportionality factor, which must satisfy the inequality $0 < c < 1$. This assumption implies that the probability of the animal's making the correct decision at trial $k + 1$ is given by

$$p_{k+1} = p_k + c(1 - p_k), \qquad 0 < c < 1$$

To determine whether this is a realistic model to employ in learning theory it is necessary to solve this equation explicitly for p_n, where n is any positive integer, and apply it to experimental data. To do so, we let $r = 1 - c$ and write this equation in the form

$$p_{k+1} = c + (1 - c)p_k = c + rp_k$$

Then we allow k to assume the values $k = 1, 2,$ and 3 to obtain the following set of relations:

$$p_2 = c + rp_1$$
$$p_3 = c + rp_2 = c + r(c + rp_1) = c + rc + r^2p_1$$
$$p_4 = c + rp_3 = c + r(c + rc + r^2p_1)$$
$$= c + rc + r^2c + r^3p_1$$

From the manner in which successive terms are built up from the preceding ones, clearly we can continue this process and, for any integer n, arrive at the formula

$$p_n = c + rc + r^2c + \cdots + r^{n-2}c + r^{n-1}p_1$$
$$= c(1 + r + r^2 + \cdots + r^{n-2}) + r^{n-1}p_1$$

Application of the formula for the sum of a geometric progression will reduce this to

$$p_n = c \frac{1 - r^{n-1}}{1 - r} + r^{n-1}p_1$$

Substituting back the value $r = 1 - c$, we obtain

$$p_n = 1 - (1 - c)^{n-1} + (1 - c)^{n-1}p_1$$

This reduces to the desired formula:

(11) $$p_n = 1 - (1 - p_1)(1 - c)^{n-1}$$

As an illustration of the use of this formula, let us suppose the constant of proportionality for an experiment of a rat running a maze is given by $c = \dfrac{1}{5}$

and the probability that it will find its way through the maze at the first trial is $p_1 = \frac{1}{20}$. Then we wish to solve these problems: (a) What is the value of p_n for $n = 2,3,4$, and 5? (b) What value does p_n approach as n becomes increasingly large?

(a) Here $p_n = 1 - \frac{19}{20}\left(\frac{4}{5}\right)^{n-1}$; hence

$$p_2 = 1 - \frac{19}{20}\frac{4}{5} = .24$$

$$p_3 = 1 - \frac{19}{20}\frac{16}{25} = .39$$

$$p_4 = 1 - \frac{19}{20}\frac{64}{125} = .51$$

$$p_5 = 1 - \frac{19}{20}\frac{256}{625} = .61$$

(b) As n becomes increasingly large $\left(\frac{4}{5}\right)^{n-1}$ approaches 0; therefore, p_n approaches 1. This implies that as the number of trials increases, the rat is almost certain to master the maze.

We can go a step further with this learning model and determine the number of mistakes the animal can be expected to make if the experiment is performed n times.

At the kth trial of the experiment the animal will make the correct decision and, hence, will make 0 mistakes, or he will make the wrong decision and, hence, will make 1 mistake. We let e_k denote the number of mistakes made at this kth trial. Thus, e_k has the value 0 or 1. Since the probability that $e_k = 0$ is p_k and the probability that $e_k = 1$ is $1 - p_k$, we can use formula (1) to calculate the expected value of the number of errors made at the kth trial. This gives

$$E[e_k] = 0 \cdot p_k + 1(1 - p_k) = 1 - p_k$$

Now the total number of mistakes that will be made if the experiment is run n times is given by

$$T_n = \sum_{k=1}^{n} e_k$$

The expected value of T_n can be obtained by calculating the expected value of the individual terms in the sum and then by summing those expected values. Although this property of expected values is intuitively plausible its proof is not simple, so we accept this property on faith. Hence

$$E[T_n] = \sum_{k=1}^{n} E[e_k]$$

$$= \sum_{k=1}^{n} (1 - p_k)$$

From formula (11), this becomes

$$E[T_n] = \sum_{k=1}^{n} (1 - p_1)(1 - c)^{k-1}$$

$$= (1 - p_1) \sum_{k=1}^{n} (1 - c)^{k-1}$$

$$= (1 - p_1)[1 + (1 - c) + (1 - c)^2 + \cdots + (1 - c)^{n-1}]$$

Application of the formula for the sum of a geometric progression then yields the formula

$$E[T_n] = \frac{1 - p_1}{c} [1 - (1 - c)^n]$$

As an illustration, let us calculate the expected number of errors that will be made by a rat of the previous example (a) if five trials are run, and (b) if a very large number of trials are run.

(a) Here $E[T_5] = \frac{19}{4} \left[1 - \left(\frac{4}{5}\right)^5\right] = 3.2$; hence, this is the average number of times a rat will fail to pass through the maze if the experiment is run five times.

(b) Since $(1 - c)^n = \left(\frac{4}{5}\right)^n$ will approach 0 when n becomes increasingly large, $E[T_n]$ will approach the value $\frac{(1 - p_1)}{c} = \frac{19}{4} = 4.75$; consequently, this is approximately the average number of failures made by a rat in a large number of runs. This implies that after a while it hardly ever makes an error.

The realism of this model could be checked by performing an experiment with a large number of rats of supposedly equal intelligence and observing whether their total number of errors for various values of n corresponds to the numbers expected from this model. In doing so, we might try different values of c or estimate the value of c by looking at the rate of learning from trial to trial.

COMMENTS

This chapter studied several types of social science problems that require the use of probability for their solution. They represent only a few of the many possible types that can be solved by applying the basic rules of probability.

The concept of expected value is particularly important in many of these problems because it represents a basis for decision making that is not universally understood or employed. It essentially asks one to use a long-range instead of a short-range point of view in making decisions. This point of view is a natural one when studying repetitive type experiments. Such experiments often give rise to independent Bernoulli trials and the binomial distribution, or to dependent trials of the Markov-chain type. Both of those models have numerous applications in the social sciences. Markov chains in particular have found much favor among mathematical psychologists for constructing learning theory models. The model that is presented in Section 7 is only one of many such models. It was selected because it is one of the simplest to explain.

EXERCISES

Section 1

1. Toss three coins simultaneously and record the number of heads obtained. Perform this experiment 100 times and then compare your experimental relative frequencies with those given by theory in Fig. 5.

2. Roll a die twice, or roll two dice simultaneously, and record the number of points obtained. Perform this experiment 100 times and compare your experimental relative frequencies with those given by theory in Fig. 6.

3. A box contains three cards consisting of the 2, 3, and 4 of hearts. Two cards are drawn from the box, with the first card returned to the box

before the second drawing. Let x denote the sum of the numbers obtained on the two cards. Use the enumeration of events technique to derive the distribution of the random variable x.

4. Work Problem 3 under the assumption that the first card is not returned to the box.

5. Use the enumeration of events technique to derive the distribution of the number of heads that will be obtained if a coin is tossed four times.

6. Use the enumeration of events technique to derive the distribution of the number of aces (ones) that will be obtained if a die is rolled three times.

Section 2

1. A roulette wheel has 37 slots numbered from 0 to 36. A player bets $2 on a certain number. If the roulette ball falls in his numbered slot he receives $70 and his $2 is returned to him. If the ball does not fall in his numbered slot he loses his $2. Calculate his expected value.

2. A roulette wheel has 37 slots numbered from 0 to 36. The numbers from 1 to 36 are evenly divided between red and black. A player bets $5 on one of those colors. If the roulette ball falls in a slot having that color, he receives $5 and his $5 is also returned to him. If the ball falls in a slot having the other color or if it falls in the 0 slot, he loses his $5. Calculate his expected value.

3. You toss a coin three times. If you get at least two heads you are permitted to roll a die and you receive in dollars the number of points that show. What can you expect to win at this game if you are permitted to play it once?

4. You toss a coin three times. If you get less than three heads you receive in dollars the number of heads obtained. If you get three heads you are permitted to roll a die once and you recieve in dollars the number of points showing plus a $2 bonus. What can you expect to win at this game?

5. Which game would you choose if given a choice: toss two dice and receive in dollars the sum of the points showing or toss four coins and receive in dollars three times the number of heads obtained?

6. Use the distribution found in Problem 3, Section 1, to calculate the expected value of that random variable.

7. Use the distribution found in Problem 4, Section 1, to calculate the expected value of that random variable.

8. Use the distribution found in Problem 6, Section 1, to calculate the expected value of that random variable.

Section 3

1. A sports promoter is contemplating taking out rain insurance for an athletic event that he is sponsoring. If it does not rain he can expect to net $10,000 on the event, but only $2000 if it does. A $7000 insurance policy will cost him $3000. Determine the probability of rain, p, such that his expectation will be the same whether or not he takes out insurance.

2. A farmer must decide when to plant a crop that is quite sensitive to frost damage. If he plants on April 10 and there is no frost thereafter, he will be able to sell at a premium and make a profit of $15,000. If there is a frost he will lose $3000 for labor and seed. By waiting two weeks, the chances of a frost are diminished; however, the later crop will yield a profit of only $3000 if no frost occurs. The probability of frost after April 10 is $\frac{8}{10}$, whereas two weeks later the probability drops to $\frac{3}{10}$. The farmer wishes to maximize his expected profit. What should he do?

3. For Problem 2, at what level of probability for frost after April 24 will the farmer be indifferent between planting early or late, assuming that the April 10 probability does not change.

4. A car owner wishes to sell his car and is contemplating spending $50 to advertise it. If the probability is .5 that he will sell it at his stipulated price of $750 without advertising and is .9 of doing so if he does advertise, should he advertise? Assume that if he does not sell it for $750 he will let a friend have it for $650.

5. Suppose that you have $10,000 to invest in either bonds or common stocks and that you would like to obtain the highest expected value of your investment at the end of five years. Your securities counselor tells you that bonds will increase in value about 25 percent over the next five years, but that the appreciation in value and dividends of stocks will

depend on economic conditions. He says that if there is deflation, they will decrease about 10 percent. If there is considerable inflation, he estimates that growth will be about 100 percent, and if the economy is fairly normal, he estimates a growth of about 15 percent. He is also willing to conjecture that the probabilities of deflation, normalcy, and inflation are .1, .7, and .2, respectively. Assuming that this advice is reasonable, should you buy bonds or stocks?

6. Each of two stockbrokers has proposed a small portfolio of four stocks and bonds for your consideration. To compare their suggestions you have asked them to give you their best estimates of the value as a percentage of today's price of these portfolios after two years under three possible economic conditions. An economist gives you his estimates of the probabilities for each of the three conditions to occur. Assembled, your data are as follows:

Economic Condition	Estimated Probabilities	Estimated Value of Securities							
		Broker A				Broker B			
		1	2	3	4	1	2	3	4
Decline	.2	30	40	70	110	90	100	108	110
No change	.5	102	103	103	110	100	120	108	110
Expansion	.3	160	230	120	110	110	140	108	110

If you wish to maximize the expected value of your portfolio, which broker should you prefer?

Section 4

1. A coin is tossed five times. By using formula (2), calculate the values of $P\{x\}$, where x denotes the number of heads, and graph $P\{x\}$ as a line chart.

2. If the probability that you will win a hand of bridge is $\frac{1}{4}$ and you play six hands, calculate the values of $P\{x\}$, where x denotes the number of wins, by means of formula (2).

3. If 20 percent of fuses are defective and a box of 10 fuses is purchased, what is the probability that at least seven of those fuses will be good?

4. In Problem 3 how many fuses would you need to purchase to have a probability of at least .9 of obtaining at least seven good fuses? Solve by calculating some guesses.

5. Would you expect the binomial distribution to be applicable to a calculation of the probability that the stock market will rise at least 20 of the days during the next month if you have a record for the last five years of the percentage of the days that it did rise? Explain.

6. Explain why it would not be strictly correct to apply the binomial distribution to a calculation of the probability that it will rain at least 10 days next January if each day in January is treated as a trial of an event, and you have a record of the percentage of rainy days in January.

7. The probability of a marksman hitting the center circle of a target is $\frac{1}{10}$. If he takes five shots at the target, what is the probability that he will hit the center circle (a) exactly once, (b) at least once?

8. An electrical device consists of five separate units connected in such a manner that it will operate successfully only if all five parts operate successfully. If the probability of successful operation for each part is .9:
   (a) What is the probability that the device will work?
   (b) What is this probability if only four of the five parts need to operate successfully?

9. A manufacturer ships his product in boxes of 10. He guarantees that not more than two out of the 10 items will be defective. If the probability that an item selected at random from his production line will be defective is $\frac{1}{10}$, what is the probability that the guarantee will be satisfied?

10. A machine normally produces 5 percent defective items. A sample of eight items is selected by chance from the production line. If the sample produces more than two defective items, the entire days production will be 100 percent inspected. What is the probability that such an inspection will occur?

11. Experience shows that only $\frac{1}{3}$ of the patients having a certain disease will recover from it under a standard treatment. A new treatment is being tried on a group of 12 patients having the disease. If the clinic

requires that at least seven of the 12 patients must recover before it will accept the new treatment as being superior:

(a) What is the probability that the new treatment will be rejected even though the recovery rate is actually $\frac{1}{2}$?

(b) What is the probability that the new treatment will be accepted even though the recovery rate is actually $\frac{1}{4}$?

12. A true-false examination has 20 statements each of which is to be marked as being true, T, or false, F. If a student has no knowledge of the subject and guesses each time, what is the probability that he will score 60 or more points if each correct answer is worth five points?

13. Two qualities of brick are purchased from a brickyard. In the top quality, 5 percent contain flaws, whereas in the second quality, 10 percent contain flaws. Equal amounts of the two qualities are ordered and received, but the quality marking is missing. If 100 bricks are selected at random from one of the two orders and eight of them possess flaws, what is the probability that the top quality order was sampled?

Section 5

1. Given the following transition matrices, calculate $p_{21}{}^{(2)}$ and $p_{12}{}^{(2)}$

$$\text{(a)} \begin{bmatrix} \frac{1}{2} & \frac{1}{2} \\ \frac{1}{3} & \frac{2}{3} \end{bmatrix}, \qquad \text{(b)} \begin{bmatrix} 0 & 0 & 1 \\ \frac{1}{2} & 0 & \frac{1}{2} \\ \frac{1}{2} & \frac{1}{2} & 0 \end{bmatrix}$$

2. Given the following transition matrices, find P^k by first calculating P^2 and P^3 and then observing the nature of the elements.

$$\text{(a)} \; P = \begin{bmatrix} 0 & 1 \\ 1 & 0 \end{bmatrix}, \qquad \text{(b)} \; P = \begin{bmatrix} 1 & 0 & 0 \\ 1 & 0 & 0 \\ 0 & 0 & 1 \end{bmatrix}$$

3. Records at a weather station show that the probability is .7 that a sunny day will be followed by a sunny day, .2 that a sunny day will be followed by a cloudy day, and .1 that a sunny day will be followed by a rainy day. Similarly, the probability is .4 that a cloudy day will be

followed by a sunny day, .4 that it will be followed by a cloudy day, and .2 that it will be followed by a rainy day. The probability for a rainy day to be followed by a sunny day is .4, by a cloudy day is .5, and by a rainy day is .1.

(a) Write down the transition matrix for this weather station.
(b) Calculate the probability that if it rains today, it will also rain two days later.
(c) Calculate the probability that if it is sunny today it will rain two days later.
(d) Calculate the probability that a sunny day will be followed by two consecutive sunny days.

4. Suppose that three companies dominate the market for a certain product and are competing against each other for a larger share of the market. Suppose, further, that market surveys have shown that buyers shift from one brand to another during a given year according to the following schedule:

A keeps 70% of its customers, but loses 15% to B and 15% to C.
B keeps 60% of its customers, but loses 25% to A and 15% to C.
C keeps 50% of its customers, but loses 20% to A and 30% to B.

(a) Write down the transition matrix that applies here.
(b) Calculate what percentage of A's customers will be with it two years from now.
(c) Calculate what percentage of A's customers will be using B's product two years from now.

5. Show that if P is a transition matrix, that is, a matrix with nonnegative elements and in which the sum of the elements of each row is 1, then P^2 is also a transition matrix.

6. One box contains n white balls and a second box contains n black balls. An experiment is conducted in which at each stage of the experiment one ball is chosen at random from the first box and placed in the second box and then one ball is chosen from the second box and placed in the first box. Hence, each box will always contain n balls at the end of each trial. After the experiment has continued for some time, assume that there are i white balls in box I. Then calculate the transition probabilities p_{ii+1}, p_{ii-1}, and p_{ii}. Use these values to form a transition matrix for this experiment, realizing that $0 \leqslant i \leqslant n$ and, therefore, that there are $n + 1$ possible states. It is unnecessary to consider box II because

the composition of box I determines that for box II. This model is often used to represent what happens to the distribution of the molecules of two different gases that are placed in a two-compartment container with the two compartments separated by a membrane and through which the gases can pass.

7. If the three companies of Problem 4 have an equal share of the market today, what share of the market will each possess (a) one year from now? (b) two years from now?

8. Work Problem 7 if today's proportions of the market possessed by the three companies are $\frac{2}{6}$, $\frac{3}{6}$, and $\frac{1}{6}$, respectively.

9. Calculate the long-run proportions of the market that will be possessed by the three companies of Problem 7. Assume that P does not change and that these three companies continue to dominate the market. How will these proportions differ from the long-run proportions that might be calculated for Problem 8?

10. A market analysis of four breakfast cereals gave the following transition matrix for the probabilities of customers shifting from one brand to another. It is assumed that all customers continue with one of these four brands.

$$P = \begin{bmatrix} .6 & .2 & .1 & .1 \\ .2 & .7 & 0 & .1 \\ .2 & .2 & .5 & .1 \\ .1 & 0 & .3 & .6 \end{bmatrix}$$

(a) Calculate P^2.
(b) If the four brands have .3, .3, .2, and .2 of the market today, what will the distribution be one year from now?
(c) What will the long-run proportions be of the market for these four brands if P does not change over time and no new competing brands enter the market?
(d) If it is not reasonable to assume that these four brands dominate the market, what might you do to correct for this possibility?

11. Assume that a gene in a chromosome has two possible forms, A and B. Let r denote the probability that a gene of type A will mutate to type B and let s be the probability that a gene of type B will mutate to type A. Assume that these probabilities hold true for each generation of off-

spring in a population and that there is no overlapping of generations, so that these probabilities may be treated as proportions of the population having the corresponding gene frequencies.

(a) Write down the transition matrix P.

(b) Calculate the long-run gene distribution for this population.

(c) Let p_{n+1} denote the probability that a gene will be of type A at the n + 1st generation. Show by adding the probabilities of the two mutually exclusive ways in which this can occur going from the nth to the n + 1st generation that

$$p_{n+1} = p_n(1-r) + (1-p_n)s = p_n k + s \qquad \text{where} \qquad k = 1 - r - s$$

Solve this difference equation for p_n by letting $n = 1,2,3, \ldots$ and observing the relationship that holds true. Simplify your formula by using the formula for the sum of a geometric progression.

(d) What can you say about the gene frequencies at the nth generation if the original gene frequency of A is $\dfrac{s}{r+s}$?

(e) Assuming that $r + s > 0$, what happens to p_n in (c) as $n \to \infty$? Compare this with the result in (b) and comment.

Section 6

1. A matches pennies with B. If A has 20 pennies and B has 30 pennies, what is the probability that A will win all of B's pennies?

2. Suppose that you are playing a game in which your chances of winning are only $\dfrac{1}{3}$. If you have \$2 and your opponent has \$6, what is the probability that you will win his \$6 if you bet: (a) \$1 each time? (b) \$2 each time?

3. Work Problem 2 if your probability is $\dfrac{1}{2}$.

4. An individual lives four blocks east of a bar and six blocks west of another bar. He can't decide which bar to patronize; therefore, he tosses a coin and agrees to walk one block east if it comes up heads and one block west if it comes up tails. He continues this coin-tossing-walking experiment until he arrives at one of the bars.

(a) Calculate the probability that he will arrive at the eastern bar.

(b) Calculate the probability in (a) if his coin is biased and the probability that it will show a head is .55.

5. Suppose that the probability A will win against B in a game is .5 and
that A has only $1 to bet. How much capital will B need to make his
probability of winning A's one dollar greater than .6?

6. Work Problem 5 if the probability .5 is changed to .6.

7. If your probability is .5 of winning a single game from an opponent who
has $20 to wager, how much capital will you need to insure a probability
of .95 of your winning his $20?

8. Work Problem 7 if your probability is .6. Use approximations in solving
this problem.

9. Suppose an individual going to Las Vegas has $70 that he is willing to
lose at gambling. Suppose also that he will be content to stop gambling
if he can win $10. He plays at a game in which his probability of win-
ning is .48. Calculate the probability that this individual will win $10
before being ruined:
(a) If he bets $1 each time.
(b) If he bets $5 each time.
(c) If he bets $10 each time.
Comment on these results with respect to strategy in a game of this
type.

10. In Problem 9, calculate the probability that the individual will win $10
before being ruined if he uses the strategy of betting $10 the first time
and then doubles his bet each time that he loses, but quits as soon as he
has won $10. Compare this probability with that of Problem 9(c) and
comment on the proper strategy to use in this situation. It can be
shown that the doubling strategy here is the best possible.

Section 7

1. In a learning situation, suppose that the proportionality constant c has
the value $\frac{1}{10}$ and that the probability of a correct decision at the first trial
is $\frac{1}{30}$.
(a) Calculate the value of p_n for $n = 2$ and $n = 3$.

(b) What value does p_n approach as $n \to \infty$?

(c) Calculate the expected number of errors that will be made in n trials of the experiment.

(d) What value does $E[T_n]$ approach as $n \to \infty$?

2. If the proportionality constant c has the value $\frac{1}{2}$, and p_1 has the value $\frac{1}{10}$, how many trials of a learning experiment must be performed before the probability is at least $\frac{3}{4}$ of making the correct decision?

3. What value of c would you choose in an animal learning experiment if an experiment was conducted with 20 animals and records showed that five of them made the correct decision on the first run and 10 of them make the correct decision on the second run?

4. Suppose that you are attempting to design a learning experiment such that the expected number of errors an animal will make in a large number of runs of the experiment will not exceed seven. If experience indicates that $c = \frac{1}{10}$ for this type of experiment, how difficult should the experiment be made?

5. Suppose that a student is learning to write programs for high-speed computing equipment and that his probability of writing a correct program after instruction is only $\frac{1}{100}$. Suppose further that his learning rate is measured by the constant c having the value $\frac{1}{4}$.

(a) Calculate the value of p_n for $n = 5$.

(b) How large would you estimate n to be before his value of p_n would exceed $\frac{1}{2}$?

(c) What value does p_n approach as $n \to \infty$?

(d) What is the expected total number of errors that he will make before mastering (theoretically) programming?

7 | Game Theory

1. INTRODUCTION

Game theory is a mathematical theory that attempts to solve the problem of how two or more competing individuals or firms should operate so as to achieve the maximum advantage for themselves. Since one competitor is likely to gain at the expense of the others, a compromise is needed to arrive at a solution that may be considered fair to all the competitors. The problem is considerably simpler and the theory much neater when there are only two individuals or firms involved; therefore, we restrict ourselves to that type of problem.

Suppose, therefore, that we have two individuals who are engaged in a competitive enterprise, such as playing a game of chess, running for a political office, or bidding on a contract to construct a factory. Only the first of these illustrations is a game in the traditional sense; however, we will call all such competitive activities *games*. Furthermore, when there are but two individuals, or firms, competing against each other we will call the activity a *two-person game*. The word individual is used hereafter to denote a competitor, regardless of whether the competitor is a person, a firm, or a larger organization.

The games that we discuss are those in which there are only a finite number of possible actions that each individual can take. In this connection, we denote the two competing individuals by R and C and their possible actions by $R_1, R_2, \ldots, R_m$, and $C_1, C_2, \ldots, C_n$, respectively. The reason for calling the two individuals R and C is that their respective possible actions, which are called *strategies*, are going to be associated with the rows and columns of a matrix.

A game consists of allowing R and C to choose one of their possible strategies, without a knowledge of the other's choice, and then on the basis of their choices giving the winner a suitable reward. The rewards are determined by the numbers found in a *payoff matrix*. If R chooses strategy R_i and C chooses strategy C_j, the payoff is the amount a_{ij} that occurs in ith row and jth column of the payoff matrix. Since R has m strategies and C has n strategies to choose from, this matrix has m rows and n columns. If $a_{ij} > 0$, R receives the amount a_{ij}, whereas if $a_{ij} < 0$, C receives the amount $|a_{ij}|$. If any amount that one player receives must be paid by the other, the game is called a *zero-sum* game. The name is quite appropriate because the sum of the winnings of R and C, whatever their choices of strategies, is always zero. We consider games of this type only; hence, we are studying *zero-sum two-person games*.

At first glance it might appear that games of the preceding type are very artificial because each competitor is permitted to choose only one strategy before the outcome of the game is decided. In a game such as chess a large number of moves may be required before a winner is determined. However, if the term strategy is understood to mean a method of playing the game, then the possible strategies $R_1, R_2, \ldots, R_m$ represent all the possible ways that R knows of playing games of chess. A strategy therefore describes which chess piece R will move at each stage of the match corresponding to what move his opponent made at the preceding stage. The number of possibilities in a game of chess even when a time limit is placed on the total number of moves that can be made is obviously extremely large. From this more general point of view, a set of strategies in any game can include a complicated set of activities of the two competitors and can, therefore, constitute a very realistic model for real-life activities. For this reason, it is customary to use the phrase *game of strategy* in describing the elements of game theory.

Since we are assuming that R has m possible strategies and C has n possible strategies, the payoff matrix corresponding to these various possibilities can be represented by the following $m \times n$ matrix:

$$
A = \begin{bmatrix}
a_{11} & \cdots & a_{1j} & \cdots & a_{1n} \\
\vdots & & \vdots & & \vdots \\
a_{i1} & \cdots & a_{ij} & \cdots & a_{in} \\
\vdots & & \vdots & & \vdots \\
a_{m1} & \cdots & a_{mj} & \cdots & a_{mn}
\end{bmatrix}
$$

Positive elements of A represent amounts won by R from C and negative elements represent losses by R to C; therefore, the elements of A can be thought of as the amounts paid to R by C. In this sense, the matrix A is a payoff matrix to R. The numbers a_{ij} are determined from the nature of the game or are the rewards agreed on by R and C.

2. OPTIMIZATION

Consider the following simple 2×3 payoff matrix in which the elements represent dollar rewards to R:

$$\begin{bmatrix} 5 & 2 & 3 \\ -5 & 1 & 6 \end{bmatrix}$$

We wish to determine which strategies R and C should choose. Suppose that R chooses R_1, which means that he chooses the first row. Then he can be certain of winning at least $2 and at most $5. If he chooses R_2, he could lose $5 but he might win as much as $6. In view of the wide range of possibilities here, it appears that R should examine more carefully his possible strategies to determine how much he can win even if C is clever or lucky enough to choose his best possible strategy each time.

If R were to choose R_1, C would certainly choose C_2 and thus would limit R to winning $2. If R were to choose R_2, C would obviously choose C_1 and thus would cause R to lose $5. The two elements that represent these two possibilities have been enclosed in circles in the matrix of Fig. 1. They represent the worst that can happen to R regardless of how clever C may be in anticipating which strategy R will select. If R wishes to protect himself best possible against these undesirable possibilities, he should choose R_1. Choosing R_1 makes it certain that R will win at least $2, and possibly more if it should happen that C does not select his best strategy. Although neither opponent knows which strategy the other will choose, one of them may outguess the other and may gain considerably thereby. The preceding analysis guards R against the occurrence of this possibility.

FIGURE 1   Strategies for R and C

Now consider which strategy C should employ. He wishes to keep the amount to be won by R as small as possible because he is required to pay R when R wins. If he reasons in the same manner as R does, he will examine each of his possible strategies to see how well he can do even if R is clever enough to anticipate which strategy C will select. Thus, if C were to select C_1, R would choose R_1 and win \$5. If C were to select C_2, R would choose R_1 and win \$2. If C were to select C_3, R would choose R_2 and win \$6. Each of the elements representing these three possibilities has been enclosed in a square in Fig. 1. They represent the worst that can happen to C regardless of how clever or lucky R may be. If C wishes to protect himself best possible against these undesirable possibilities, he should choose C_2 because that minimizes the amount that he will need to pay to R.

The preceding analysis, which assumes that R wishes to maximize his winnings and C wishes to minimize his losses, leads to the conclusion that R should choose strategy R_1 and C should choose strategy C_2. If the game is played with these two strategies, R will win \$2 from C. The number 2 is called the *value* of this game. It represents the amount that R expects to win and C expects to pay if R and C play their best defensive strategies.

If the best strategies for R and C do not yield a common circle-square element of the payoff matrix, the game does not have a solution, and therefore does not have a value. Games of this type are studied in Section 4.

3. RANDOMIZED STRATEGIES

The preceding game was played under the assumption that it was to be played only once. Hence, once R and C have selected their strategies, the result of the game is determined. Suppose, however, that a sequence of games is to be played and that R and C are permitted to change their strategies from game to game. How should they proceed then? Since each contestant assumes that the other is just as intelligent as he is, they dare not develop a systematic pattern of choosing strategies for fear of this being discovered; hence, it would be advisable for them to choose their successive strategies by means of some random scheme. This can be accomplished by having R select a set of probabilities $p_1, p_2, \ldots, p_m$ that will determine the relative frequencies with which he wishes his strategies $R_1, R_2, \ldots, R_m$ to be played. Similarly, C is permitted to select a set of probabilities $q_1, q_2, \ldots, q_n$ that will determine the relative frequencies with which he wishes his strategies $C_1, C_2, \ldots, C_n$ to be played.

As an illustration, in the game of the preceding section R might choose $p_1 = \frac{2}{3}$ and $p_2 = \frac{1}{3}$, and C might choose $q_1 = \frac{2}{4}$, $q_2 = \frac{1}{4}$, and $q_3 = \frac{1}{4}$. Each time the game is to be played, R will use a game of chance that yields P_1 twice as frequently as P_2 to choose one of those strategies. This could be done, for example, by drawing a card from a set of three cards that contains two aces and one 2. Similarly, C could draw a card from a set of four cards that contains two aces, one 2, and one 3.

Strategies that are selected by chance according to a set of probabilities are called *randomized strategies*. They include the one-play strategies that were obtained in the preceding section because it is merely necessary to choose $p_1 = 1$, $p_2 = 0$, and $q_1 = 0$, $q_2 = 1$, $q_3 = 0$ to arrive at the strategies R_1 and C_2 that were determined for that game. If a player uses a set of probabilities in which one of them is 1 and all the rest are 0, he is said to be using a *pure strategy*, otherwise he is using a *mixed strategy*.

Even though a game is to be played only once, and this is the natural situation in most business-type games, it may be that one of the competitors can do better than the other by employing randomization in choosing his strategy. Therefore, we take a fresh look at our earlier one-play games in studying randomized games to see if improvements are possible by using randomization.

Since the payoffs will vary from game to game because they will depend on chance, we look at the average payoff in a long sequence of games. This is equivalent to looking at the expected value of the payoff. Now the expected value of the payoff in the ith row and jth column of the payoff matrix A is merely $a_{ij}p_iq_j$ because the row and column choices are made independently, and therefore the probability of R winning the amount a_{ij} is the product of the ith row and jth column probabilities.

As an illustration of a randomized payoff matrix, consider the matrix of Section 2 if the probabilities are $p_1 = \frac{2}{3}$, $p_2 = \frac{1}{3}$, $q_1 = \frac{1}{2}$, $q_2 = \frac{1}{4}$, and $q_3 = \frac{1}{4}$. Letting B denote this matrix, we find that

$$
B = \begin{bmatrix} 5 \cdot \frac{2}{3} \cdot \frac{1}{2} & 2 \cdot \frac{2}{3} \cdot \frac{1}{4} & 3 \cdot \frac{2}{3} \cdot \frac{1}{4} \\ -5 \cdot \frac{1}{3} \cdot \frac{1}{2} & 1 \cdot \frac{1}{3} \cdot \frac{1}{4} & 6 \cdot \frac{1}{3} \cdot \frac{1}{4} \end{bmatrix}
$$

The expected payoff to R is the sum of these individual expected payoff values. Their sum is found to be $2\frac{1}{4}$; therefore, R would do slightly better under these two sets of randomized strategies than under the original nonrandomized version of the game. It may well be, however, that C chose a poor set of probabilities here. The purpose of this example is to illustrate how expected payoffs are calculated; it is not intended to illustrate good randomized strategies.

After a set of probabilities has been selected by each of R and C and the corresponding expected payoff matrix is calculated, the game is completely determined and could be played by a machine that selects successive pairs of strategies according to the probabilities $p_1, \ldots, p_m$ and $q_1, \ldots, q_n$.

The interesting question now is: How should R and C choose their probabilities? For example, is it possible for R to do better than C in the preceding illustrative game if his probabilities are chosen properly, regardless of what probabilities C chooses? In the preceding illustration R did do better but perhaps C could have prevented this with a better set of probabilities. The answer to this question is as follows: Independent of what probabilities R selects, C can find a set of probabilities such that the expected value of the payoff to R will not exceed \$2. In addition, regardless of what probabilities C may select, R can find a set of probabilities such that he can be assured of averaging at least \$2. Thus neither R nor C can gain in this particular problem by using mixed strategies rather than pure strategies, provided that both R and C employ their best defensive randomized strategies.

4. NONDETERMINED GAMES

The illustration used in Section 2 has the property that there is a common element in the payoff matrix, namely 2, that represents what R expects to win and C expects to pay. Suppose, however, that our payoff matrix is the following one:

$$\begin{bmatrix} 5 & 1 & 3 \\ -5 & 2 & 6 \end{bmatrix}$$

If, as previously, we place circles around elements chosen by C, corresponding to each choice of R, and squares around elements chosen by R, corresponding to each choice of C, we obtain

According to the reasoning used in Section 2, R should choose strategy R_1 because that yields the larger of the two encircled elements and assures R that he will win at least \$1 regardless of how C plays. Similarly, C should choose C_2 because that yields the smallest of the three elements inside squares and assures C that he will need to pay out at most \$2 regardless of how R plays. But now we have the paradox that C would have been willing to pay out \$2 but is required to pay out only \$1 if these best defensive strategies are employed by both R and C. Thus in this game there is not a unique element that represents how much R expects to win and at the same time the amount that C expects to pay R.

For games of this type we can no longer speak of the value of the game. By definition, a game has a value if, and only if, there is an element of the payoff matrix that is the largest of the payoffs that R can receive if C uses his best strategy and also is the smallest of the payoffs that R can be assured of receiving. The value of the game is the value of that common element. It is customary to speak of the element that produces the value of a game as being a *saddle point*. The name arises in studying surfaces in calculus where a saddle point is the point that is lowest on the saddle in the direction of the horse's head and is highest in the direction of the rider's legs. It is the smallest of the large values in one direction and the largest of the small values in the other direction. This corresponds to the property of the value of a game being the smallest element in its row and the largest element in its column. To find a saddle point in a payoff matrix, it suffices to place circles around the smallest element in each row and then to check to see whether one of those encircled elements is also the largest element in its column. There is usually only one such element. If there were two such elements they would necessarily have the same value.

If a game does have a value, it is said to be determined, otherwise it is called a *nondetermined game*. In the literature of game theory the more technical phrases *strictly determined* and *nonstrictly determined* are used to describe games of this kind. The game that is being used as an illustration in this section is an example of a nondetermined game. Now just as in the case of the determined game of Section 2, we may ask whether R or C can do better in this game by employing randomized strategies. For a determined game we discovered that nothing was gained by employing randomized strategies, provided that both R and C used their best randomized strategies. But since R will not be receiving as much as C would have been willing to pay, it may well be that R can do better by employing a randomized strategy here. Hence, the question arises: What happens to a non-determined game if randomized strategies are employed? The answer to that question is given by the following remarkable theorem, which is the basic theorem of game theory. The theorem is too difficult to prove here; therefore, it is only stated and applied.

Theorem. *For every matrix game there exist optimal randomized strategies for R and C and a number v such that if R employs his optimal strategy his expected payoff will be $\geq v$ for every strategy of C, and such that if C employs his optimal strategy the expected payoff to R will be $\leq v$ for every strategy of R. The number v is called the value of the game.*

From this theorem it follows that every game has a solution when randomized strategies are employed. Hence, if we have a game for which the payoff matrix contains a saddle point, the solution is given by the pure strategies of R and C corresponding to the saddle point, and the value of the game is the value of the saddle point element. If, however, there is no saddle point, we employ randomized strategies, and we are assured that there exist strategies for R and C, which are called optimal randomized strategies, that yield an expected payoff v which plays the same role for randomized strategies that the value of the game does for pure strategies.

5. CALCULATION OF THE OPTIMAL STRATEGIES

In view of the preceding theorem it is theoretically possible to solve any matrix game. The problem remains, of course, of obtaining the solution in a systematic practical way. Fortunately, any game theory problem can be converted over to a linear programming problem to which the simplex method may be applied. The purpose of this section is to show how this can be done.

Let $A = [a_{ij}]$ be any $m \times n$ payoff matrix that does not have a saddle point. Assume that all the elements are positive. This can be realized by adding a sufficiently large positive constant to all the elements of the original payoff matrix. It is easily shown that this merely increases the value of the game by this constant amount without changing the optimal strategies. Let v be the value of the randomized game for the matrix A. It is easy to show that, when the elements of A are all positive, then, v must also be positive.

Let

$$P = [p_1, p_2, \ldots, p_m]$$

and

$$Q = \begin{bmatrix} q_1 \\ q_2 \\ \cdot \\ \cdot \\ \cdot \\ q_n \end{bmatrix}$$

be any randomized strategies selected by R and C, respectively. Then the amount that R can expect to win can be expressed as a product of the following matrices:

$$PAQ$$

This is merely a compact matrix way of writing out the sum of all the products of the form $a_{ij} p_i q_j$.

If C chooses his optimal randomized strategy, which will be denoted by Q^*, then it must be true from our theorem that

(1) $$PAQ^* \leqslant v$$

for every probability vector P. In particular, this inequality must hold true for the vectors $P_1 = [1,0, \ldots ,0]$, $P_2 = [0,1, \ldots ,0]$, $\ldots$,$P_m = [0,0, \ldots ,1]$. But, with the understanding that the q's on the right are the components of the optimal vector Q^*, calculations give

$$P_1 A Q^* = [1,0, \ldots ,0] \begin{bmatrix} a_{11} & \cdots & a_{1n} \\ \cdot & & \cdot \\ \cdot & & \cdot \\ \cdot & & \cdot \\ a_{m1} & \cdots & a_{mn} \end{bmatrix} \begin{bmatrix} q_1 \\ \cdot \\ \cdot \\ \cdot \\ q_n \end{bmatrix}$$

$$= [a_{11}, a_{12} \ldots a_{1n}] \begin{bmatrix} q_1 \\ \cdot \\ \cdot \\ \cdot \\ q_n \end{bmatrix}$$

$$= a_{11} q_1 + a_{12} q_2 + \cdots + a_{1n} q_n$$

Similarly

$$P_2AQ^* = a_{21}q_1 + a_{22}q_2 + \cdots + a_{2n}q_n$$

$$P_mAQ^* = a_{m1}q_1 + a_{m2}q_2 + \cdots + a_{mn}q_n$$

In view of inequality (1), it follows from these calculations that

$$a_{11}q_1 + a_{12}q_2 + \cdots + a_{1n}q_n \leqslant v$$
$$a_{21}q_1 + a_{22}q_2 + \cdots + a_{2n}q_n \leqslant v$$

$$a_{m1}q_1 + a_{m2}q_2 + \cdots + a_{mn}q_n \leqslant v$$

Since $v > 0$, we may divide both sides of these inequalities by v to arrive at a set of equivalent inequalities. After doing so, we simplify the notation by letting $x_i = \dfrac{q_i}{v}$. Then we obtain the inequalities

(2)

$$a_{11}x_1 + a_{12}x_2 + \cdots + a_{1n}x_n \leqslant 1$$
$$a_{21}x_1 + a_{22}x_2 + \cdots + a_{2n}x_n \leqslant 1$$

$$a_{m1}x_1 + a_{m2}x_2 + \cdots + a_{mn}x_n \leqslant 1$$

Since Q^* is a probability vector, it follows from the definition of the x's that we also have the relation

(3)
$$x_1 + \cdots + x_n = \frac{q_1 + \cdots + q_n}{v} = \frac{1}{v}$$

Now if the probabilities chosen by C are optimal, they must minimize v. To find these optimal probabilities, it therefore suffices to find the values of $x_1, \ldots, x_n$ that minimize v, subject to the inequalities given in (2). But in view of equation (3), this is equivalent to finding the values of $x_1, \ldots, x_n$ that maximize the function $f = x_1 + \cdots + x_n$ subject to the restrictions given in (2). Thus the problem of finding the optimal strategy for C has been converted to the linear programming problem of maximizing

$$f = x_1 + x_2 + \cdots + x_n$$

subject to the restrictions

$$
\begin{aligned}
a_{11}x_1 + \cdots + a_{1n}x_n &\leq 1 \\
a_{21}x_1 + \cdots + a_{2n}x_n &\leq 1
\end{aligned}
$$

(4)

$$
\begin{aligned}
a_{m1}x_1 + \cdots + a_{mn}x_n &\leq 1 \\
x_1 \geq 0,\ x_2 \geq 0,\ \ldots\ ,x_n &\geq 0
\end{aligned}
$$

The same type of reasoning shows that if we let $y_i = \dfrac{p_i}{v}$, the problem of finding an optimal set of probabilities for R is equivalent to the linear programming problem of minimizing the function

$$
g = y_1 + y_2 + \cdots + y_m
$$

subject to the restrictions

$$
\begin{aligned}
a_{11}y_1 + a_{21}y_2 + \cdots + a_{m1}y_m &\geq 1 \\
a_{12}y_1 + a_{22}y_2 + \cdots + a_{m2}y_m &\geq 1
\end{aligned}
$$

(5)

$$
\begin{aligned}
a_{1n}y_1 + a_{2n}y_2 + \cdots + a_{mn}y_m &\geq 1 \\
y_1 \geq 0,\ y_2 \geq 0,\ \ldots\ ,y_m &\geq 0
\end{aligned}
$$

Notice that each inequality in (5) uses as the coefficients of the y's a column of the payoff matrix A, whereas each inequality in (4) uses as the coefficients of the x's a row of the payoff matrix.

As a first illustration, let us solve the game discussed in Section 4 and for which there is not a determined solution without randomization. The payoff matrix is

(6)
$$
\begin{bmatrix} 5 & 1 & 3 \\ -5 & 2 & 6 \end{bmatrix}
$$

Although the elements are not all positive here, it seems clear that the value of the game will be positive; therefore, it is not necessary to add a positive constant, such as 6, to all the elements to insure that $v > 0$.

To find an optimal strategy for C, we maximize

$$
f = x_1 + x_2 + x_3
$$

subject to

$$5x_1 + x_2 + 3x_3 \leqslant 1$$
$$-5x_1 + 2x_2 + 6x_3 \leqslant 1$$
$$x_1 \geqslant 0,\ x_2 \geqslant 0,\ x_3 \geqslant 0$$

The simplex method of solving this problem then proceeds as follows:

⑤	1	3	1	0		1
−5	2	6	0	1		1
−1	−1	−1	0	0		0

①	$\frac{1}{5}$	$\frac{3}{5}$	$\frac{1}{5}$	0		$\frac{1}{5}$
−5	2	6	0	1		1
−1	−1	−1	0	0		0

1	$\frac{1}{5}$	$\frac{3}{5}$	$\frac{1}{5}$	0		$\frac{1}{5}$
0	③	9	1	1		2
0	$-\frac{4}{5}$	$-\frac{2}{5}$	$\frac{1}{5}$	0		$\frac{1}{5}$

1	$\frac{1}{5}$	$\frac{3}{5}$	$\frac{1}{5}$	0		$\frac{1}{5}$
0	①	3	$\frac{1}{3}$	$\frac{1}{3}$		$\frac{2}{3}$
0	$-\frac{4}{5}$	$-\frac{2}{5}$	$\frac{1}{5}$	0		$\frac{1}{5}$

1	0	0	$\frac{2}{15}$	$-\frac{1}{15}$		$\frac{1}{15}$
0	1	3	$\frac{1}{3}$	$\frac{1}{3}$		$\frac{2}{3}$
0	0	2	$\frac{7}{15}$	$\frac{4}{15}$		$\frac{11}{15}$

The solution is now seen to be $x_1 = \dfrac{1}{15}$, $x_2 = \dfrac{2}{3}$, $x_3 = 0$, and the maximum value of f is $\dfrac{11}{15}$.

To obtain an optimal strategy for R, we minimize

$$g = y_1 + y_2$$

subject to

$$5y_1 - 5y_2 \geqslant 1$$
$$y_1 + 2y_2 \geqslant 1$$
$$3y_1 + 6y_2 \geqslant 1$$
$$y_1 \geqslant 0, \; y_2 \geqslant 0$$

We observe that the third inequality is automatically satisfied if the second one is; therefore, we may delete the third inequality. Since this problem involves only two variables, it is easily solved by geometrical methods. Hence, we graph the two lines $5y_1 - 5y_2 = 1$ and $y_1 + 2y_2 = 1$ and determine the feasibility region as shown in Fig. 2.

Since the lines of the family $y_1 + y_2 = g$ have slope -1, the line with the smallest g that intersects the feasibility region is the line denoted by l in Fig. 1. It intersects the feasibility region in the point that is the intersection of the lines $5y_1 - 5y_2 = 1$ and $y_1 + 2y_2 = 1$. This point has the coordinates $y_1 = \dfrac{7}{15}$ and $y_2 = \dfrac{4}{15}$. The minimum value of $g = y_1 + y_2$ is therefore $\dfrac{11}{15}$, which was also found earlier to be the maximum value of $f = x_1 + x_2 + x_3$.

We now use the relationship between the x's and q's, and between the y's and the p's to obtain optimal strategies for R and C. From the solution of

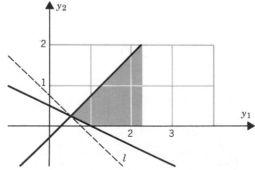

FIGURE 2   The feasibility region for y_1 and y_2.

the maximizing problem we have

$$\frac{q_1}{v} = \frac{1}{15}, \qquad \frac{q_2}{v} = \frac{2}{3}, \qquad \text{and} \qquad \frac{q_3}{v} = 0$$

Since the maximum value of f is $\frac{11}{15}$, it follows from (3) that the value of v is $\frac{15}{11}$; hence, by applying this to the preceding equations, we find that the optimal strategy for C is

$$q_1 = \frac{1}{11}, \qquad q_2 = \frac{10}{11}, \qquad \text{and} \qquad q_3 = 0$$

To obtain the optimal strategy for R, we first write down the solution to the minimizing problem:

$$\frac{p_1}{v} = \frac{7}{15} \qquad \text{and} \qquad \frac{p_2}{v} = \frac{4}{15}$$

Since $v = \frac{15}{11}$, the optimal strategy for R is

$$p_1 = \frac{7}{11} \qquad \text{and} \qquad p_2 = \frac{4}{11}$$

This optimal randomized solution should be compared with the results obtained for the nondetermined game discussed in Section 4 that were based on pure strategies. In that game, R would have selected strategy R_1 and C would have selected strategy C_2. This would have yielded R a payoff of $1, in spite of the fact that C was prepared to pay as much as $2. In the randomized game, R can expect to win $\frac{15}{11} = 1\frac{4}{11}$ dollars. Thus, randomization is beneficial to R for this game. If R is willing to use the expected value of a payoff table as a basis for making a decision, even though it may be a one-event decision only, he should employ an optimal randomized strategy.

It is instructive to find the optimal randomized strategies for a game that is determined under pure strategies and show, as was stated previously, that the value of the game will be the same. As an illustration, consider the game of Section 2 for which the payoff matrix is

(7)
$$\begin{bmatrix} 5 & 2 & 3 \\ -5 & 1 & 6 \end{bmatrix}$$

Proceeding as before, we maximize the function

$$f = x_1 + x_2 + x_3$$

subject to the restrictions

$$5x_1 + 2x_2 + 3x_3 \leqslant 1$$
$$-5x_1 + x_2 + 6x_3 \leqslant 1$$
$$x_1 \geqslant 0,\ x_2 \geqslant 0,\ x_3 \geqslant 0$$

The simplex solution then proceeds as follows:

⑤	2	3	1	0		1
-5	1	6	0	1		1
-1	-1	-1	0	0		0

①	$\frac{2}{5}$	$\frac{3}{5}$	$\frac{1}{5}$	0		$\frac{1}{5}$
-5	1	6	0	1		1
-1	-1	-1	0	0		0

1	$\left(\frac{2}{5}\right)$	$\frac{3}{5}$	$\frac{1}{5}$	0		$\frac{1}{5}$
0	3	9	1	1		2
0	$-\frac{3}{5}$	$-\frac{2}{5}$	$\frac{1}{5}$	0		$\frac{1}{5}$

$\frac{5}{2}$	①	$\frac{3}{2}$	$\frac{1}{2}$	0		$\frac{1}{2}$
0	3	9	1	1		2
0	$-\frac{3}{5}$	$-\frac{2}{5}$	$\frac{1}{5}$	0		$\frac{1}{5}$

$\frac{5}{2}$	1	$\frac{3}{2}$	$\frac{1}{2}$	0		$\frac{1}{2}$
$-\frac{15}{2}$	0	$\frac{9}{2}$	$-\frac{1}{2}$	1		$\frac{1}{2}$
$\frac{3}{2}$	0	$\frac{5}{10}$	$\frac{5}{10}$	0		$\frac{1}{2}$

The solution is now seen to be $x_1 = 0$, $x_2 = \dfrac{1}{2}$, $x_3 = 0$, and the maximum value of f is $\dfrac{1}{2}$.

The minimizing problem is that of minimizing

$$g = y_1 + y_2$$

subject to the restrictions

$$5y_1 - 5y_2 \geqslant 1$$
$$2y_1 + y_2 \geqslant 1$$
$$3y_1 + 6y_2 \geqslant 1$$
$$y_1 \geqslant 0,\ y_2 \geqslant 0$$

As previously, the last restriction may be deleted because it is satisfied if the second one is. This problem is readily worked by graphical methods. Fig. 3 gives the solution. The line l with slope -1 minimizes $y_1 + y_2 = g$. It intersects the feasibility region in the point $y_1 = \dfrac{1}{2}$, $y_2 = 0$.

Converting over to q and p values we have

$$\frac{q_1}{v} = 0, \quad \frac{q_2}{v} = \frac{1}{2}, \quad \text{and} \quad \frac{q_3}{v} = 0$$

together with

$$\frac{p_1}{v} = \frac{1}{2} \quad \text{and} \quad \frac{p_2}{v} = 0$$

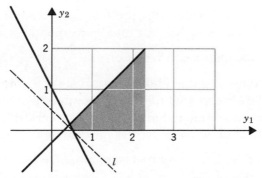

FIGURE 3 The feasibility region for y_1 and y_2.

Since the maximum value of f is $\frac{1}{2}$, the value of v is 2; hence, the optimal strategies for R and C are, respectively,

$$p_1 = 1 \quad \text{and} \quad p_2 = 0$$

and

$$q_1 = 0, \quad q_2 = 1, \quad \text{and} \quad q_3 = 0$$

This, of course, is the solution obtained before without randomization.

Although the optimal randomized strategies reduce to the optimal pure strategies here, it is possible to have a game for which there exist optimal pure strategies but for which the optimal randomized strategies do not reduce to pure strategies. This can occur when there is more than one saddle point. The value of the game, however, will be the same.

There exists a theorem in linear programming, called the duality theorem, that makes it unnecessary to solve both the maximizing and the minimizing problem to obtain the optimal strategies for R and C. It is sufficient to solve one of those problems by the simplex method because the solution to the other problem can be read off the final matrix configuration. Since the theorem is difficult to prove and the statement of the theorem gives no indication of why we can read off both solutions after the final calculations have been made, we merely apply the consequences of the theorem.

This theorem assures us that if we have solved the maximization problem, we can find the solution to the minimization problem in the last row of our final configuration, corresponding to the columns representing the slack variables. The same property holds if we interchange maximization and minimization.

As a first illustration, consider the problem in Section 5 associated with matrix (6). The elements in the last row, corresponding to the two slack variables, of the final matrix configuration are $\frac{7}{15}$ and $\frac{4}{15}$. According to the duality theorem, these are the minimizing values of y_1 and y_2 for the function $g = y_1 + y_2$. We observe that these are the values that we obtained from Fig. 2 by solving the minimization problem geometrically.

As a second illustration, consider the pure strategy game of matrix (7). The final simplex calculations for that game produced the numbers $\frac{5}{10}$ and

0 in the last row, corresponding to the two slack variables. These are the values that were obtained for y_1 and y_2 in Fig. 3.

Since the simplex technique calculations are essentially cut in half by use of the consequences of the duality theorem, we obviously should use that property when solving any matrix game. This is done in the next section where we solve a 3×3 matrix game.

In view of the preceding results, we should first check a payoff matrix to see whether there exists a saddle point. If there does, optimal strategies exist as pure strategies, and we can read them off the payoff table. If a saddle point does not exist, we introduce randomized strategies and find the optimal such strategies by solving two linear programming problems with a single simplex calculation.

6. AN ILLUSTRATION

Two companies, R and C, that manufacture the same product are bidding on a contract that has been put up for bids. Each company has three bids that it can submit. One bid will yield a high profit, another a modest profit, and the third a small profit. The two sets of possible bids are not the same numerically; therefore, one company always wins the contract. The following matrix represents the relative values to the two companies in certain monetary units corresponding to their possible bids. In this bidding, R submits a slightly lower bid than C at the same level but loses the bid to C if C bids at a lower profit level.

$$
\begin{array}{cc}
 & \text{C} \\
 & \text{High \ Modest \ Small}
\end{array}
$$

$$
\text{R}\quad
\begin{array}{c}
\text{High} \\
\text{Modest} \\
\text{Small}
\end{array}
\begin{bmatrix}
4 & -1 & -2 \\
-1 & 4 & -1 \\
-2 & 2 & 3
\end{bmatrix}
$$

This payoff matrix does not possess a saddle point; therefore, we introduce randomized strategies and solve for the optimal strategies by linear programming techniques. Since the positive terms are larger numerically than the negative terms, the value of v is undoubtedly positive; therefore, it is not necessary to add a positive constant to all terms to insure that $v > 0$.

We first use the simplex method to maximize

$$f = x_1 + x_2 + x_3$$

subject to the restrictions

$$4x_1 - x_2 - 2x_3 \leqslant 1$$
$$-x_1 + 4x_2 - x_3 \leqslant 1$$
$$-2x_1 + 2x_2 + 3x_2 \leqslant 1$$
$$x_1 \geqslant 0, \ x_2 \geqslant 0, \ x_3 \geqslant 0$$

The simplex technique proceeds as follows:

④	−1	−2	1	0	0	1
−1	4	−1	0	1	0	1
−2	2	3	0	0	1	1
−1	−1	−1	0	0	0	0

①	$-\frac{1}{4}$	$-\frac{1}{4}$	$\frac{1}{4}$	0	0	$\frac{1}{4}$
−1	4	−1	0	1	0	1
−2	2	3	0	0	1	1
−1	−1	−1	0	0	0	0

1	$-\frac{1}{4}$	$-\frac{1}{2}$	$\frac{1}{4}$	0	0	$\frac{1}{4}$
0	$\frac{15}{4}$	$-\frac{3}{2}$	$\frac{1}{4}$	1	0	$\frac{5}{4}$
0	$\frac{3}{2}$	②	$\frac{1}{2}$	0	1	$\frac{3}{2}$
0	$-\frac{5}{4}$	$-\frac{3}{2}$	$\frac{1}{4}$	0	0	$\frac{1}{4}$

1	$-\frac{1}{4}$	$-\frac{1}{2}$	$\frac{1}{4}$	0	0	$\frac{1}{4}$
0	$\frac{15}{4}$	$-\frac{3}{2}$	$\frac{1}{4}$	1	0	$\frac{5}{4}$
0	$\frac{3}{4}$	①	$\frac{1}{4}$	0	$\frac{1}{2}$	$\frac{3}{4}$
0	$-\frac{5}{4}$	$-\frac{3}{2}$	$\frac{1}{4}$	0	0	$\frac{1}{4}$

$$
\begin{array}{cccccc|c}
1 & \dfrac{1}{8} & 0 & \dfrac{3}{8} & 0 & \dfrac{1}{4} & \dfrac{5}{8} \\[2mm]
0 & \boxed{\dfrac{39}{8}} & 0 & \dfrac{5}{8} & 1 & \dfrac{3}{4} & \dfrac{19}{8} \\[2mm]
0 & \dfrac{3}{4} & 1 & \dfrac{1}{4} & 0 & \dfrac{1}{2} & \dfrac{3}{4} \\[2mm]
\hline
0 & -\dfrac{1}{8} & 0 & \dfrac{5}{8} & 0 & \dfrac{3}{4} & \dfrac{11}{8}
\end{array}
$$

$$
\begin{array}{cccccc|c}
1 & \dfrac{1}{8} & 0 & \dfrac{3}{8} & 0 & \dfrac{1}{4} & \dfrac{5}{8} \\[2mm]
0 & \boxed{1} & 0 & \dfrac{5}{39} & \dfrac{8}{39} & \dfrac{6}{39} & \dfrac{19}{39} \\[2mm]
0 & \dfrac{3}{4} & 1 & \dfrac{1}{4} & 0 & \dfrac{1}{2} & \dfrac{3}{4} \\[2mm]
\hline
0 & -\dfrac{1}{8} & 0 & \dfrac{5}{8} & 0 & \dfrac{3}{4} & \dfrac{11}{8}
\end{array}
$$

$$
\begin{array}{ccccccc|c}
1 & 0 & 0 & \dfrac{14}{30} & -\dfrac{1}{39} & \dfrac{9}{39} & \dfrac{22}{39} \\[2mm]
0 & 1 & 0 & \dfrac{5}{39} & \dfrac{8}{39} & \dfrac{6}{39} & \dfrac{19}{39} \\[2mm]
0 & 0 & 1 & \dfrac{11}{39} & \dfrac{1}{39} & \dfrac{30}{39} & \dfrac{15}{39} \\[2mm]
\hline
0 & 0 & 0 & \dfrac{25}{39} & \dfrac{1}{39} & \dfrac{30}{39} & \dfrac{56}{39}
\end{array}
$$

The solution is therefore $x_1 = \dfrac{22}{39}$, $x_2 = \dfrac{19}{39}$, $x_3 = \dfrac{15}{39}$, and the maximum value

of f is $\dfrac{56}{39}$.

The minimizing problem that needs to be solved is that of minimizing

$$g = y_1 + y_2 + y_3$$

subject to the restrictions

$$4y_1 - y_2 - 2y_3 \geqslant 1$$
$$-y_1 + 4y_2 + 2y_3 \geqslant 1$$
$$-2y_1 - y_2 + 3y_3 \geqslant 1$$
$$y_1 \geqslant 0, \; y_2 \geqslant 0, \; y_3 \geqslant 0.$$

But, as we pointed out in the preceding section, the solution to this problem can be read off the last row of the final configuration that produced the solution of the maximizing problem. Hence, the solution is given by

$$y_1 = \frac{25}{39}, \quad y_2 = \frac{1}{39}, \quad \text{and} \quad y_3 = \frac{30}{39}$$

We now calculate the two sets of optimal probabilities by using the fact that $v = \dfrac{39}{56}$. Thus

$$q_1 = x_1 v = \frac{22}{56}, \quad q_2 = x_2 v = \frac{19}{56}, \quad q_3 = x_3 v = \frac{15}{56}$$
$$p_1 = y_1 v = \frac{25}{56}, \quad p_2 = y_2 v = \frac{1}{56}, \quad p_3 = y_3 v = \frac{30}{56}$$

From these results it appears that R should confine most of his bidding to high profit or low profit, whereas C should make all three types of bids, but should favor the high profit bid slightly more and the low profit bid slightly less than the modest profit bid. It also follows from this analysis that R can expect to average $\dfrac{39}{56}$ units of money if these optimal strategies are used.

If we had analyzed the game without the use of randomization, we would have arrived at the following configuration:

$$\begin{bmatrix} \boxed{4} & \boxed{-1} & -2 \\ \boxed{-1} & \boxed{4} & -1 \\ \boxed{-2} & 2 & \boxed{3} \end{bmatrix}$$

If R and C had chosen their best conservative pure strategies, R would have been willing to settle for a 1 unit loss, whereas C would have been willing to pay three units to R. That is a large difference and certainly justifies the use of randomization here.

COMMENTS

This chapter explained the basic principles of game theory as they apply to zero-sum two-person games. A basic theorem of game theory was introduced to guarantee that every randomized game possesses a solution. It was shown that the solution of any such game can be obtained by solving a maximization and a minimization problem of linear programming. Another basic theorem, called the duality theorem, was introduced to guarantee that the solution of both of those problems can be obtained by solving only the maximization problem.

Although game theory appears to be a promising way of treating many practical competitive business situations, it has had only limited application thus far because most business problems are too complex to satisfy its basic assumptions. More complicated versions of the theory exist and are being developed. Perhaps they will become more successful in solving realistic problems. Meanwhile, the ideas of game theory are well worth knowing because they give one a broader point of view in analyzing problems.

EXERCISES

Section 2

1. Determine the best strategies for R and C for each of the following payoff matrices. State the amount that R can be assured of winning. State the maximum amount that C will be required to pay. Does the game have a solution?

   (a) $\begin{bmatrix} 5 & 7 \\ 3 & 0 \end{bmatrix}$
   (b) $\begin{bmatrix} 2 & 0 \\ 3 & 1 \end{bmatrix}$

   (c) $\begin{bmatrix} 5 & 0 \\ -3 & 1 \end{bmatrix}$
   (d) $\begin{bmatrix} 1 & 2 \\ -3 & -4 \\ -2 & -1 \end{bmatrix}$

   (e) $\begin{bmatrix} 0 & 3 & -2 \\ 1 & -2 & -1 \\ 2 & 2 & 1 \end{bmatrix}$
   (f) $\begin{bmatrix} 0 & -1 & -2 \\ 1 & -3 & 3 \\ 2 & 4 & 1 \end{bmatrix}$

2. R and C match pennies. If they both show heads, R receives $2 from C. If they both show tails, R receives $3 from C. If R shows a head and C a

tail, C receives $1 from R. If R shows a tail and C a head, C receives $3 from R. What is the payoff matrix for this game?

3. R and C each choose one of the numbers 1, 2, and 3. If they choose the same number, R receives from C the amount of the chosen number in dollars. If they choose different numbers, R pays C the amount of the number chosen by R in dollars. What is the payoff matrix for this game?

4. R and C each choose one of the numbers 1, 2, and 3. If they choose the same number, R receives from C the amount of the chosen number in dollars. If they choose different numbers, R pays C the amount in dollars obtained by subtracting C's number from R's number. What is the payoff matrix for this game?

5. R and C play a game in which they simultaneously hold up one or two fingers. If they both hold up one finger, C pays $2 to R. If they both hold up two fingers, C pays $4 to R. If they hold up a different number of fingers, R pays $3 to C. What is the payoff matrix?

6. A more complicated version of the game in Problem 5 allows each player to hold up from one to five fingers. If the total number of fingers showing is divisible by three there is no payoff. If this division has a remainder of one, C must pay R an amount in dollars equal to the total number of fingers showing. If the division has a remainder of two, R must pay C an amount in dollars equal to the total number of fingers showing. What is the payoff matrix?

7. What are the best strategies for R and C in Problem 2? Does the game have a value?

8. What are the best strategies for R and C in Problem 3? Does the game have a value?

9. What are the best strategies for R and C in Problem 4? Does the game have a value?

10. What are the best strategies for R and C in Problem 5? Does the game have a value?

11. What are the best strategies for R and C in Problem 6? Does the game have a value?

Section 3

1. Calculate the expected payoff to R for each of the following payoff matrices if the randomized strategies for R and C are those shown.

(a) $\begin{bmatrix} 2 & 1 \\ 3 & 3 \end{bmatrix}$ $\begin{bmatrix} 3 & 2 \\ 6 & 0 \end{bmatrix}$ $\begin{bmatrix} \frac{1}{2} \\ \frac{1}{2} \end{bmatrix}$

(b) $\begin{bmatrix} \frac{1}{5} & \frac{4}{5} \end{bmatrix}$ $\begin{bmatrix} -2 & 1 \\ -1 & 2 \end{bmatrix}$ $\begin{bmatrix} \frac{1}{3} \\ \frac{2}{3} \end{bmatrix}$

(c) $\begin{bmatrix} \frac{1}{3} & \frac{1}{3} & \frac{1}{3} \end{bmatrix}$ $\begin{bmatrix} 0 & 1 & 2 \\ -1 & 0 & -2 \\ 1 & 3 & 1 \end{bmatrix}$ $\begin{bmatrix} \frac{1}{2} \\ 0 \\ \frac{1}{2} \end{bmatrix}$

2. For each part of Problem 1 determine whether R would do better or worse with the given randomized strategies than with his best pure strategy.

3. Calculate the expected payoff to R for the following payoff matrix if the randomized strategies for R and C are those shown. Compare this with what R could expect to win if he employed his best pure strategy.

$$\begin{bmatrix} \frac{1}{2} & \frac{1}{3} & \frac{1}{6} \end{bmatrix} \begin{bmatrix} 2 & -1 & 0 & -2 \\ -1 & 0 & 2 & 1 \\ 0 & 1 & -1 & 1 \end{bmatrix} \begin{bmatrix} \frac{1}{4} \\ \frac{1}{4} \\ 0 \\ \frac{1}{2} \end{bmatrix}$$

Section 4

1. For what values of a and b will the following game be determined?

$$\begin{bmatrix} a & 0 \\ 0 & b \end{bmatrix}$$

2. Show that the following game will be determined if the elements of the first row are larger than the corresponding elements of the second row.

$$\begin{bmatrix} a & b & c & d & e \\ r & s & t & u & v \end{bmatrix}$$

3. In each of the following games determine whether the payoff matrix can be reduced to a smaller matrix because one of the players would be foolish to use one of his possible strategies. Explain your reasoning in each case.

(a) $\begin{bmatrix} 2 & 6 \\ 3 & 4 \end{bmatrix}$ (b) $\begin{bmatrix} -1 & 2 \\ 4 & 3 \end{bmatrix}$,

(c) $\begin{bmatrix} 2 & -2 & 3 \\ 1 & 3 & 2 \end{bmatrix}$ (d) $\begin{bmatrix} 1 & 0 \\ -3 & 3 \\ -4 & 2 \end{bmatrix}$

4. For each of the matrices in Problem 3 determine the value of the game, provided that it has a value under pure strategies.

5. Reduce the following payoff matrices to smaller matrices by deleting any strategies that would never be used by a player. If the resulting game has a value under pure strategies, find it.

(a) $\begin{bmatrix} 4 & 5 & -1 \\ 5 & 2 & 3 \\ 2 & -1 & -2 \end{bmatrix}$ (b) $\begin{bmatrix} 1 & 4 & 5 & -2 \\ 2 & -1 & 1 & 2 \\ 2 & 3 & 4 & 3 \\ -1 & 0 & 3 & 4 \end{bmatrix}$

6. Show that if a payoff matrix has two saddle points, the corresponding elements must have the same value.

Section 5

1. For each of the following payoff matrices, determine the optimal randomized strategies for R and C. Use geometrical methods for solving the linear programming problems.

(a) $\begin{bmatrix} 4 & -2 \\ 2 & 3 \end{bmatrix}$ (b) $\begin{bmatrix} 3 & 4 \\ 0 & 5 \end{bmatrix}$

2. For each of the following payoff matrices, determine the optimal randomized strategies for R and C. Use geometrical methods for solving any two-variable linear programming problem.

(a) $\begin{bmatrix} 2 & 1 & -1 \\ -1 & 1 & 3 \end{bmatrix}$ (b) $\begin{bmatrix} 2 & -1 \\ 1 & 0 \\ 0 & 2 \end{bmatrix}$

3. Find the optimal randomized strategies for R and C for the following payoff matrix.

$$\begin{bmatrix} 4 & 3 & -2 \\ 0 & 2 & 1 \\ -3 & 0 & 2 \end{bmatrix}$$

4. Find the optimal randomized strategies for R and C for the following payoff matrix.

$$\begin{bmatrix} 2 & 2 & -1 \\ 4 & 0 & -4 \\ -1 & 2 & 4 \end{bmatrix}$$

5. Show that if each element of a payoff matrix is positive, the value of the game will be positive.

6. Show that if each element of a payoff matrix is increased by the amount $c > 0$, the optimal randomized strategies will be the same as before and that the value of the game will be increased by the amount c.

Section 6

1. A children's game consists in each of two players simultaneously saying the word stone, scissors, or paper. According to the rules of the game, stone beats scissors, scissors beats paper, and paper beats stone. If both players say the same word the game is a tie and there is no payoff. If one player chooses a word that beats the other's choice, he wins one point. Construct the payoff matrix and find the optimal randomized strategies. Add 1 to each element of the payoff matrix to insure that $v > 0$ before solving the game.

2. The following matrix represents the expected return on investments corresponding to three future economic conditions. What percentages of each type of investment would you select if you wished to maximize your expected return independent of what the future economic conditions turn out to be? What probabilities for those economic conditions will be least favorable to your investments if you use your calculated percentages?

	Recession	Normal	Inflation
Stocks	−60	40	80
Bonds	40	30	50
Banks	20	30	40

8 | The Derivative

One of the basic problems in the various sciences is that of measuring the rate at which some function is changing over time, or with respect to some other variable. For example, a knowledge of such rates is needed to make predictions about the future size of a school population, the future demand for electricity, or the amount of inflation as a function of the money supply. Although our motivating interest is that of determining how to measure the rate at which a function is changing, we approach the problem from a geometrical point of view. After the geometry of the problem is explained, we easily treat practical rate problems.

Although our ultimate objective is to solve practical rate problems, this chapter is concerned only with the mathematical development of the tools needed to solve such problems. The application of these tools to practical problems is considered in Chapter 9. Students should therefore behave like mathematicians when studying this chapter and wait patiently for the applications that occur later.

1. THE SLOPE OF A CURVE

The slope of a straight line was defined in Section 4, Chapter 1, to be the ratio

$$m = \frac{y_2 - y_1}{x_2 - x_1}$$

where (x_1, y_1) and (x_2, y_2) are the coordinates of any two points on the line. In this section we extend the notion of slope so that it may be applied to more general curves.

236

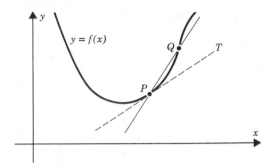

FIGURE 1   Finding the slope of a curve.

Let f be a function of x that is being studied and consider the curve obtained by graphing the equation $y = f(x)$. Let P denote a point on this curve,
as shown in Fig. 1. We wish to define what is meant by the slope of this
curve at the point P.

We choose a second point on the curve, call it Q, that lies to the right of P
and draw a line passing through P and Q. Since any well-behaved curve
will look like a straight line over a sufficiently small segment of it, the line
segment connecting P and Q will fit the curve quite well between P and Q
if Q is sufficiently close to P. We capitalize on this geometric intuition and
define the slope of the curve at P to be the value that the slope of the line
passing through P and Q approaches as Q approaches P along the curve. Furthermore, we define the line that passes through P and that has this
resulting slope as the tangent line to the curve at P. This line is labeled T in
Fig. 1. Hereafter, we use the phrase "limiting value" to describe the value
that the slope approaches as Q approaches P.

Although we chose the point Q to the right of P, we could have chosen it
to the left of P and proceeded in the same manner. If a different limiting
value for the slope had been obtained, we would have been in a quandry as
to how to define the slope at such a point. As a result, we insist that the
same limiting value must be obtained for both sides before our definition is
applicable.

For the purpose of illustrating how the slope of a curve is calculated, consider the particular curve $y = x^2$. Suppose that we wish to find the slope at
the point P whose coordinates are $(1,1)$. We choose a point Q whose x coordinate is $1 + h$, where h is some positive or negative number. The y coordinate of Q is then given by $y = (1 + h)^2$. The geometry of this problem is
shown in Fig. 2.

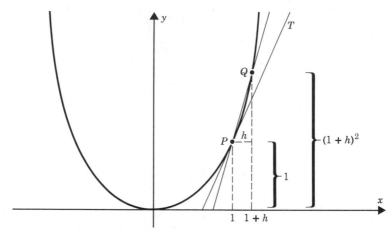

FIGURE 2   Finding the slope of a parabola.

In terms of the notation used previously, the points P and Q are the points (x_1,y_1) and (x_2,y_2), respectively. For this problem we have $(x_1,y_1) = (1,1)$ and $(x_2,y_2) = (1+h,(1+h)^2)$. The slope of the line passing through P and Q is therefore given by

$$\frac{y_2 - y_1}{x_2 - x_1} = \frac{(1+h)^2 - 1}{(1+h) - 1} = \frac{2h + h^2}{h} = 2 + h$$

When the point Q approaches the point P the number h will approach the value 0; consequently, the slope of the line through P and Q, which is given by $2 + h$, will approach the value 2. This will be true whether h is a positive number or a negative number. By definition, therefore, the slope of the parabola $y = x^2$ at the point $(1,1)$ is 2. The corresponding tangent line at this point is shown in Fig. 2 as the line labeled T. From the formula for the equation of a straight line given in Section 4, Chapter 1, it follows that the equation of T is $y - 1 = 2(x - 1)$, or equivalently, $y = 2x - 1$.

The preceding technique can be extended to produce a formula that will give the slope of this parabola at an arbitrary point on it. Toward this objective, let (x_0,y_0) denote an arbitrary point P on the parabola. Since the equation of the parabola is $y = x^2$, this point must have coordinates (x_0,x_0^2). A neighboring point Q on the parabola, whether to the right or left of P, will have coordinates that can be expressed in the form $(x_0 + h,(x_0 + h)^2)$. Consequently, the slope of the line passing through P and Q is given by

$$\frac{y_2 - y_1}{x_2 - x_1} = \frac{(x_0 + h)^2 - x_0^2}{(x_0 + h) - x_0} = \frac{2hx_0 + h^2}{h} = 2x_0 + h$$

Since x_0 is a fixed number and h approaches 0 as Q approaches P along the curve, the limiting value of this slope is $2x_0$. This limiting value is the same whether h is positive or negative. The slope of the parabola at the point on it which has the x coordinate x_0 is therefore $2x_0$. It should be noted that this formula immediately yields the particular slope that was calculated previously for $x_0 = 1$.

2. THE DERIVATIVE

The preceding section introduced the concept of the slope of a curve. We use this geometrical concept to motivate a more general algebraic concept called the derivative.

Let f be any function of x. Choose any two points on the x axis. They may be represented by x_0 and $x_0 + h$. The values of f at these two points are $f(x_0)$ and $f(x_0 + h)$, respectively. Then form the quotient

$$\frac{f(x_0 + h) - f(x_0)}{(x_0 + h) - x_0} = \frac{f(x_0 + h) - f(x_0)}{h}$$

Although we are not relying on geometry in the present discussion, we should observe that this quotient is the slope of the line passing through the points P and Q on the curve $y = f(x)$, where P and Q have the x coordinates x_0 and $x_0 + h$, respectively. This is shown in Fig. 3.

If this quotient approaches a limiting value as h approaches 0, the limiting value is called the *derivative* of the function f at x_0. This value, when it exists, is denoted by the symbol $f'(x_0)$. Just as in defining the slope of a curve, it is necessary that the same limiting value be obtained whether

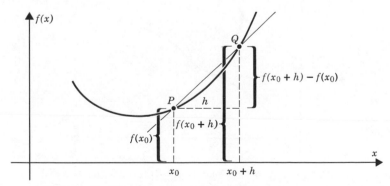

FIGURE 3   The geometry of the derivative.

h approaches 0 through positive values or negative values. We can summarize these comments as follows:

Definition. *The derivative of the function f at the point x_0 is denoted by $f'(x_0)$ and is given by the formula*

(1)
$$f'(x_0) = \lim_{h \to 0} \frac{f(x_0 + h) - f(x_0)}{h}$$

The symbol $\lim\limits_{h \to 0}$ is a brief way of representing the limiting value of whatever follows it as h approaches 0 through both positive and negative values. It is read "the limit as h approaches 0 of ———," where the blank is to be filled in with the expression that follows the symbol $\lim\limits_{h \to 0}$.

In view of the geometrical motivation for this algebraic definition, it follows that the derivative of a function f at a point x_0 gives the slope of the curve $y = f(x)$ at the point where $x = x_0$. This, however, is merely the geometrical meaning of the derivative. In later sections, other interpretations of the derivative are introduced.

If $f'(x_0)$ exists, we say that the function f is *differentiable* at the point x_0. If $f'(x)$ exists for all values of x in some interval, we say that f is differentiable in that interval. The function $f(x) = x^2$ is an example of a function that is differentiable for all values of x. Such a function is called a differentiable function. Most of the functions that we use as models for solving problems in the social sciences are differentiable, at least over some interval.

From a geometrical point of view, a function is differentiable in an interval if its graph is such that it possesses a tangent line at every point on it in that interval.

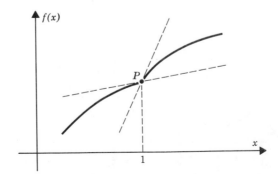

FIGURE 4   A function that is not differentiable at $x = 1$.

The function f whose graph is shown in Fig. 4 is an example of a function that is not differentiable throughout its indicated domain because it does not possess a tangent line at the point P where $x = 1$. The limiting tangent lines at the point P are quite different, depending on whether the point Q is chosen to the right or to the left of P before taking the limit. Thus, there does not exist a unique tangent line at P, and therefore the function is not differentiable at the point $x = 1$.

3. LIMITS

In the preceding section the derivative of a function is defined as the limiting value of a certain quotient. For the illustration of that section it is clear how that limiting value is obtained. For more complicated functions, however, it is not so obvious how the limit is to be calculated. In this section we take a closer look at the meaning of this word and then list rules that assist us in calculating complicated limits.

Consider a function f and its graph. Let a be a fixed value of x. Then we wish to give meaning to the expression

$$(2) \qquad\qquad \lim_{x \to a} f(x) = L$$

which is read "the limit of $f(x)$ as x approaches a is L."

In view of our previous discussion of the limit symbol as it was used to define the derivative, we interpret this expression to mean that *the value of $f(x)$ approaches the number L as the value of x approaches the number a.* Furthermore, we require that $f(x)$ approach L whether x approaches a from the left side or the right side of a. Since we are concerned only with what happens to $f(x)$ as x approaches a, we say nothing about the value of $f(x)$ at $x = a$. The function $f(x)$ need not even be defined at $x = a$ in our definition of limit.

As an illustration, consider once more the function $f(x) = x^2$ and let $a = 2$. Then we can write x in the form $x = 2 + h$ and calculate

$$f(x) = f(2 + h) = (2 + h)^2 = 4 + 4h + h^2$$

Since the value of $f(x)$ approaches the number 4 when h approaches 0, which is equivalent to having x approach 2, it follows from our definition that

$$\lim_{x \to 2} f(x) = 4$$

and, hence, that $L = 4$.

The preceding result would be true even if we defined f by the formula $f(x) = x^2$ for $x \neq 2$ and to have the value 6 for $x = 2$. The graph of this modified function is shown in Fig. 5. It is clear from this graph, and our preceding calculations, that $f(x)$ approaches the value 4 from both sides of $x = 2$ and that our limiting process does not depend on the value of $f(x)$ at $x = 2$.

Suppose now that $f(a)$ exists and that $f(a) = L$ in the preceding definition. Then (2) will assume the form

(3) $$\lim_{x \to a} f(x) = f(a)$$

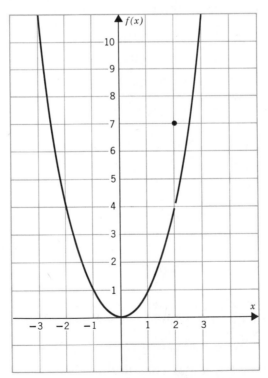

FIGURE 5   The graph of a discontinuous function.

When this is true, the function f is said to be *continuous* at the point a. If a function is continuous at all points in its domain, it is called a *continuous function*. Most of the functions that are encountered in applied problems are continuous. For example, the function $f(x) = x^2$ is continuous at every point on the x axis because, if x_0 is any point, we can write $x = x_0 + h$ and calculate

$$f(x_0 + h) = (x_0 + h)^2 = x_0^2 + 2hx_0 + h^2$$

But the value of $f(x_0 + h)$ can be made arbitrarily close to x_0^2 by choosing h sufficiently close to 0; consequently, since $f(x_0) = x_0^2$, it follows that

$$\lim_{x \to x_0} f(x) = f(x_0)$$

Similar algebraic techniques will show that $f(x) = x^n$, where n is any positive integer, is a continuous function.

A continuous function on an interval is often described geometrically by stating that its graph can be drawn without lifting one's pencil off the paper. The graph in Fig. 5, for example, cannot be drawn in this manner because of the jump occurring at $x = 2$.

There is an interesting relationship between the property of a function being continuous at a point and possessing a derivative at that point. It may be expressed as follows:

Theorem 1. *If a function possesses a derivative at a point x_0 it must be continuous at that point.*

The requirement that a function possess a derivative at a point, or throughout an interval, is therefore a stronger restriction on it than that it be continuous at the point, or throughout the interval. Figure 4 is an illustration of a function that is continuous at a point $(x = 1)$ but does not possess a derivative there.

We do not prove Theorem 1; however, we need to use it in a later section. In that section we derive a set of formulas that enable us to differentiate some fairly complicated functions. To derive those formulas we also need several properties of limits. These properties, which we do not demonstrate but which intuitively seem very plausible, can be demonstrated by a systematic application of the proper algebraic techniques applied to a more precise definition of limit. They may be summarized as follows:

(4) *If* $\lim\limits_{x \to a} f(x) = L$ *and* $\lim\limits_{x \to a} g(x) = M,$ *then*

(a) $\lim\limits_{x \to a} [f(x) + g(x)] = L + M$

(b) $\lim\limits_{x \to a} cf(x) = cL$

(c) $\lim\limits_{x \to a} f(x)g(x) = LM$

(d) $\lim\limits_{x \to a} \dfrac{f(x)}{g(x)} = \dfrac{L}{M},$ *provided that* $M \neq 0$

4. THE DERIVATIVE OF $f(x) = x^n$

Consider the special function $f(x) = x^n$, where n is some positive integer. We wish to calculate the derivative of this function at a point $x = x_0$. In Section 2 we calculated the derivative of this function for the special case in which $n = 2$. Now we wish to do so for the general case. Applying the definition given in (1), we obtain

$$f'(x_0) = \lim_{h \to 0} \frac{(x_0 + h)^n - x_0^n}{h}$$

But from the binomial expansion formula we know that $(x_0 + h)^n$ can be expanded as follows:

$$(x_0 + h)^n = x_0^n + nx_0^{n-1}h + \frac{n(n-1)}{2} x_0^{n-2}h^2 + \cdots + h^n$$

Hence

$$\frac{(x_0 + h)^n - x_0^n}{h} = \frac{1}{h} \left[nx_0^{n-1}h + \frac{n(n-1)}{2} x_0^{n-2}h^2 + \cdots + h^n \right]$$

$$= nx_0^{n-1} + \frac{n(n-1)}{2} x_0^{n-2}h + \cdots + h^{n-1}$$

Therefore

$$f'(x_0) = \lim_{h \to 0} \left[nx_0^{n-1} + \frac{n(n-1)}{2} x_0^{n-2}h + \cdots + h^{n-1} \right]$$

Now we apply properties (4)(a) and (b) of the preceding section to obtain

$$f'(x_0) = \lim_{h \to 0} nx_0^{n-1} + \frac{n(n-1)}{2} x_0^{n-2} \lim_{h \to 0} h + \cdots + \lim_{h \to 0} h^{n-1}$$

Since the terms on the right, except for the first one, contain h to succes- sively higher powers, all those terms have zero limits. The first term on the right is a constant; therefore, its limiting value is that constant value. As a result, we obtain

$$f'(x_0) = n x_0^{n-1}$$

This is the derivative of $f(x) = x^n$ at the point $x = x_0$. If we omit the subscript on x we obtain a general formula that is valid for any desired value of x. This result may be summarized in the form of a theorem as follows:

Theorem 2. *The function* $f(x) = x^n$, *where* n *is any positive integer, has a derivative which is given by the formula*

(5) $$f'(x) = n x^{n-1}$$

Although this formula has been shown to be valid only for positive in- teger values of n, it can be shown to be valid for all real number exponents. Thus n may be a negative integer, or a fraction, or a negative fraction, or even a number such as π.

The following examples illustrate the type of functions that can be dif- ferentiated by means of formula (5).

Example 1. If $f(x) = x^4$, then $f'(x) = 4x^3$.

Example 2. If $f(x) = x^{-3}$, then $f'(x) = -3x^{-4}$.

Example 3. If $f(x) = x^{3/2}$, then $f'(x) = \dfrac{3}{2} x^{1/2}$.

Example 4. If $f(x) = x^{-1/2}$, then $f'(x) = -\dfrac{1}{2} x^{-3/2}$.

Before applying this formula further, we consider a particular closely related formula and derive some useful differentiation formulas.

Let $f(x) = c$, where c is any constant. Then since the graph of this func- tion is a horizontal straight line, it follows from the geometrical interpreta- tion of the derivative as the slope of the curve $y = f(x)$ that $f'(x) = 0$. It is easy to demonstrate this algebraically, however. We calculate

$$\frac{f(x_0 + h) - f(x_0)}{h} = \frac{c - c}{h} = 0$$

Since this quotient is 0 for all values of h except $h = 0$, its limit must be 0, as was to be shown.

5. THE DERIVATIVE OF SUMS, PRODUCTS, AND QUOTIENTS

Suppose that we wish to calculate the derivative of the function $f(x) = 2x + x^2$. We could apply definition (1) directly; however, in view of our properties of limits and the fact that a derivative is merely a special kind of limit, it would appear that we might be able to break the problem down into simpler components. Consider, therefore, the problem of calculating the derivative of the function r, where

$$r(x) = f(x) + g(x)$$

For the purpose of saving energy, hereafter in deriving differentiation formulas we dispense with the zero subscript on x_0. Therefore, in all such derivations it is to be understood that x has a fixed value in the domain of the function, which previously has been denoted by x_0, and that h is the only variable involved in taking limits. With this understanding, we proceed to calculate the quotient

$$\frac{r(x+h) - r(x)}{h} = \frac{[f(x+h) + g(x+h)] - [f(x) + g(x)]}{h}$$

$$= \frac{f(x+h) - f(x)}{h} + \frac{g(x+h) - g(x)}{h}$$

Then applying definition (1), we obtain

$$r'(x) = \lim_{h \to 0} \frac{r(x+h) - r(x)}{h}$$

$$= \lim_{h \to 0} \left[\frac{f(x+h) - f(x)}{h} + \frac{g(x+h) - g(x)}{h} \right]$$

Application of property (4)(a) allows us to calculate the separate limits on the right and sum them. Hence

$$r'(x) = \lim_{h \to 0} \frac{f(x+h) - f(x)}{h} + \lim_{h \to 0} \frac{g(x+h) - g(x)}{h}$$

If f and g both possess derivatives, this gives

$$r'(x) = f'(x) + g'(x)$$

We have arrived at the result that the derivative of the sum of two functions is equal to the sum of the derivatives of the two functions. This result may be summarized in the form of a theorem as follows:

Theorem 3. *If f and g are differentiable functions of x, their sum r = f + g is differentiable and has the derivative given by the formula*

$$r'(x) = f'(x) + g'(x)$$

This theorem enables us to calculate the derivative of $f(x) = 2x + x^2$ by calculating the derivative of $2x$ and x^2 separately and adding the results. Before doing so, we need another property of derivatives. Consider, therefore, the problem of calculating the derivative of the function $r = cf$, where c is some constant. Application of definition (1) yields

$$r'(x) = \lim_{h \to 0} \frac{r(x+h) - r(x)}{h}$$

$$= \lim_{h \to 0} \frac{cf(x+h) - cf(x)}{h}$$

$$= \lim_{h \to 0} c \frac{f(x+h) - f(x)}{h}$$

But from property (4)(b) we are permitted to factor out the constant c and obtain

$$r'(x) = c \lim_{h \to 0} \frac{f(x+h) - f(x)}{h} = cf'(x)$$

This demonstrates the validity of the next theorem:

Theorem 4. *If f is a differentiable function of x, the function r = cf is differentiable and has the derivative given by the formula*

$$r'(x) = cf'(x)$$

The two preceding theorems, together with Theorem 2, enable us to calculate the derivative of $f(x) = 2x + x^2$. Theorem 3 permits us to calculate the derivative of $2x$ and the derivative of x^2 and add the results. Because of Theorem 4 we are permitted to factor out the 2 in $2x$ and calculate the derivative of x, after which we restore the factor 2. Since the derivative of x and of x^2 are both obtained from the formula of Theorem 2, we obtain

$$f'(x) = 2 + 2x$$

As another illustration, let $f(x) = 4x^3 - 7x^4$. Applying Theorems 3 and 4 in succession, and then applying Theorem 2, we obtain

$$f'(x) = 4 \cdot 3x^2 - 7 \cdot 4x^3 = 12x^2 - 28x^3$$

Now suppose that we had a function of the form

$$f(x) = (x^4 + 3x^2 - x + 2)(x^5 - 3x^3 + 4x^2 + 7x)$$

We could multiply the two expressions in parentheses and then apply the preceding theorems to obtain $f'(x)$. It would be nice, however, if we had a formula that tells us how to differentiate the product of two functions, because that might require less work. Such a formula does exist. One's intuition is incorrect here, however, because the derivative of a product of two functions is not, except for some very special functions, equal to the product of the derivatives of the two functions. The proper formula is given by the following theorem.

Theorem 5. *If f and g are differentiable functions of x, their product $r = fg$ is differentiable and has the derivative given by the formula*

(6) $$r'(x) = f(x)g'(x) + g(x)f'(x)$$

To demonstrate this formula, we first calculate the quotient needed to apply definition (1):

(7) $$\frac{r(x+h) - r(x)}{h} = \frac{f(x+h)g(x+h) - f(x)g(x)}{h}$$

Then we add and subtract $f(x+h)g(x)$ to the numerator on the right to express it in the form

$$\frac{1}{h}\left[f(x+h)g(x+h) - f(x+h)g(x) + f(x+h)g(x) - f(x)g(x)\right]$$

Performing some algebra, we can simplify this as follows:

(8) $$f(x+h)\frac{g(x+h) - g(x)}{h} + g(x)\frac{f(x+h) - f(x)}{h}$$

We are now ready to take limits as $h \to 0$. Assuming that it exists, the limit of the left side of (7) is $r'(x)$. The limit of the right side of (7) is the same as the limit of (8), which can be computed by applying properties (4) concerning the limits of sums and products. Application of those properties gives

$$r'(x) = \lim_{h \to 0} f(x+h) \lim_{h \to 0} \frac{g(x+h) - g(x)}{h} + \lim_{h \to 0} g(x) \lim_{h \to 0} \frac{f(x+h) - f(x)}{h}$$

From Theorem 1 we know that a differentiable function is necessarily continuous. Since f is assumed to be differentiable, it is continuous and it therefore follows from (3) that $\lim_{h \to 0} f(x+h) = f(x)$. That takes care of the first limit on the right. The second limit on the right is $g'(x)$ because g is also assumed to be differentiable. The third limit is clearly $g(x)$ because $g(x)$ is a constant when x has a fixed value, as it has here. The last limit is $f'(x)$ in view of f's differentiability. By combining these results, we obtain the formula given in our theorem:

$$r'(x) = f(x)g'(x) + g(x)f'(x)$$

As a first illustration, we choose a simple problem so that we can check the answer without any difficulty. Let $r(x) = x^2 \cdot x^3$. Here $f(x) = x^2$ and $g(x) = x^3$. Formula (6) and the rule for differentiating x^n then give

$$r'(x) = x^2 \cdot 3x^2 + x^3 \cdot 2x = 5x^4$$

Since $r(x) = x^2 \cdot x^3 = x^5$, we can differentiate $r(x)$ directly to obtain $5x^4$, thereby checking our formula result.

As a more complicated illustration, consider the following function:

$$r(x) = (x^2 + x + 1)(x^2 - x + 1)$$

Applying formula (6) and our earlier rules, we obtain

$$r'(x) = (x^2 + x + 1)(2x - 1) + (x^2 - x + 1)(2x + 1)$$

By multiplying out the parentheses and collecting terms, we find that

$$r'(x) = 4x^3 + 2x$$

Suppose that we had multiplied out the parentheses in $r(x)$ before differentiating it. This would have given

$$r(x) = x^4 + x^2 + 1$$

Differentiation of this gives

$$r'(x) = 4x^3 + 2x$$

The equivalence of these two answers is a check on our work.

Although there is no appreciable saving in time in using the product formula on the preceding illustrations, it will be found that the formula is very useful for treating the more complicated functions that we study later.

Division is multiplication by reciprocals; however, we cannot use the preceding multiplication rule directly to obtain a rule for division because we do not have a rule for differentiating reciprocal functions. We therefore derive the following rule which enables us to differentiate quotients.

Theorem 6. *If f and g are differentiable functions of x, the quotient $r = \dfrac{f}{g}$ is differentiable and has the derivative given by the formula*

$$(9) \qquad r'(x) = \frac{g(x)f'(x) - f(x)g'(x)}{g^2(x)}$$

This formula is valid for any value of x for which $g(x) \neq 0$.

To derive this formula, we first calculate the quotient needed to apply definition (1):

$$\frac{r(x+h) - r(x)}{h} = \frac{1}{h}\left[\frac{f(x+h)}{g(x+h)} - \frac{f(x)}{g(x)}\right]$$

$$= \frac{g(x)f(x+h) - f(x)g(x+h)}{hg(x)g(x+h)}$$

Then we add and subtract the quantity $f(x)g(x)$ to the numerator on the right side to express it in these equivalent forms:

$$\frac{r(x+h) - r(x)}{h} = \frac{g(x)f(x+h) - f(x)g(x) - [f(x)g(x+h) - f(x)g(x)]}{hg(x)g(x+h)}$$

$$= \frac{g(x)\dfrac{f(x+h) - f(x)}{h} - f(x)\dfrac{g(x+h) - g(x)}{h}}{g(x)g(x+h)}$$

Finally, we take the limit of these expressions by applying properties (4) and using Theorem 1 to justify that $\lim\limits_{h \to 0} g(x+h) = g(x)$. This gives the desired result:

$$r'(x) = \frac{g(x)f'(x) - f(x)g'(x)}{g(x)^2}$$

As a simple illustration for which the answer is easily checked, let $r(x) =$

$\frac{x^4}{x^2}$. Then, by applying formula (9), we obtain

$$r'(x) = \frac{x^2 \cdot 4x^3 - x^4 \cdot 2x}{(x^2)^2} = \frac{2x^5}{x^4} = 2x$$

Since $r(x) = \frac{x^4}{x^2} = x^2$, we can differentiate directly to obtain $2x$.

As a more complicated illustration, suppose that we wish to calculate the value of $r'(1)$ where

$$r(x) = \frac{x^5 + x^2 - x + 1}{x^2 + 1}$$

Application of formula (9) yields

$$r'(x) = \frac{(x^2 + 1)(5x^4 + 2x - 1) - (x^5 + x^2 - x + 1)(2x)}{(x^2 + 1)^2}$$

Then

$$r'(1) = \frac{2 \cdot 6 - 2 \cdot 2}{(2)^2} = 2$$

If we were to calculate this value without using the quotient formula, we would need to divide the numerator of $r(x)$ by the denominator by long division techniques and then hope to differentiate the resulting quotient. This would be a lengthy process compared to the one based on the quotient formula. Furthermore, unless the division produced a polynomial without a remainder, we would not be able to differentiate the outcome of the long division.

The following examples are intended to illustrate the types of problems that can be handled by means of our basic differentiation theorems. The calculations are made without comment under the assumption that the student can supply the justification for the various steps. In each case the problem is to find $f'(x)$.

Example 1. $f(x) = \frac{4}{3} x^3 - \frac{6}{7} x^7$

$$f'(x) = \frac{4}{3} \cdot 3 \cdot x^2 - \frac{6}{7} \cdot 7 \cdot x^6 = 4x^2 - 6x^6$$

Example 2. $f(x) = (x^3 + x)(x^2 - 1)$

$$f'(x) = (x^3 + x)2x + (x^2 - 1)(3x^2 + 1) = 5x^4 - 1$$

Example 3. $f(x) = \dfrac{x^2}{x^2+1}$

$$f'(x) = \frac{(x^2+1)2x - x^2(2x)}{(x^2+1)^2} = \frac{2x}{(x^2+1)^2}$$

Example 4. $f(x) = 2x^4 - 3x^{-2}$

$$f'(x) = 8x^3 + 6x^{-3} = 8x^3 + \frac{6}{x^3}$$

In this example we used the fact that the formula for differentiating x^n is valid for negative exponents as well as for positive exponents.

Example 5. $f(x) = \dfrac{(x+2)(x^2+1)}{x-2}$

$$f'(x) = \frac{(x-2)[(x+2)2x + (x^2+1)] - (x+2)(x^2+1)}{(x-2)^2}$$

$$= \frac{2x^3 - 4x^2 - 8x - 4}{(x-2)^2}$$

6. NOTATION

Thus far we have used the symbol $f'(x)$ to denote the derivative of the function f at an arbitrary point x. There are, however, several other symbols that are also used to represent the derivative. One of the most commonly used is $\dfrac{df}{dx}$. The letter d represents differentiation, with the d in the numerator indicating that the function f is to be differentiated and the denominator indicating that the differentiation is with respect to the variable x. If f had been a function of some other variable, say t, the derivative of f with respect to the variable t would have been written $\dfrac{df}{dt}$. Symbols like these are to be treated as single entities; they are not fractions.

When the function $f(x)$ is represented by the letter y, that is when we write $y = f(x)$, our customary notation for the derivative becomes y . The new symbol introduced in the preceding paragraph then assumes the form $\dfrac{dy}{dx}$. This form possesses an advantage over y' in that it makes clear that the basic variable is x. This new notation, particularly the symbol $\frac{dy}{dx}$, possesses certain advantages in calculating the derivative of a complicated expression. An illustration of this advantage will be given in the next section.

In some applications of calculus it is useful to study how the derivative function $f'(x)$ changes as x changes. This leads naturally to studying the derivative of $f'(x)$. Since we are denoting the derivative of a function by means of a prime symbol, we should place another prime on $f(x)$ to represent the derivative of the derivative of $f(x)$. Hence, we should write $f''(x)$ here. The derivative of the derivative of a function is called the second derivative of the function. If we use the letter y to represent the function $f(x)$ and wish to use the new symbol introduced in the preceding paragraph, then we would use the symbol $\frac{d^2y}{dx^2}$ to represent the second derivative. Higher order derivatives are designated in a similar manner. Thus, $f'''(x)$ is a symbol that tells us to first differentiate $f(x)$, then to differentiate the result of the first differentiation, and finally to differentiate the result of the second differentiation. We could also represent this third derivative by the symbol $\frac{d^3y}{dx^3}$.

The calculation of higher order derivatives presents no problems. For example, suppose that we wish to calculate $f'''(x)$ where $f(x) = x^5 - 2x^3 + 5x + 6$.

We proceed as follows:

$$f'(x) = 5x^4 - 6x^2 + 5$$
$$f''(x) = 20x^3 - 12x$$
$$f'''(x) = 60x^2 - 12$$

7. THE CHAIN RULE

A function of the type $y = (x^3 + x^2 + x + 1)^5$ would be very difficult to differentiate because with our present rules for differentiation it would require us to multiply the expression in parentheses by itself four times before differentiating. Fortunately, there exists a differentiation rule that enables us to circumvent most of this algebra. This rule is called the *chain rule*. We show how it works on the preceding problem before considering its general formulation.

Suppose that we introduce a new variable u by letting $u = x^3 + x^2 + x + 1$. Then our function assumes the simple form of $y = u^5$. If we treat y as a function of u, it is easy to differentiate it with respect to u because $\dfrac{dy}{du} = 5u^4$. Now, it can be shown that the derivative of y with respect to x can be obtained by first differentiating y with respect to u and then multiplying the result by the derivative of u with respect to x. That is,

$$\frac{dy}{dx} = \frac{dy}{du}\frac{du}{dx}$$

Since $\dfrac{du}{dx} = 3x^2 + 2x + 1$ here, this rule gives

$$\frac{dy}{dx} = 5u^4(3x^2 + 2x + 1)$$

$$= 5(x^3 + x^2 + x + 1)^4(3x^2 + 2x + 1)$$

The preceding problem is a special case of the following type of problem. Let f and g be two differentiable functions and let a new function y be constructed from them by means of

$$y = f(g(x))$$

The function y, which, of course, is a function of x, is called a *composite function*. It is said to be the *composition* of f and g. To obtain the value of y for any specified value of x, we first calculate the value of g for the selected value of x and then calculate the value of f for the value of g that was just obtained. For example, in the earlier numerical illustration if we wish to calculate the value of y for $x = 1$, we first calculate the value $g(1) = 1^3 + 1^2 + 1 + 1 = 4$ and then calculate the value of $y = f(4) = 4^5 = 1024$. The rule for differentiating general composite functions then proceeds as follows.

Introduce a new variable u by letting $u = g(x)$. Then the function y assumes the simple form $y = f(u)$. Now treat y as a function of u and differentiate it with respect to u, obtaining $\dfrac{dy}{du} = f'(u)$. Then differentiate the function $u = g(x)$ with respect to x, obtaining $\dfrac{du}{dx} = g'(x)$. Finally, multiply these two derivatives together to obtain the derivative of y with respect to x. This rule is summarized in the form of a theorem.

Theorem 7. *If f and g are differentiable functions, the composite function defined by $y = f(g(x))$ is differentiable and has the derivative given by the formula*

(10)
$$\frac{dy}{dx} = \frac{dy}{du}\frac{du}{dx} = f'(u)g'(x)$$

where $u = g(x)$.

The proof of this theorem is quite lengthy; therefore, we do not give it here. We do, however, apply it to a number of problems to indicate how useful it is for differentiating complicated functions.

Example 1. Let $y = (x^2 + 2x)^{-2}$. We can write this as the composite function $y = f(g(x))$ by letting $u = g(x) = x^2 + 2x$ and letting $f(u) = u^{-2}$. Then formula (10) states that

$$\frac{dy}{dx} = f'(u)g'(x) = -2u^{-3}(2x + 2) = -\frac{4x + 4}{(x^2 + 2x)^3}$$

This problem could have been solved without the chain rule by writing

$$y = \frac{1}{(x^2 + 2x)^2} = \frac{1}{x^4 + 4x^3 + 4x^2}$$

and then applying the quotient rule to obtain

$$\frac{dy}{dx} = \frac{(x^4 + 4x^3 + 4x^2) \cdot 0 - 1 \cdot (4x^3 + 12x^2 + 8x)}{(x^4 + 4x^3 + 4x^2)^2}$$

$$= -\frac{4x(x^2 + 3x + 2)}{x^4(x^2 + 4x + 4)^2}$$

$$= -\frac{4x(x + 2)(x + 1)}{x^4(x + 2)^4}$$

$$= -\frac{4x + 4}{(x^2 + 2x)^3}$$

The chain rule is certainly much simpler for this problem.

Example 2. Let $y = \left(\frac{x + 1}{x - 1}\right)^4$. We can write this as the composite function $y = f(g(x))$ by letting $u = g(x) = \frac{x + 1}{x - 1}$ and letting $f(u) = u^4$. Then formula (10) gives

$$\frac{dy}{dx} = f'(u)g'(x) = 4u^3 g'(x)$$

$$= 4u^3 \left[\frac{(x-1) - (x+1)}{(x-1)^2} \right]$$

$$= -8 \frac{(x+1)^3}{(x-1)^5}$$

Example 3. Let $y = \dfrac{(x^2+1)^2}{(x^2+4)^3}$. In differentiating a complicated function such as this, we do not make a formal substitution of u for x^2+1 or for x^2+4 because that would require two substitutions. Instead, we make a mental substitution for each of those functions and apply the chain rule to each part when we meet it. Thus, we apply the quotient formula and proceed as follows:

$$\frac{dy}{dx} = \frac{(x^2+4)^3 \dfrac{d}{dx}(x^2+1)^2 - (x^2+1)^2 \dfrac{d}{dx}(x^2+4)^3}{(x^2+4)^6}$$

The calculation of the two derivatives in the numerator is then carried out by applying the chain rule mentally to each part, letting $u = x^2 + 1$ for the first part and letting $u = x^2 + 4$ for the second part. We obtain

$$\frac{dy}{dx} = \frac{(x^2+4)^3 2(x^2+1)2x - (x^2+1)^2 3(x^2+4)^2 2x}{(x^2+4)^6}$$

$$= \frac{(x^2+1)(10x - 2x^3)}{(x^2+4)^4}$$

The preceding example is typical of how the chain rule is applied after one becomes familiar with the technique. It is not necessary to write down the substitution u each time. This is done mentally instead of formally on paper. However, until a student has worked a fair number of problems he would do well to write out the solutions formally to guard against possible errors.

The following examples are additional illustrations of the kind of functions that can be differentiated by means of the chain rule and the earlier rules for differentiation. The choice of the substitution $u = g(x)$ is usually quite obvious.

Example 4. $f(x) = \dfrac{(x^2+1)^3}{x}$. Choose $u = x^2 + 1$; then

$$f'(x) = \frac{x \cdot 3(x^2+1)^2 2x - (x^2+1)^3}{x^2} = \frac{(5x^2-1)(x^2+1)^2}{x^2}$$

Example 5. $f(x) = (x^2 + x)(1 - x)^2$. Only one substitution is necessary here. It is made mentally when needed.

$$f'(x) = (x^2 + x)2(1 - x)(-1) + (1 - x)^2(2x + 1) = (1 - x)(1 - x - 4x^2)$$

Example 6. $f(x) = (x - 4)^{-2}(x + 1)^{-1}$. Two substitutions are needed here. They give

$$\begin{aligned} f'(x) &= (x - 4)^{-2}(-1)(x + 1)^{-2} + (x + 1)^{-1}(-2)(x - 4)^{-3} \\ &= (x - 4)^{-3}(x + 1)^{-2}(2 - 3x) \end{aligned}$$

8. IMPLICIT DIFFERENTIATION

Functions in the social sciences are sometimes defined implicitly by means of equations. The equation $y(1 + x) = 33,600$ that occurred in Section 6, Chapter 1, is an example of a function y that is defined implicitly by means of an equation. We can solve this equation for y in terms of x to obtain the explicit functional form:

$$y = \frac{33,600}{1 + x}$$

It may occur, however, that an equation which defines y implicitly as a function of x cannot be solved explicitly for y in terms of x, except numerically; therefore, we may not be able to calculate the derivative of y with respect to x by means of our differentiation formulas for explicit functions. It clearly would be very helpful to have a method for calculating $\frac{dy}{dx}$ without the necessity of solving for y in an implicit relationship.

As an illustration, the implicit functional relationship given by the equation $y^5 + y^2 = x^2 + 2$ cannot be changed into an explicit functional relationship because this fifth-degree equation cannot be solved for y in terms of x, except numerically. In spite of this fact, it is possible to calculate the derivative of y as a function of x in this situation, although the result may also involve an implicit relationship.

Before we discuss this complicated function we look at the first illustration: $y(1 + x) = 33,600$. Since we can solve this equation and obtain y as a function of x, we can substitute this value into the left side of this equation to make it a function of x only, which happens to have the constant value 33,600 because the right side does. We can differentiate both sides of this equation with respect to x to obtain a valid equation. Using the product rule, we obtain

$$y + (1 + x)\frac{dy}{dx} = 0$$

As a result,

$$\frac{dy}{dx} = -\frac{y}{1 + x}$$

Thus, we obtain the derivative of our function without the necessity of expressing it in explicit form.

The explicit version of this function is, of course,

$$y = \frac{33{,}600}{1 + x}$$

and its derivative is

$$\frac{dy}{dx} = -\frac{33{,}600}{(1 + x)^2}$$

To illustrate the calculation of $\frac{dy}{dx}$ by the two methods for a particular value of x, let us choose $x = 4$. The implicit version requires us to first calculate $y = \frac{33{,}600}{1 + 4} = 6720$ and then to calculate $\frac{dy}{dx} = -\frac{6720}{1 + 4} = -1344.$

The explicit version requires only the calculation of $\frac{dy}{dx} =$

$-\frac{33{,}600}{(1 + 4)^2} = -1344.$

By replacing y by $\frac{33{,}600}{(1 + x)}$ in the implicit formula for $\frac{dy}{dx}$, it will be seen that these two formulas for $\frac{dy}{dx}$ are equivalent for all values of x for which the function is defined.

Now let us consider the second, more complicated, problem. Assuming that we could solve the preceding equation $y^5 + y^2 = x^2 + 2$ for y in terms of x, we could then substitute this function of x back into our equation and obtain an identity in x. That is, we would have a function of x on both sides of the equation and this equation would be satisfied for every value of x for which the relationship between y and x is assumed to hold true. If two differentiable functions of x are equal for, say, an interval of x values, they must possess the same derivative for values of x in that interval. Hence, we can differentiate both sides of the preceding equation with respect to x and obtain a valid equation. We now apply this reasoning to the relationship $y^5 + y^2 = x^2 + 2$.

To differentiate y^5 with respect to x, we apply the chain rule. Since there will be confusion in attempting to apply formula (10) directly to this problem because of the different roles played by y; let us write formula (10) in the form

$$\frac{dz}{dx} = \frac{dz}{dy}\frac{dy}{dx}$$

where z is the function to be differentiated and y plays the role of u in that earlier formula. Thus, letting $z = y^5$, we obtain

$$\frac{dz}{dx} = 5y^4 \frac{dy}{dx}$$

To differentiate y^2, we use the same procedure as for y^5. This gives

$$2y \frac{dy}{dx}$$

The right side of our equation causes no difficulty. Combining these calculations, we obtain

$$5y^4 \frac{dy}{dx} + 2y \frac{dy}{dx} = 2x$$

Factoring out $\frac{dy}{dx}$ on the left side, we get

$$[5y^4 + 2y] \frac{dy}{dx} = 2x$$

The solution of this equation for $\frac{dy}{dx}$ yields the desired result:

$$\frac{dy}{dx} = \frac{2x}{5y^4 + 2y}$$

Now if we wish to calculate the derivative at a given point, we can substitute the selected x value into the original equation and solve for y numerically. With this pair of x and y values we then substitute into the expression just obtained for $\dfrac{dy}{dx}$ to obtain the desired value of the derivative at that point. Thus, we circumvent the necessity of solving for y in terms of x to obtain an explicit function of x to differentiate. Since the latter operation is impossible here anyway, the implicit technique is quite an improvement, to say the least.

The following are some additional illustrations on the technique of implicit differentiation. In each case the problem is to calculate $\dfrac{dy}{dx}$.

Example 1. $x^3 + y^3 = x$.

$$3x^2 + 3y^2\,\frac{dy}{dx} = 1, \quad \frac{dy}{dx} = \frac{1 - 3x^2}{3y^2}$$

Example 2. $xy = x + y$.

$$x\,\frac{dy}{dx} + y = 1 + \frac{dy}{dx}$$

$$\frac{dy}{dx} = \frac{1 - y}{x - 1}$$

Example 3. $x^{1/3} + y^{1/3} = 1$.

$$\frac{1}{3}\,x^{-2/3} + \frac{1}{3}\,y^{-2/3}\,\frac{dy}{dx} = 0$$

$$\frac{dy}{dx} = -\frac{x^{-2/3}}{y^{-2/3}} = -\frac{y^{2/3}}{x^{2/3}}$$

In this example we use the fact that the formula for differentiating x^n is valid for fractional exponents as well as for integer exponents.

Example 4. $x^2 y + xy^2 = 2$.

$$x^2\,\frac{dy}{dx} + y(2x) + x(2y)\,\frac{dy}{dx} + y^2 = 0$$

$$\frac{dy}{dx} = -\frac{y^2 + 2xy}{x^2 + 2xy}$$

Example 5. $\sqrt{x} + \sqrt{y} = \dfrac{x}{2}$. This is equivalent to $x^{1/2} + y^{1/2} = \dfrac{x}{2}$. Hence

$$\frac{1}{2} x^{-1/2} + \frac{1}{2} y^{-1/2} \frac{dy}{dx} = \frac{1}{2}$$

$$\frac{dy}{dx} = \frac{1 - x^{-1/2}}{y^{-1/2}} = \frac{1 - \dfrac{1}{\sqrt{x}}}{\dfrac{1}{\sqrt{y}}} = \frac{\sqrt{y}}{\sqrt{x}} \, (\sqrt{x} - 1)$$

COMMENTS

This chapter introduced the concept of the derivative of a function and studied some of its fundamental properties. This material should enable us to solve some of the interesting types of problems that led to the introduction of calculus. Thus this chapter is a necessary preliminary to the next chapter, which considers several of the useful applications of calculus.

EXERCISES

Section 1

1. Using the graph of Fig. 2, Section 1, and drawing tangent lines to it at the corresponding points, estimate the slope of the curve at the points $\left(\frac{1}{2}, \frac{1}{4}\right)$ and $\left(\frac{1}{4}, \frac{1}{16}\right)$. Compare your answer with the exact answer given by the formula derived in Section 1.

2. (a) Graph the curve $y = x^3$ for $0 \leqslant x \leqslant 2$.
   (b) Sketch a tangent line to the graph at the point P whose coordinates are $(1,1)$ and estimate its slope by inspection.
   (c) Choose a point Q with coordinates $(1 + h, (1 + h)^3)$ and calculate the slope of the chord between the points P and Q. Then calculate the limiting value of this slope as $h \to 0$ to obtain the exact answer to (b).

3. Derive a formula for the slope of the curve $y = x^3$ at an arbitrary point (x_0, y_0) on it. Use your result to solve (c) of Problem 2.

4. (a) Graph the curve $y = \dfrac{1}{x}$ for $0 < x \leqslant 4$.

(b) Sketch a tangent line to the graph at the point P whose coordinates are $\left(2, \frac{1}{2}\right)$ and estimate its slope by inspection.

(c) Choose a point Q with coordinates $\left(2 + h, \frac{1}{2+h}\right)$ and calculate the slope of the chord between the points P and Q. Then calculate the limiting value of this slope as $h \rightarrow 0$ to obtain the exact answer to (b).

5. Derive a formula for the slope of the curve $y = \frac{1}{x}$ at an arbitrary point (x_0, y_0) on it. Use your result to solve Problem 4(c).

6. (a) Graph the curve $y = \sqrt{x}$ for $0 \leqslant x \leqslant 4$.
   (b) Sketch a tangent line to the graph at the point P whose coordinates are $(1,1)$ and estimate its slope by inspection.
   (c) Choosing $x_0 = 1$ and $x_0 + h = 1.3, 1.2, 1.1, 1.05$ and 1.01, use square root tables to find the value of $\sqrt{x_0 + h}$ for those values of $x_0 + h$. Then calculate the slope of the chord PQ where P is the point $(1,1)$ and Q is the point $(1 + h, \sqrt{1+h})$ on the curve. From these values, guess the limiting value of the slope as $h \rightarrow 0$. Compare your guess with your estimate in (b).

7. Work Problem 6 for the curve $y = \sqrt[3]{x}$.

Section 2

In all of the following problems assume that the limiting value of a function of h as $h \rightarrow 0$ is the value obtained by setting $h = 0$ wherever it occurs in that function, provided that a zero value is not obtained in any denominators.

1. Use the definition of the derivative to calculate the derivative of $f(x) = \frac{x^2 + 1}{x}$ at $x = 1$.

2. Use the definition of the derivative to calculate the derivative of $f(x) = \frac{1}{x^2}$ at $x = 2$.

3. Use the definition of the derivative to calculate the derivative of $f(x) = \frac{x - 1}{x + 1}$ at $x = 0$.

4. Use the definition of the derivative to calculate the derivative of $f(x) = \dfrac{x^2}{x+1}$ at $x = x_0$.

5. Find the equation of the tangent to the curve $y = x^3$ at the point $(1,1)$. Use the answer to Problem 2, Section 1.

6. Find the equation of the tangent to the curve $y = \dfrac{1}{x}$ at the point $\left(2, \dfrac{1}{2}\right)$. Use the answer to Problem 4, Section 1.

7. Find the equation of the tangent to the curve $y = \dfrac{1}{x^2}$ at the point $\left(2, \dfrac{1}{4}\right)$. Use the answer to Problem 2.

Section 3

1. Let $f(x) = x$ for $0 \leqslant x \leqslant 1$, $f(x) = 2 - x$ for $1 < x \leqslant 2$, and $f(x) = 0$ elsewhere.
   (a) Graph $f(x)$.
   (b) What is the value of $\lim\limits_{x \to 1} f(x)$?

   (c) Are there any values of a for which $\lim\limits_{x \to a} f(x)$ does not exist?

   (d) Where is f continuous?

2. Let $f(x) = x$ for $0 \leqslant x \leqslant 1$, $f(x) = 3 - x$ for $1 < x \leqslant 2$ and $f(x) = 0$ elsewhere.
   (a) Graph $f(x)$.
   (b) What is the value of $\lim\limits_{x \to 1} f(x)$?

   (c) Are there any values of a for which $\lim\limits_{x \to a} f(x)$ does not exist?

   (d) Where is f continuous?

3. Let $f(x) = 1$ if x is an integer and $f(x) = 0$ for all other values of x.
   (a) Graph $f(x)$.
   (b) What is the value of $\lim\limits_{x \to 2} f(x)$?

   (c) Are there any values of a for which $\lim\limits_{x \to a} f(x)$ does not exist?

   (d) Where is f continuous?

4. For what values of x does the function $f(x)$ of Problem 1 fail to have a derivative? Is $f(x)$ continuous at those points?

5. For what values of x does the function $f(x)$ of Problem 2 fail to have a derivative? Is $f(x)$ continuous at those points?

6. For what values of x does the function $f(x)$ of Problem 3 fail to have a derivative? Is $f(x)$ continuous at those points?

7. In each part draw the graph of a function for which

   (a) $\lim_{x \to 3} f(x) = 2$.

   (b) $\lim_{x \to 3} f(x) = 2$ but for which $f(3) \neq 2$.

   (c) $\lim_{x \to 3} f(x) = 2$ and $f(3) = 2$ but for which $f'(3)$ does not exist.

8. Why does the function $f(x) = \dfrac{1}{x}$ not have a derivative at $x = 0$?

9. Does the function $f(x) = \sqrt{1 - x^2}$ whose graph is a semicircle have a derivative at all points where it is defined? Explain.

10. In each case determine whether the function $f(x)$ is continuous, or differentiable, or both, at $x = 0$.

    (a) $f(x) = \begin{cases} x^2, & x \leqslant 0 \\ x^3, & x > 0 \end{cases}$

    (b) $f(x) = \begin{cases} |x|, & x \neq 0 \\ 1, & x = 0 \end{cases}$

11. Explain why the function $f(x) = \sqrt{x}$ does not possess a derivative at $x = 0$.

Section 5

1. Use the theorems on differentiation to calculate $f'(x)$ if
    (a) $f(x) = 4x^3 - 3x^4$
    (b) $f(x) = (x^2 + 1)(x^4 + x)$

    (c) $f(x) = \dfrac{2x + 1}{2x - 1}$
    (d) $f(x) = \dfrac{1}{x^2 + 1}$

    (e) $f(x) = x^2 + 3x + \dfrac{1}{x^2}$
    (f) $f(x) = (2x^4 - 1)(5x^3 + 6x)$

    (g) $f(x) = \dfrac{4}{x^2} + \dfrac{6}{x^4}$
    (h) $f(x) = \dfrac{x}{x^2 + x}$

2. Use the theorems on differentiation to calculate $f'(x)$ if

(a) $f(x) = \dfrac{x^6}{6} - x^4$

(b) $f(x) = \dfrac{3}{8x^4}$

(c) $f(x) = \dfrac{2x - 1}{2 - 3x}$

(d) $f(x) = (2x^2 + 1)(x^3 + 3x)$

(e) $f(x) = \dfrac{x}{1 + 2x^2}$

(f) $f(x) = \sqrt{7}\,(x^3 - x^{-2})$

(g) $f(x) = (x^{-2} + 1)(x + x^2)$

(h) $f(x) = \dfrac{(2x + 1)(3x - 1)}{x + 5}$

3. Use the theorems on differentiation to calculate $f'(x)$ if

(a) $f(x) = 3x^4 - 5x^2 + 7$

(b) $f(x) = \dfrac{x^3 - 3x + 4}{3}$

(c) $f(x) = \dfrac{6}{5x^5}$

(d) $f(x) = \dfrac{x}{x - 1}$

(e) $f(x) = \dfrac{4x}{2x^2 + 1}$

(f) $f(x) = \dfrac{x^2}{x + 1}$

(g) $f(x) = \dfrac{x - 1}{x^2}$

(h) $f(x) = \dfrac{x^3 - 8}{x^3 + 8}$

4. Find the equation of the tangent to the curve $y = \dfrac{1}{x + 1}$ at the point $\left(1, \dfrac{1}{2}\right)$.

5. Find the equation of the tangent to the curve $y = \dfrac{8}{x^2 + 4}$ at the point $(2,1)$.

6. Find the equation of the tangent to the curve $y = \dfrac{x - 1}{x + 1}$ at the point $(-3,2)$.

7. For each of the following functions, determine the values of x for which $f'(x) = 0$.

(a) $f(x) = x^3 - 3x + 12$

(b) $f(x) = x^4 - 2x^2$

(c) $f(x) = \dfrac{x}{x^2 + 4}$

(d) $f(x) = \dfrac{x}{(x + 1)^2}$

8. For what value of x does the tangent to the curve $y = x^2 + 3x + 2$ have the slope 5?

9. Find the values of x for which the tangent to the curve $y = 2x^3 - 9x^2 + 12x + 10$ is horizontal.

10. Find the equation of the tangent line (or lines) to the curve $y = \frac{1}{3}(x^3 - 3x^2 + 6x + 6)$ which is parallel to the line $2x - y + 5 = 0$.

11. Show that the equation of the tangent to the curve whose equation is $xy = 1$ at the point (x_0, y_0) lying on it is given by the formula $y_0 x + x_0 y = 2$.

12. If r, s, and t are differentiable functions of x, show that the derivative of the function $w = rst$ is given by the formula $w' = r'st + rs't + rst'$.

13. Use the formula of Problem 12 to calculate the derivative of $f(x) = (x^2 + 4)(x^3 - x)(x^4 + x)$.

14. Given $f(x) = x^{-k}$, where k is a positive integer, write $f(x) = \frac{1}{x^k}$ and apply the quotient rule of differentiation to show that the formula for differentiating x^n is also valid when n is a negative integer.

Section 7

1. Calculate $f'(x)$ if $f(x) = (1 + x^2)^2$
   (a) Without using the chain rule.
   (b) Using the chain rule. Compare your answers.

2. Calculate $f'(x)$ using the chain rule if
   (a) $f(x) = (3x^2 + x)^6$
   (b) $f(x) = 3x^2 + (x^2 + x)^2$
   (c) $f(x) = \frac{(x + 1)^5}{x}$
   (d) $f(x) = \frac{1}{(x^2 + 2)^2}$
   (e) $f(x) = \left(\frac{x - 7}{x + 2}\right)^2$
   (f) $f(x) = \left(\frac{x^2 + 1}{x - 1}\right)^2$

3. Calculate $f'(x)$ using the chain rule if
   (a) $f(x) = (x^2 + 4x - 3)^3$
   (b) $f(x) = (x^2 + 2)(x - 1)^2$
   (c) $f(x) = \frac{x^2 - 4x}{(x + 1)^2}$
   (d) $f(x) = \frac{(x^2 + 4)^2}{(x^2 + 2)}$
   (e) $f(x) = (x + 1)^{-1}(2x + 1)^{-2}$
   (f) $f(x) = [(x^2 + 1)^2 + x]^2$

4. Calculate $f'(x)$ using the chain rule if
   (a) $f(x) = (x^2 + 3)^{-2}$
   (b) $f(x) = x(x^2 + 4)^2$
   (c) $f(x) = \frac{1}{(x^2 + 3x)^4}$
   (d) $f(x) = \frac{(x + 1)^2}{(x - 1)^2}$
   (e) $f(x) = (x^2 - 4)^{-1}(x + 2)^{-2}$
   (f) $f(x) = \frac{(x + 1)^2(x + 3)^2}{(x - 2)^3}$

5. Calculate $f'(x)$ using the chain rule if
   (a) $f(x) = (x+3)^3$
   (b) $f(x) = (x-2)^{-4}$
   (c) $f(x) = (x^2-1)^2$
   (d) $f(x) = (1-x^3)^{-2}$
   (e) $f(x) = \dfrac{(x^2+3)^2}{x-1}$
   (f) $f(x) = (x^{-2}+4)^2(x^2+1)^{-1}$

6. Find the equation of the tangent to the curve $y = x(x^2+4)^2$ at the point $(0,0)$.

7. Find the equation of the tangent to the curve $y = \dfrac{2}{(4-x)^2}$ at the point $(3,2)$.

8. Find the equation of the tangent (or tangents) to the curve $y = \dfrac{x^2}{x+2}$ that is parallel to the line $5x - 9y + 3 = 0$.

9. If $r(x) = [f(x)]^m[g(x)]^n$ and f and g are differentiable functions of x, use the chain rule to derive a formula for $r'(x)$.

Section 8

1. In each case calculate $\dfrac{dy}{dx}$ by using implicit differentiation techniques.

   (a) $x^2 + y^2 = 4xy$
   (b) $\dfrac{1}{x} + \dfrac{1}{y} = 1$
   (c) $x^2 + y^2 = x^2y^2$
   (d) $x^3y + xy^3 = 3$
   (e) $x + y = \dfrac{y}{x}$
   (f) $x^4 - y^4 = x^2y^3$

2. In each case calculate $\dfrac{dy}{dx}$ by using implicit differentiation techniques.

   (a) $\sqrt{x} + \sqrt{y} = 1$
   (b) $x^2 + xy^3 = y$
   (c) $\dfrac{x+y}{x-y} = x+1$
   (d) $\dfrac{y}{x-y} = x^2 + 2$
   (e) $\dfrac{y-1}{y+1} = \dfrac{x+1}{x-1}$
   (f) $\sqrt{y}\,x^2 + \sqrt{x}\,y^2 = 1$

3. Find the equation of the tangent to the curve $x^3 + y^3 = 9$ at the point $(2,1)$.

4. Given $y^2 - x^2 = 16$, calculate $\dfrac{dy}{dx}$ by (a) implicit differentiation, and (b) explicit differentiation. Assume that $y > 0$. Show that your answers are equivalent.

5. Given $x^2 - y^2 = 9$, calculate $\dfrac{dy}{dx}$ by (a) implicit differentiation, and (b) explicit differentiation. Assume that $y > 0$. Show that your answers are equivalent.

6. Find the equation of the tangent to the curve $x^2 + 4y^2 - 2x - 8y = 12$ at the point $(2, -1)$.

7. Show that the equation of the tangent to the circle $x^2 + y^2 = r^2$ at the point (x_0, y_0) lying on it is given by $x_0 x + y_0 y = r^2$.

8. If $y = x^4 + 1$, $x > 0$, calculate $\dfrac{dx}{dy}$, where x is the inverse function of y by implicit differentiation and also by solving for x in terms of y and using explicit differentiation.

9. If $y = \sqrt{2x - 1}$, $x > \dfrac{1}{2}$, calculate $\dfrac{dx}{dy}$, where x is the inverse function of y, by implicit differentiation and also by solving for x in terms of y and using explicit differentiation.

10. Calculate $f'(x)$ using the chain rule if

   (a) $f(x) = (4x + 3)^{3/2}$

   (b) $f(x) = \sqrt{\dfrac{x - 1}{x + 1}}$

   (c) $f(x) = \dfrac{\sqrt{x^2 - 1}}{x}$

   (d) $f(x) = (2x)^{2/3} + (3x)^{1/3}$

   (e) $f(x) = \sqrt{3x} + \dfrac{1}{\sqrt{3x}}$

   (f) $f(x) = \sqrt[3]{\dfrac{x - 2}{x + 2}}$

   (g) $f(x) = \sqrt{1 + \sqrt{x}}$

   (h) $f(x) = \left(\dfrac{x^3}{1 + x^2}\right)^{2/3}$

11. If $[f(x)]^m + [g(y)]^n = c$, where m and n are integers and c is a constant, use implicit differentiation and the chain rule to derive a formula for $\dfrac{dy}{dx}$.

12. Write $y = x^{m/n}$ in the implicit form $y^n = x^m$. Then differentiate both sides with respect to x and solve for $\dfrac{dy}{dx}$. From your answer show that the formula for differentiating x^n applies to this more general type of exponent also.

9 | Applications of the Derivative

In this chapter we use differential calculus to assist us in the solution of certain types of problems that arise in the social sciences. We restrict ourselves to three classes of problems. The first class, which is geometrical in nature, is concerned with using the derivative to help graph functions. The second class is concerned with using the derivative to determine where a function assumes its maximum, or minimum, value. The third class involves measuring the rate of change of various functions.

The discussion in this chapter is based on a number of useful theorems. Most of these theorems will not be proved because that requires more time and effort than seems justified in this brief treatment of calculus; however, in a few instances a sketch of a proof based on geometric intuition is given.

1. CURVE SKETCHING

In studying mathematical models for a social science problem it is very instructive to look at the graphs of the functions that are being contemplated as models for the problem. Many of the functions that are used as models have been obtained by inspecting empirical data and plotting the data. For example, a manufacturer can construct a cost function, which we denote by $C(x)$, by keeping a record of how much it costs him to manufacture x units of his product for various values of x. If he plots the values of $C(x)$ against x and draws a smooth curve through the resulting points, he will have a geometrical representation of his cost function. It may then be possible to obtain an algebraic representation of $C(x)$ that can be studied further by means of calculus techniques.

269

Conversely, if we are given a particular function and wish to study its properties to determine whether it may serve, for example, as a cost function, a graph of the function is very useful. It is usually much easier to observe general properties of a function visually from its graph than to study those properties algebraically.

In view of the importance of studying functions geometrically, this section is devoted to explaining how calculus techniques can be used to assist in sketching the graphs of functions.

In the theorems that follow it is often necessary to distinguish between a *closed interval*, denoted by [a,b], and the corresponding *open interval*, denoted by (a,b). The closed interval [a,b] includes the end points of the interval as well as all the points between a and b, whereas the open interval (a,b) does not include the end points of the interval. Since we are dealing with one-dimensional intervals on the x axis, there should be no confusion here with the earlier notation in which (a,b) represented the coordinates of a point in two dimensions. The theorems that follow could be simplified somewhat by using only closed intervals; however, since we do not prove many of these theorems, it is just as well to state them in their traditional more general form.

The first theorem that is needed in the application of calculus to curve sketching may be expressed as follows:

Theorem 1. *If the function f is continuous in [a,b] and f'(x) = 0 for all x in (a,b), then f(x) has the same value for all x in [a,b].*

This theorem seems very plausible because $f'(x) = 0$ for all x in (a,b) implies that the graph of $f(x)$ will have a horizontal tangent at every point inside that interval. This, together with the fact that the graph of $f(x)$ is unbroken because of the continuity of f, requires that the graph be a hori-

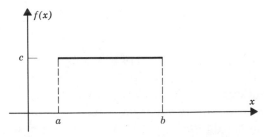

FIGURE 1   The graph of a function for which $f'(x) = 0$.

zontal straight line and, therefore, that the function have a constant value.

Figure 1 illustrates the geometry of this theorem. The preceding geometrical explanation is not a proof of the theorem; it only suggests that our intuition readily agrees with the truth of the theorem.

A consequence of Theorem 1 is another theorem that is used on numerous occasions in later sections. It may be stated as follows:

Theorem 2. *If f and g are continuous in [a,b] and f'(x) = g'(x) for all x in (a,b) then*

$$f(x) = g(x) + c, \qquad a \leqslant x \leqslant b$$

where c is some constant.

This theorem is easily demonstrated algebraically. Let $r(x) = f(x) - g(x)$. From the equality of the derivatives of f and g it follows that

$$r'(x) = f'(x) - g'(x) = 0, \qquad a < x < b$$

Since the difference of two continuous functions is a continuous function, r is continuous in $[a,b]$; hence, by Theorem 1, $r(x)$ has the same value for all x in $[a,b]$. Let $r(x) = c$ be this constant value. Then

$$r(x) = f(x) - g(x) = c, \qquad a \leqslant x \leqslant b$$

Or

(1) $$f(x) = g(x) + c, \qquad a \leqslant x \leqslant b$$

This result shows that if two continuous functions have the same derivative at all points in an interval, they can differ from each other throughout that interval only by an additive constant. This relationship is illustrated graphically in Fig. 2.

The next theorem is concerned with how to determine where a function is increasing and where it is decreasing. Geometrically, this means determining where the graph is rising and where it is falling. We say that a function f is *increasing* at a point x_0 if there is a number $h_0 > 0$ such that

$$f(x_0 + h) > f(x_0)$$

and

$$f(x_0 - h) < f(x_0)$$

for all values of h satisfying $0 < h < h_0$. This technical language implies that the value of $f(x)$ must be smaller than $f(x_0)$ for values of x to the left of

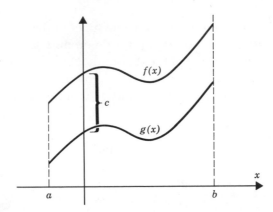

FIGURE 2   Graphs of two functions having the same derivative.

x_0, but close to x_0, and must be larger than $f(x_0)$ for values of x to the right of x_0, but close to x_0. If a function is increasing at all points of an interval (a,b), we say that it is *increasing throughout the interval* (a,b).

The derivative of a function is useful in determining where the function is increasing and where it is decreasing. This usage is given by the following theorem.

Theorem 3. *If $f'(x) > 0$ for all x in (a,b), then $f(x)$ is increasing throughout the interval (a,b).*

Let x be any point in (a,b). Since $f'(x) > 0$, it follows that

$$\lim_{h \to 0} \frac{f(x+h) - f(x)}{h} > 0$$

This can hold only if the quotient is positive for sufficiently small values of $|h|$. But if $h > 0$, this requires that $f(x+h) - f(x) > 0$ for sufficiently small values of h, which in turn implies that

$$f(x+h) > f(x)$$

for sufficiently small positive values of h. Similarly, if $h < 0$, it is necessary that $f(x+h) - f(x) < 0$ for sufficiently small values of $|h|$, which in turn implies that

$$f(x+h) < f(x)$$

for $h < 0$ and sufficiently small values of $|h|$. In view of our definition of

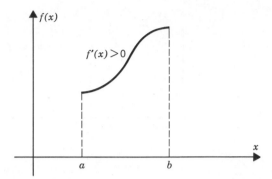

FIGURE 3   A graph of an increasing function.

an increasing function, these inequalities show that f is increasing at the point x. Since x was any point in (a,b), f must be increasing at every point in (a,b) and, hence, throughout the interval.

The geometrical interpretation of this theorem is that the graph of $f(x)$ is rising between a and b, as illustrated in Fig. 3.

There is, of course, a corresponding theorem for the case in which $f'(x) < 0$ for all values of x in (a,b) and which states that f is decreasing throughout the interval. Geometrically, this implies that the graph of $f(x)$ is falling throughout the interval.

We shall work several examples to illustrate how the derivative can be used to assist in the graphing of functions.

Example 1. Let $f(x) = x^2 - x$ and consider its graph in the interval $0 \leqslant x \leqslant 2$. This function is clearly continuous and differentiable in this interval and will have a smooth graph in that interval. Here $f'(x) = 2x - 1$; consequently, $f'(x) > 0$ will hold wherever $2x - 1 > 0$. This inequality is satisfied for $\frac{1}{2} < x < 2$; hence, by Theorem 3, f is increasing throughout the interval from $x = \frac{1}{2}$ to $x = 2$. Similarly, $f'(x) < 0$ holds for $2x - 1 < 0$ and, hence, for the interval $0 < x < \frac{1}{2}$. Therefore, by a theorem corresponding to Theorem 3, f is decreasing throughout the interval from $x = 0$ to $x = \frac{1}{2}$. A graph of $f(x)$ is now easily constructed by finding a few points on it and using the information just obtained. For this purpose we calculate

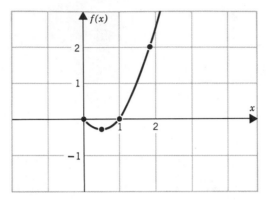

FIGURE 4   A graph of $f(x) = x^2 - x$.

$$f(0) = 0, \qquad f\left(\tfrac{1}{2}\right) = -\tfrac{1}{4}, \qquad f(1) = 0, \qquad f(2) = 2$$

A graph of this function is shown in Fig. 4.

Example 2. Let $f(x) = 4x^3 - 5x^2 - 2x - 1$.

This function is clearly continuous and differentiable for all values of x, but we graph it over the interval $[-1,2]$ only. We first compute $f'(x)$, which is

$$f'(x) = 12x^2 - 10x - 2 = 2(6x^2 - 5x - 1)$$
$$= 2(x - 1)(6x + 1)$$

We then determine the values of x that make $f'(x) = 0$. These are called the *critical points*. They are useful in determining the intervals over which f is increasing or decreasing. Here the critical points are $x = 1$ and $x = -\tfrac{1}{6}$. If $x < -\tfrac{1}{6}$, $f'(x)$ will be positive because the factor $x - 1$ will be negative and the factor $6x + 1$ will also be negative. Hence, by Theorem 3, f is increasing from $x = -1$ to $x = -\tfrac{1}{6}$. For $-\tfrac{1}{6} < x < 1$, $f'(x)$ will be negative because the factor $x - 1$ will be negative but the factor $6x + 1$ will now be positive. Hence, by the analogue of Theorem 3, f is decreasing in this interval. Finally, $f'(x)$ will be positive for $x > 1$ because both factors will be positive then. Therefore, f is increasing from $x = 1$ to $x = 2$. To determine the points where the function changes from increasing to decreasing and then to increasing, we calculate the functional values at the critical points.

They are

$$f\left(-\frac{1}{6}\right) = -\frac{89}{108} \quad \text{and} \quad f(1) = -4$$

To obtain a little more accuracy in the graph of $f(x)$, it is useful to locate a few more points on it. Hence, we also calculate the values

$$f(-1) = -8, \quad f(0) = -1, \quad f(2) = 7$$

After plotting the preceding five points and using the increasing-decreasing information that we obtained previously, we can easily graph f in the interval $[-1,2]$. The graph is shown in Fig. 5. The interval $[-1,2]$ was

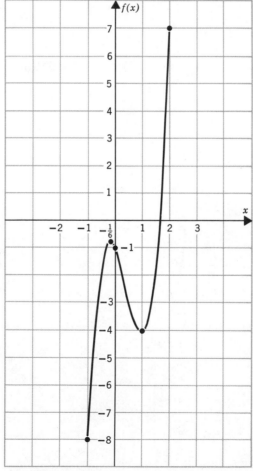

FIGURE 5   A graph of $f(x) = 4x^3 - 5x^2 - 2x - 1$.

chosen because the functional values become extremely large in absolute value for values of x outside that interval.

Example 3. Graph $f(x) = x^4 - 2x^3$ in the interval $[-1,2]$. $f'(x) = 4x^3 - 6x^2 = 0$ gives $2x^2(2x - 3) = 0$. The critical points are $x = 0$ and $x = \frac{3}{2}$. Since $x^2(2x - 3) > 0$ if $x > \frac{3}{2}$, it follows that f is increasing from $x = \frac{3}{2}$ to $x = 2$. Since $x^2(2x - 3) < 0$ if $x < \frac{3}{2}$, except when $x = 0$, f is decreasing from $x = -1$ to $x = 0$, and then it decreases again from $x = 0$ to $x = \frac{3}{2}$. Calculations give

$$f(-1) = 3, \quad f(0) = 0, \quad f(1) = -1, \quad f\left(\frac{3}{2}\right) = -\frac{27}{16}, \quad f(2) = 0.$$

A graph of the function based on all this information is given in Fig. 6.

Example 4. Graph $f(x) = \frac{1}{4} x^4 - \frac{1}{3} x^3 - x^2 + 2$ in $[-2,3]$. $f'(x) = x^3 - x^2 - 2x = 0$ gives $x(x - 2)(x + 1) = 0$. The critical points are $x = -1$, $x = 0$,

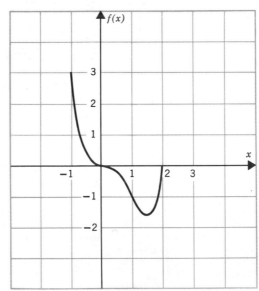

FIGURE 6   A graph of $f(x) = x^4 - 2x^3$.

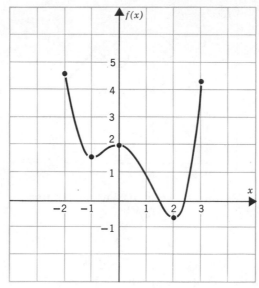

FIGURE 7   A graph of $f(x) = \dfrac{1}{4}x^4 - \dfrac{1}{3}x^3 - x^2 + 2$.

and $x = 2$. Since $x(x - 2)(x + 1) < 0$ for $-2 < x < -1$ and for $0 < x < 2$, and $x(x - 2)(x + 1) > 0$ for $-1 < x < 0$ and for $2 < x < 3$, the graph is falling in the first two intervals and is rising in the second two intervals. Calculations give

$$f(-2) = \frac{14}{3}, \qquad f(-1) = \frac{19}{12}, \qquad f(0) = 2, \qquad f(2) = -\frac{2}{3}, \qquad f(3) = \frac{17}{4}$$

A graph of the function based on all this information is given in Fig. 7.

2. FINDING MAXIMA AND MINIMA

In this section we study the basic problem of how to determine where a function assumes its maximum, or minimum, value. This is the second of the three classes of problems that we outlined at the beginning of this chapter. Although this is not a geometrical problem, it is helpful to look at it from a geometrical point of view.

In studying problems of this type, we introduce a new term called a local maximum. If f has the property that its value at x_0 satisfies the inequality

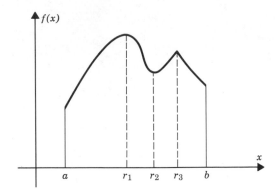

FIGURE 8   Local maxima and minima.

$$f(x_0) \geqslant f(x)$$

for all values of x close to x_0, then f is said to have a *local maximum* at x_0. That is, the value of $f(x)$ at x_0 must be at least as large as its value in the immediate vicinity of x_0. A corresponding statement with the inequality reversed defines a *local minimum.* The word local aptly describes the situation because we are concerned only with how the function behaves in the local vicinity of x_0.

The geometry of local maxima and minima is illustrated in Fig. 8 which gives the graph of a function that has two local maxima, located at r_1 and r_3, and one local minimum, located at r_2. The local maximum at r_1 is also an *absolute maximum* of the function in the interval $[a,b]$. The *absolute minimum* of the function occurs at $x = a$.

Not all functions possess a maximum, or minimum, value in an interval. However, if we restrict ourselves to continuous functions, then it can be shown that a function does possess both a maximum and a minimum value in an interval. A famous theorem in mathematics guarantees this property. It may be stated as follows:

Theorem 4. *If f is continuous in $[a,b]$, it will take on a maximum, and a minimum, value in that interval.*

If we had not required f to be continuous in the closed interval $[a,b]$, there would be no assurance that the function possessed a maximum, or minimum, value. For example, the function defined by

$$f(0) = 0, \qquad f(1) = 1, \qquad f(x) = \frac{1}{x}, \qquad 0 < x < 1$$

is not continuous in the closed interval [0,1]. From its graph, which is partly shown in Fig. 9, it is clear that this function does not possess a maximum value in the interval [0,1].

The next theorem is the one that enables us to locate the maximum, or minimum, value of a function f. Although the theorem itself does not require f to be continuous throughout a closed interval, we assume that such is the case when we apply the theorem, otherwise we might be searching for something that does not exist.

Theorem 5. *If f possesses a derivative at every point of an interval (a,b) and if it assumes its absolute maximum value at a point c somewhere inside this interval, then $f'(c) = 0$.*

If $f'(c)$ were not equal to zero, it would have a positive or a negative value. Suppose that $f'(c) > 0$. Then by the arguments used in Theorem 3, f must be increasing at $x = c$; therefore, it could not have a maximum at that point. Similarly, if it were true that $f'(c) < 0$, f would be decreasing at $x = c$; therefore, it could not have a maximum at that point. Thus it must be true that $f'(c) = 0$.

Geometrically, this theorem merely states that if the graph of $f(x)$ has a tangent line at every point inside an interval that contains a maximum point, then the tangent line at the maximum point must be horizontal. This is illustrated in Fig. 8 where it is clear that the tangent line at $x = r_1$ will be horizontal if that function possesses a derivative at all points in the immediate vicinity of $x = r_1$.

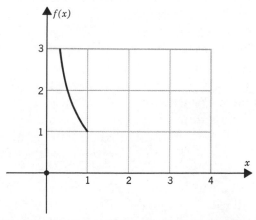

FIGURE 9   A function that does not possess a maximum value.

There is, of course, a corresponding theorem with the word maximum replaced by the word minimum.

We are now in a position to proceed in a systematic manner to find the maximum, or minimum, value of a function f in a closed interval $[a,b]$. We assume that f is continuous in the interval; therefore, by Theorem 4 it must possess a maximum and a minimum somewhere in the interval. Let us concentrate on the maximum value only. The same procedure would be followed in searching for a minimum. This maximum value must occur either at one of the end points of the interval or inside the interval. If it occurs inside the interval and if f possesses a derivative at all points inside the interval, then Theorem 5 assures us that it must occur where $f'(x) = 0$. Hence, it must occur at one of the critical points. In this case our procedure is clear. We set $f'(x) = 0$ and solve the equation for the critical points. We then calculate the value of f at each of the critical points and also at the end points of the interval. The largest of these functional values will be the absolute maximum value of the function and the point where it occurs produces the value of x that we were seeking.

If f does not possess a derivative at all points inside the interval, we must also look at the points where the derivative fails to exist and calculate the value of f at such points. If none of these is larger than the largest value obtained under the procedure just explained, our answer is the same. If, however, the largest of these values exceeds the largest value obtained before, then the maximum is this new largest value.

The geometry of this procedure is nicely illustrated by means of Fig. 8. Suppose that we wish to find where f assumes its maximum value in $[a,b]$. We first solve the equation $f'(x) = 0$ to find the critical points. They are seen to be $x = r_1$ and $x = r_2$. Then we calculate the value of $f(x)$ for $x = a$, $x = b$, $x = r_1$ and $x = r_2$. The largest of these is $f(r_1)$; therefore, if f possessed a derivative at all points inside the interval, our solution would be given by $x = r_1$. Since $f'(x)$ does not exist for $x = r_3$, we calculate $f(r_3)$ and find that $f(r_3) < f(r_1)$; therefore, we retain our earlier tentative solution, namely, $x = r_1$, and conclude that f is maximized at this point.

If we were seeking a minimizing point we would proceed in the same manner and arrive at the conclusion that $f(a)$ was smaller than any of the values $f(r_1)$, $f(r_2)$, $f(r_3)$, and $f(b)$ and, therefore, that the minimum value was assumed at the point $x = a$.

The functions that occur in applied problems are usually continuous and differentiable for all values of x in their domain; therefore, points such as r_3 in Fig. 8 seldom arise in problems of this kind. Consequently, the procedure usually consists in finding the critical points and then calculating the value at the critical points and the boundary points to determine which yields the largest (or smallest) value of the function. For such problems it therefore follows that the absolute maximum (or minimum) of a function will be either a local maximum (or minimum) or it will be located at a boundary point of the interval.

As illustrations of how calculus can be used to solve the problem of maximizing, or minimizing, a function in a practical situation, consider the following problems and their solutions.

Example 1. A packaging firm wishes to make a closed box that has a square base and that will have 27 cubic feet of space in its interior. The material for making the box costs 2 cents per square foot. What are the dimensions of the box that will minimize the cost for material?

To solve this problem, let x be the length of a side of the square base and let y be the height of the box in feet. These dimensions are illustrated in Fig. 10. The problem is to minimize the cost function C. The total area is given by $2x^2 + 4xy$ because the top and bottom have area x^2 each and there are four sides of area xy each. Since each square foot of material costs 2 cents, the total cost in cents is given by

$$C = 2[2x^2 + 4xy]$$

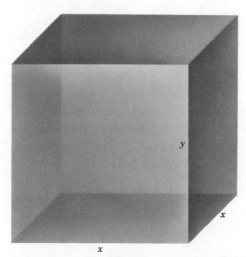

FIGURE 10   The geometry of a minimization problem.

The requirement that the total volume shall be 27 cubic feet yields the restriction equation

$$x^2 y = 27$$

We first use the restriction equation to solve for y in terms of x and substitute it into the cost equation to reduce C to a function of x only. Since $y = \dfrac{27}{x^2}$, we obtain

$$C(x) = 2\left[2x^2 + 4x\,\frac{27}{x^2}\right]$$

Or

$$C(x) = 4x^2 + \frac{216}{x}$$

This function is clearly continuous and differentiable everywhere except at $x = 0$. Since x and y represent the dimensions of a box, negative values for these variables are ignored. Furthermore, because of the restriction $x^2 y = 27$, x cannot assume the value zero. In view of these considerations, our interval $[a,b]$ should consist of a segment of the positive x axis. We proceed with the technique explained in the preceding paragraph even though we have not as yet completely specified the interval $[a,b]$ of Theorem 5 for this problem. This is done later.

It is convenient to write $C(x)$ in the form

$$C(x) = 4x^2 + 216x^{-1}$$

Then

$$C'(x) = 8x - 216x^{-2}$$

$$= 8x - \frac{216}{x^2}$$

Setting $C'(x) = 0$ for the purpose of finding the critical points, and multiplying through by x^2, we obtain the equation

$$8x^3 - 216 = 0$$

Or

$$x^3 = \frac{216}{8} = 27$$

Or

$$x = 3$$

The value of the cost function at this critical point is given by $C(3) = 4(3)^2 +$ $\frac{216}{3} = 108$. Now the question remains as to whether this is an absolute minimum, a local minimum, or possibly neither.

An inspection of the cost function shows that $C(x)$ becomes infinite as x approaches 0 or as x becomes infinite. Thus $C(x)$ clearly cannot attain its minimum value for a small positive or a large positive value of x. Hence, we choose an interval $[a,b]$ to be one where a is a small positive value and b is a large positive value. Then clearly $C(x)$ will not assume its minimum value at the end points of this interval. Since $C(x)$ is a continuous function in this closed interval, it must assume its minimum value inside this interval. Therefore, because of Theorem 5, it follows that the minimum must occur where $f'(x) = 0$. Since there is only one critical point, given by $x = 3$, this value of x produces the desired minimum. The dimensions of the box that produce minimum material costs are therefore given by $x = 3$ and $y = 3$; hence, a cubical box is the cheapest kind. The geometry of this function is shown in Fig. 11.

Example 2. The manufacturer of a certain product has found from experience that his manufacturing costs can be expressed by means of the function

$$C(x) = 60x$$

where x is the number of units produced. That is, each unit costs \$60 to manufacture. He has also found from experience that the revenue he receives for selling x units can be approximated by the quadratic function

$$R(x) = -3x^2 + 480x$$

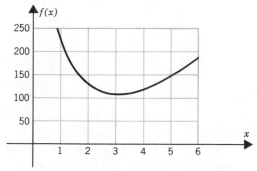

FIGURE 11   A graph of $C(x) = 4x^2 + 216x^{-1}$

This function seems to imply that the revenue increases as x increases, at least for small values of x, but that it decreases and becomes negative as x becomes very large. Producing too many units for the market could easily lead to a lower total revenue because of the sales costs and the inability to sell all the units produced.

The problem is to determine how many units should be produced to maximize total profit. Letting $P(x)$ denote the profit function, we have that

$$P(x) = R(x) - C(x) = -3x^2 + 480x - 60x$$
$$= -3x^2 + 420x$$

The functions in this problem are assumed to hold for x in some interval of the form $[a,b]$ where a and b are positive numbers and b is fairly large. Although x must be a positive integer, because it represents the number of units produced, we ignore this fact and treat it as an ordinary variable of calculus. As a result of this assumption, $P(x)$ may be treated as a continuous function that is differentiable for all positive values of x. We first calculate

$$P'(x) = -6x + 420$$

Solving the equation $P'(x) = 0$, we obtain the single critical point $x = 70$. The value of $P(x)$ at $x = 70$ is given by $P(70) = -3(70)^2 + 420(70) = 14{,}700$. Since $P(x)$ is a quadratic function with a negative coefficient for its x^2 term, its graph is a parabola that opens downward; hence, $x = 70$, which is the x coordinate of the vertex of the parabola, must produce a maximum. The manufacturer should therefore plan on producing 70 units if he wishes to maximize his profit.

Problems of the preceding type can be characterized in the following manner. If there is a cost function $C(x)$ and a revenue function $R(x)$, then the value of x that maximizes profit will be a value that satisfies the equation

$$P'(x) = R'(x) - C'(x) = 0$$

Hence, it will be a value of x that satisfies the equation

$$R'(x) = C'(x)$$

For realistic cost and revenue functions the maximum will almost always occur at a critical point obtained in this manner.

In most practical problems the domain over which the maximum of a function is being sought is the positive x axis and the conditions of the problem insure that a maximum exists. For such problems the maximum invariably occurs at a critical point. It is only when the interval over which the maximum is desired is restricted in some manner that boundary values need to be calculated. Thus in the first of the preceding illustrations we could have argued from practical considerations that the minimum could not be attained for small or large values of x and, therefore, that the single critical value $x = 3$ had to produce a minimum.

Example 3. Determine the values of x in the interval $[-3,1]$ where $f(x)$ assumes its maximum and minimum values if

$$f(x) = x^3 + 2x^2 - 4x + 2$$

$f'(x) = 3x^2 + 4x - 4 = 0$ gives $(3x - 2)(x + 2) = 0$. The critical points are $x = \frac{2}{3}$ and $x = -2$. Calculations give $f(-3) = 5$, $f(-2) = 10$, $f\left(\frac{2}{3}\right) = \frac{14}{27}$, $f(1) = 1$; hence $x = -2$ produces the maximum value and $x = \frac{2}{3}$ produces the minimum value of the function in this restricted interval.

Example 4. A page in a book is to have $1\frac{1}{2}$-inch margins at the top and bottom and 1-inch margins at the two sides. If it is desired to have 24 square inches of printed area, what page dimensions will yield the smallest page?

Let x and y be the width and height, respectively, of the printed area part of the page. Then the total area is given by

$$f = (x + 2)(y + 3)$$

where x and y must satisfy $xy = 24$. Elimination of y gives

$$f(x) = (x + 2)\left(\frac{24}{x} + 3\right) = 30 + 3x + \frac{48}{x}$$

Hence, $f'(x) = 3 - \frac{48}{x^2}$. Solving $f'(x) = 0$ we obtain the critical value $x = 4$. This value obviously minimizes $f(x)$; therefore, the page should be 6 inches wide and 9 inches high.

Example 5. An assembly plant orders x units of a product needed in its operation each time that it places an order. The yearly cost of placing orders and maintaining an inventory for the product is given by

$$f(x) = 4000 + 4x + \frac{10,000}{x}$$

What size order should be placed each time if the yearly cost is to be minimized?

As before, we treat $f(x)$ as a continuous and differentiable function of x even though from a practical viewpoint x must be an integer. To find the critical points, we solve the equation

$$f'(x) = 4 - \frac{10,000}{x^2} = 0$$

The only positive root of this equation is $x = 50$. This value obviously minimizes $f(x)$. If the solution had not been an integer, we could have replaced it by the nearest integer value to obtain a practical solution to the problem.

3. RATE PROBLEMS

In Section 1 the derivative was used to study from a geometrical point of view how a function is changing as the basic variable x is changing. In this section we examine the same problem from an algebraic point of view as it applies to certain problems of interest in the social sciences.

There are many functions that can be treated as functions of time. Thus the growth of an industry or a biological group, the depreciation of a machine, or the amortization of a debt, may all be treated as functions of time. Many functions, however, change very little with time but are directly influenced by other variables. For example, the profit in manufacturing a certain article may depend heavily on the number of units of the article that are being produced. Whether the basic variable is time or some other variable, interest often centers on determining the rate of change of the function with respect to the basic variable. In this section we study such problems. First, we consider some problems where time is the basic variable, after which we look at functions of other basic variables.

Suppose that a switching locomotive is traveling along a straight railroad track near a station according to the formula

$$(2) \qquad\qquad s(t) = 20t^2 - 80t$$

where s represents the distance in feet measured from left to right along the track from the station and t represents the time elapsed in minutes from when the locomotive was at the station. We wish to study the motion of this locomotive over the time interval [0,8]. It is assumed that the formula holds true for this eight-minute period only.

First, let us determine how fast the locomotive is traveling at some specified time point, say when $t = 5$. At that time the locomotive is at a distance of $s(5) = 20(5)^2 - 80(5) = 100$ feet from the station. A similar calculation for $t = 6$ gives $s(6) = 240$; hence, the locomotive traveled a distance of 140 feet during that one-minute time interval. Since the average speed of an object over a given time interval is the distance traveled divided by the elapsed time, the average speed of the locomotive from time $t = 5$ to $t = 6$ is 140 feet per minute. This, however, is only an average speed, and it does not represent the speedometer speed of the locomotive at $t = 5$. To arrive at a measure of what we believe should represent the speedometer, or instantaneous, speed at $t = 5$ we calculate the average speed for a small time interval starting at $t = 5$ and observe what happens to it as the interval becomes increasingly small. Hence, we calculate the value of s for $t = 5$ and for $t = 5 + h$, where h is a small positive number, and take the difference in the s values to find out how far the locomotive traveled in those h minutes. We found earlier that $s(5) = 100$. We use formula (2) to obtain

$$s(5 + h) = 20(5 + h)^2 - 80(5 + h)$$
$$= 100 + 120h + 20h^2$$

The average speed over the time interval $(5, 5 + h)$ is therefore given by

$$\frac{s(5 + h) - s(5)}{h} = \frac{100 + 120h + 20h^2 - 100}{h}$$

$$= 120 + 20h$$

As h approaches 0, this quantity approaches 120; hence, we conclude that the instantaneous speed of the locomotive at the time $t = 5$ is 120 feet per minute. Since this is merely the derivative of the function $s(t)$ with respect to the variable t, evaluated at $t = 5$, we could have started with formula (2) and calculated

(3) $$\frac{ds}{dt} = 40t - 80$$

and then evaluated it at $t = 5$ to obtain $\dfrac{ds}{dt} = 120$.

In studying the rate of change of a moving object it is customary to use the word speed, or velocity, to represent this rate of motion. Velocity differs from speed in that it is negative if the object is traveling in a negative direction (backward), whereas speed measures only how fast it travels regardless of direction. A locomotive that is traveling backward at 5 miles per hour has a speed of 5 miles per hour but a velocity of -5 miles per hour. Since a derivative may be positive or negative, the quantity $\dfrac{ds}{dt}$ measures the velocity of the locomotive rather than merely its speed. In view of the preceding discussion, we define the *instantaneous velocity* at a given point of an object moving along a straight line as the value of the derivative $\dfrac{ds}{dt}$ at that point.

The motion of the locomotive may now be described by means of its distance function (2) and its velocity function (3). Setting $\dfrac{ds}{dt} = 0$ in (3), we obtain $t = 2$, which tells us that the locomotive is at rest when $t = 2$. At time $t = 0$ we have $s = 0$ and $\dfrac{ds}{dt} = -80$, which means that the locomotive is at the station but backing away from it at a speed of 80 feet per minute when we first started observing it. Two minutes later it has come to a stop. Since $s(2) = -80$, the locomotive is then 80 feet to the left of the station. For $t > 2$, the value of $\dfrac{ds}{dt}$ is positive and increases as t becomes larger; hence, the locomotive is speeding up as it starts moving along the track in the positive (right) direction. At the end of the time interval [0,8], the locomotive is $s(8) = 640$ feet from the station and has a velocity of 240 feet per minute. The movement of the locomotive during the time interval [0,8] can be represented schematically by the curve shown in Fig. 12.

In studying the motion of an object it is often useful to determine how the

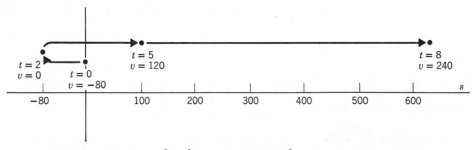

FIGURE 12   A graphical representation of a movement over time.

velocity of the object is changing with respect to time. Since the derivative measures the rate of change of a function, this suggests that we differentiate the velocity function with respect to time. The derivative of the velocity of an object moving along a straight line is called the acceleration of the object. Since the velocity is the derivative of the distance function $s(t)$, it follows that the acceleration function is given by the second derivative of $s(t)$; hence

$$a(t) = \frac{d^2s}{dt^2}$$

As an illustration, to find the acceleration of the train in the preceding problem we proceed as follows:

$$s(t) = 20t^2 - 80t$$

$$v(t) = \frac{ds}{dt} = 40t - 80$$

$$a(t) = \frac{dv}{dt} = \frac{ds^2}{dt^2} = 40$$

Thus the train has a constant acceleration of 40 units. Since s is in feet and t is in minutes, the units of acceleration are in feet per minute-per minute. The acceleration of an object is caused by a force acting on the body; hence, there is a constant force in the positive direction acting on the train to cause it to travel in the positive direction.

Simple problems related to projectiles are examples of problems for which the acceleration is constant and is due to the force of gravity acting on the projectile. For these problems it is known that the acceleration due to gravity is -32 feet per second-per second. The negative sign indicates that our distance function $s(t)$, which measures how high the projectile is above ground level, measures positive distances upward and the force of gravity pulling the projectile back to earth produces a negative upward acceleration.

Although the word velocity is appropriate when discussing the rate with which a distance function is changing with respect to time, for more general situations we merely speak of the derivative as measuring the rate of change of the function with respect to time. As an illustration of a situation for which the word velocity is inappropriate, consider the following problem.

Suppose the value of a car depreciates in such a manner that its value in dollars t years after it was purchased is given by the formula

$$f(t) = 50 + \frac{2000}{t+1}$$

Here our basic variable is time. By means of this formula and differentiation we can answer various questions concerning the rate at which the
value of the car is depreciating. For example, what is the rate of depreciation when the car is three years old? How does the rate of depreciation at
the end of two years compare with that at the end of three years?

To answer such questions we first calculate $f'(t)$ by using the chain rule
on $f(t)$ when $f(t)$ is expressed in the form

$$f(t) = 50 + 2000(t+1)^{-1}$$

This gives

$$f'(t) = -2000(t+1)^{-2} = -\frac{2000}{(t+1)^2}$$

For $t = 3$, we have

$$f'(3) = -\frac{2000}{(3+1)^2} = -125$$

Hence, the car is decreasing in value at the rate of \$125 per year at the end
of three years. For $t = 2$, we obtain

$$f'(2) = -\frac{2000}{9} = -222\frac{2}{9}$$

The rate of depreciation at the end of two years is therefore almost \$100
more per year than at the end of three years.

Since the derivative gives the limiting value of an average rate of change
over a short time interval, the statement that the car depreciates \$125 per
year is to be interpreted as the rate of change at precisely the end of the
third year and not as an average over the third year or over any other one-
year time interval containing the end of three years. If we wished to find
the amount that the car had depreciated during the third year we would
have calculated $f(2) - f(3)$, which turns out to be 166.

As an illustration of the type of problems that involve rates of change
with respect to variables other than time, consider the following typical
problems in economic theory.

Let x denote the number of units of a product being produced in a factory. Then the total cost, C, of producing these units is some function of x. Similarly, if the price of one unit is p dollars and if all units produced can be sold, the total revenue, R, is a function of x and is given by $R(x) = px$. If the price per unit were fixed, R would be a linear function of x; however, it is more realistic to assume that the price to charge per unit is a function of the number of units produced, in which case R will be a more complicated function of x.

Suppose that we are given the total cost and revenue functions, $C(x)$ and $R(x)$, and that we wish to see what happens to them if we increase production by one unit, say, from x to $x + 1$. Then the increase in cost and the increase in revenue due to this one additional item produced are given by $C(x + 1) - C(x)$ and $R(x + 1) - R(x)$, respectively. If, however, we were to calculate

$$\frac{C(x + h) - C(x)}{h} \quad \text{and} \quad \frac{R(x + h) + R(x)}{h}$$

we would obtain the average change in these two functions as a result of producing h additional items. For $h = 1$ these reduce to our earlier differences. If, now, we allow h to approach 0, we obtain the instantaneous rate of change of these functions, which are approximations of the earlier changes as a result of increasing production by one unit. The derivatives that result from taking these limiting values are given the following special names.

$$\text{Marginal cost:} \qquad \frac{dC}{dx}$$

$$\text{Marginal revenue:} \qquad \frac{dR}{dx}$$

The variable x here is assumed to represent the number of units that can be sold. It, therefore, represents the demand for the product and not necessarily the amount that the manufacturer would like to produce.

As a numerical illustration of the preceding formulas, suppose that the cost function is a quadratic function:

$$C(x) = ax^2 + bx + c$$

Suppose, further, that the demand function is a linear function of the price, in which case the price is a linear function of the demand. Since x denotes demand here, we are assuming that $p(x) = \alpha x + \beta$. Then the revenue function assumes the form

$$R(x) = xp = \alpha x^2 + \beta x$$

The marginal cost and the marginal revenue are then given by the formulas

$$\frac{dC}{dx} = 2ax + b$$

and

$$\frac{dR}{dx} = 2\alpha x + \beta$$

Since profit is given by $P = R - C$ and since, as we observed in Section 2, the maximum value of a profit function normally occurs at a critical point, it follows that the maximum profit will occur for x satisfying the equation

$$P'(x) = R'(x) - C'(x) = 0$$

But this implies that the maximum profit will occur at that value of x which makes the marginal cost equal the marginal revenue.

Another related concept involving derivatives that has useful application for certain problems of economics is the *elasticity* of a function. If x represents the demand for a product and p represents the price per unit, the elasticity of demand is defined as follows.

Let p have a fixed value and let $x(p)$ be the corresponding demand. Allow p to increase by an amount h. Then the increase in demand will be $x(p + h) - x(p)$ and the relative increase in demand with respect to the original demand will be

$$\frac{x(p + h) - x(p)}{x(p)}$$

Similarly, since the increase in the price is h, the relative increase in the price with respect to the original price will be

$$\frac{h}{p}$$

We take the ratio of these two relative changes in the function and the variable to obtain

$$\frac{x(p + h) - x(p)}{x(p)} \cdot \frac{p}{h} = \frac{x(p + h) - x(p)}{h} \cdot \frac{p}{x(p)}$$

Finally, we take the limit of this ratio as $h \to 0$ to obtain the *elasticity of demand*. Hence

(4)
$$Elasticity\ of\ demand = \frac{p}{x}\frac{dx}{dp}$$

The elasticity of any other function can be defined in a similar manner. In general, if we are given $y = f(x)$, the elasticity of y with respect to x is given by the limit, as $h \to 0$, of the ratio

(5)
$$\frac{\dfrac{f(x+h)-f(x)}{f(x)}}{\dfrac{(x+h)-x}{x}} = \frac{x}{f(x)}\frac{f(x+h)-f(x)}{h}$$

This limit gives the following general formula:

$$Elasticity\ of\ f(x) = \frac{x}{f(x)}\ f'(x)$$

The elasticity of a function is a useful tool for studying the relative rates of change of two related variables independent of the units in which they are measured. For example, in calculating the elasticity of demand, it is irrelevant whether the demand, x, is in actual units or units of thousands, or whether the price, p, is in dollars or cents. Both numerator and denominator in (5) are independent of the units of measurement used for f and x, respectively; consequently, the limiting value of (5) must be independent of such units.

As a numerical illustration, let us calculate the elasticity of demand for a quadratic demand function, which may be written in the form $x(p) = ap^2 + bp + c$. We first calculate

$$\frac{dx}{dp} = 2ap + b$$

Then we apply formula (4) to obtain

$$\frac{p(2ap+b)}{ap^2+bp+c} = \frac{2ap^2+bp}{ap^2+bp+c}$$

The following examples are additional illustrations of rate problems.

Example 1. A rocket is fired straight up. Its distance in feet above the ground at the end of t seconds is given by the formula $y = 640t - 16t^2$.

Describe its motion.

$$y'(t) = 640 - 32t$$
$$y''(t) = -32$$

Hence, its velocity is 0 when $t = \dfrac{640}{32} = 20.$ At that time the rocket will have traveled a distance of $y(20) = 6400$ feet. Hence, the rocket starts upward with a velocity of 640 feet per second and travels upward for 20 seconds to reach a maximum height of 6400 feet, after which it falls to the ground. It takes a total of 40 seconds to come back to earth, because $y = 0$ when $640t - 16t^2 = 16t(40 - t) = 0.$ Its acceleration has the constant value of $-32.$

Example 2. The cost function for producing x units of a product is given by

$$C(x) = 600 + 2x + \frac{x^2}{400}$$

What is the marginal cost when producing 400 units? What is the average cost per unit? What production level will minimize the average cost per unit?

The marginal cost function is

$$C'(x) = 2 + \frac{x}{200}$$

For $x = 400$ this function has the value 4; hence, 4 is the desired marginal cost. From a practical point of view this implies that production costs increase \$4 per unit when 400 units are produced, assuming that costs are measured in dollars.

The average cost function is

$$A(x) = \frac{C(x)}{x} = \frac{600}{x} + 2 + \frac{x}{400}$$

The function $A(x)$ will be minimized when x is a solution of the equation

$$A'(x) = -\frac{600}{x^2} + \frac{1}{400} = 0$$

The solution is $x = \sqrt{240,000} \doteq 490;$ hence, this production level will minimize the average cost per unit.

Example 3. The price-demand function for a commodity is given by

$$p = \frac{400}{x + 2}$$

What is the revenue function? What is the marginal revenue when 18 units are produced? What is the elasticity of demand when 18 units are produced?

The revenue function is

$$R(x) = xp = \frac{400\,x}{x + 2}$$

The marginal revenue function is

$$R'(x) = 400 \left[\frac{x + 2 - x}{(x + 2)^2} \right] = \frac{800}{(x + 2)^2}$$

At $x = 18$, this function has the value 2.

The demand function is obtained by solving the price-demand function for x in terms of p; hence, it is given by

$$x = \frac{400}{p} - 2$$

The elasticity of demand is therefore given by

$$\frac{p}{x}\frac{dx}{dp} = \frac{p}{\frac{400}{p} - 2}\left(-\frac{400}{p^2} \right) = \frac{400}{2p - 400}$$

For $x = 18$ the value of p is 20; hence, the elasticity of demand when $x = 18$ is $-\frac{10}{9}$.

COMMENTS

Of the three problems studied in this chapter, the one on how to determine where a function is maximized or minimized is undoubtedly the most important. Many business problems, for example, are concerned with how to maximize profits or minimize costs. The illustration of determining how often to place orders to minimize inventory costs is typical of these problems. We are able to study only a few of the many such practical problems that can be solved with the help of calculus. Following chapters consider other problems of this type, some of which involve several variables. From the few illustrations given, however, it should be evident that calculus is a powerful tool for solving many important practical problems of the social sciences. We made only a modest beginning in that direction here.

EXERCISES

Section 1

1. Given $f(x) = 2x^4 - 4x^3$.
   (a) Where is its graph rising?
   (b) Where is its graph falling?
   (c) Where is the tangent line horizontal?
   (d) Sketch the graph using the preceding information.

2. Given $f(x) = \dfrac{1}{1+x^2}$.

   (a) Where is its graph rising?
   (b) Where is its graph falling?
   (c) Where is the tangent line horizontal?
   (d) Sketch the graph using the preceding information.

3. Suppose that the supply of a commodity, y, is related to the price, x, by the equation $y = a + b \sqrt{x-c}$, where $a > 0$, $b > 0$, and $c > 0$ are given constants. Show that the supply curve is increasing for all values of $x > c$. Sketch the curve for $x \geqslant c$. Choose convenient units on the axes.

4. Suppose that the price, p, of a commodity is related to the demand, x, for the commodity by the equation $p = \dfrac{a}{x} - c$, where $a > 0$ and $c > 0$ are given constants. Show that the revenue curve, defined by $R = xp$, is decreasing as demand increases. Sketch the curve for $x > 0$.

5. In a certain industry the total cost of producing x units of a product is given by the function $C(x) = \sqrt{ax + b}$, where $a > 0$ and $b > 0$ are given constants. Show that the average cost curve, defined by $A(x) = \dfrac{C(x)}{x}$, is decreasing as x increases. Sketch the curve for $x > 0$.

6. Find the critical points, if any, for each of the following functions.

(a) $f(x) = 2x^2 - x^4$

(b) $f(x) = (x - 1)^4$

(c) $f(x) = 2x^3 - 2x^2 - 16x + 1$

(d) $f(x) = (x^2 - 4)^{2/3}$

(e) $f(x) = \dfrac{x}{x^2 - 4}$

(f) $f(x) = \sqrt{x} + \dfrac{1}{\sqrt{x}}$

(g) $f(x) = x^2 - \dfrac{1}{x^2}$

(h) $f(x) = \dfrac{x^2 + 1}{x^2 - x}$

(i) $f(x) = \dfrac{x + 1}{x^2 - 5x + 4}$

(j) $f(x) = \sqrt{x^2 - 3x + 2}$

7. Graph each of the following functions by first finding its critical points, determining where it is increasing and decreasing, and plotting a few points. Indicate its domain.

(a) $y = x^4 + 4x$

(b) $y = x^3 + x^2 - x + 1$

(c) $y = 2x + \dfrac{1}{2x}$

(d) $y = 4x^4 - x^2 + 4$

(e) $y = x\sqrt{2 - x}$

(f) $y = (x - 1)^2(x + 2)^2$

(g) $y = x\sqrt{4 - x^2}$

(h) $y = x^{4/3} + 4x^{1/3}$

8. Graph each of the following functions by first finding its critical points, determining where it is increasing and decreasing, and plotting a few points. Indicate its domain.

(a) $y = (x - 1)^4$

(b) $y = x^4 - 2x^2$

(c) $y = \sqrt{4 - x^2}$

(d) $y = x^4 - 8x^2 + 16$

(e) $y = x^3 + 3x^2 - 9x$

(f) $y = \sqrt{x} + \dfrac{1}{\sqrt{x}}$

(g) $y = (x^2 - 4)^{3/2}$

(h) $y = \dfrac{x^2}{\sqrt{x + 5}}$

Section 2

1. Find the maximum and minimum values of each of the following functions in the specified interval.

(a) $f(x) = \dfrac{x^2 - 27}{x - 6}$ $[0,5]$

(b) $f(x) = (x - 1)^3$ $[-1,3]$

(c) $f(x) = 3x^4 - 4x^3 - 12x^2$ $[-1,3]$

(d) $f(x) = \dfrac{x^3}{x^4 + 27}$ $[-4,4]$

(e) $f(x) = \dfrac{x^2 + 4}{x + 2}$ $[0,4]$

(f) $f(x) = 15x^7 - 63x^5 + 70x^3$ $[-\sqrt{2}, \sqrt{2}]$

2. Suppose that the sales revenue function, S, and the total cost function, C, as functions of the number of units produced, x, are given by

$$S(x) = -2x^2 + 60x \quad \text{and} \quad C(x) = 20x$$

Find the value of x that will maximize the profit $P = S - C$.

3. A firm has found from experience that its profit as a function of the amount produced, x, is given by

$$P(x) = -\frac{x^3}{3} + 625x - 2000, \quad 0 \leqslant x \leqslant 40$$

Find the value of x that maximizes the profit and determine the amount of profit for that amount of production.

4. Given the following sales revenue function, S, and the total cost function, C, as functions of the number of units sold, x, find the value of x that maximizes the profit $P = S - C$.

$$S(x) = -\frac{6}{10^6} x^3 + \frac{18}{10^3} x^2 - 2x + 100, \; x > 100$$

$$C(x) = \frac{2}{10^2} x^2 - 24x + 11{,}000$$

5. A company can make a profit of $16 on each unit of a product it manu-factures if it does not produce more than 300 units per week. Its profit decreases 4 cents per unit over 300. How many units should it produce each week to maximize its total weekly profit?

6. A firm's revenue and cost functions are given by

$$R(x) = -\frac{11}{54} x^2 + \frac{235}{108} x - \frac{28}{27}, \, 2 \leqslant x \leqslant 5$$

and

$$C(x) = \frac{3}{4} x + 1, \, 2 \leqslant x \leqslant 5$$

Here x represents the output in thousands of units. Find the value of x that (a) minimizes the cost, (b) maximizes the revenue, and (c) max-imizes the profit.

7. Let x denote the demand for a product, p the price per unit of the product, $C(x)$ the total cost of producing x units, and $R(x)$ the total revenue obtained from selling x units. It is assumed that the price p is a function of the number of units sold; hence, $p = p(x)$. If P denotes the profit function,

$$P(x) = R(x) - C(x) = xp(x) - C(x)$$

Find conditions on these functions that must be satisfied at $x = x_0$ if x_0 is to maximize $P(x)$.

8. Find the dimensions of the rectangle with maximum area if its perimeter is to be 200 inches.

9. Find the dimensions of the most economical open box with a square bottom if the volume is to be 4 cubic feet and the material for the bottom costs twice as much as for the sides.

10. A rectangular field for grazing cattle is to be fenced off along the bank of a river with no fence needed along the river. If the fencing material costs \$2 a foot, find the dimensions of the field with maximum area that can be enclosed with \$2000 worth of fencing.

11. A factory is located on the opposite side of a river from a power station and one mile farther down the river. The river is $\frac{1}{4}$ mile wide. If it costs \$6 a foot to run a cable over land and \$10 a foot under water, what is the most economical way of running a cable from the power station to the factory?

12. An open tin box is to be made from a rectangular piece of tin of dimensions 5 by 8 inches by cutting out equal size squares from each corner and folding up the sides. What size squares should be cut out if the box is to possess maximum volume?

13. Find the maximum and minimum values of the function $y = f(x)$ defined implicitly by the relationship

   (a) $\sqrt{x} + \sqrt{y} = 2$, $[0,4]$ (b) $x^2 + y^2 + xy = x$, $y > 0$

14. The specific weight of water, s, is given by the formula

$$s = 1 + 5.3 \times 10^{-5}t - 6.53 \times 10^{-6}t^2 + 1.4 \times 10^{-8}t^3, \quad 0° < t < 100°$$

where t is the temperature in degrees centigrade. At what temperature will water have the maximum specific weight?

15. The rate of a chemical reaction, v, is given by the formula $v = kx(a - x)$, where x is the amount of the product, a is the amount of material at the beginning of the reaction, and k is a positive constant. At what concentration x will the rate of reaction be a maximum?

16. Two airports are 10 miles apart. Large jets fly from one but not from the other. As a result, the noise from it is four times as loud as from the other airport. The noise level at any point away from a source is assumed to be proportional to the strength of the source and inversely proportional to the distance from the source. Under these assumptions, determine where between the two airports an office should be located to minimize the total noise level from the two airports.

17. Work Problem 16 if the noise level is assumed to be proportional to the strength of the source and inversely proportional to the square of the distance from the source.

18. Let A denote the fixed yearly demand for a raw product that a manufacturing company needs for its production and let x denote the number of units it orders of this raw product each time it places an order. Let $C(x)$ and $O(x)$ denote the carrying and ordering costs, respectively. Assume that $C(x)$ is proportional to x; hence, $C(x) = ax$, where a is some constant. Assume that $O(x)$ is proportional to the number of orders placed during the year. Since $\dfrac{A}{x}$ give the number of such orders, it follows that $O(x) = b\,\dfrac{A}{x}$, where b is some constant. The total cost T is therefore given by

$$T(x) = C(x) + O(x) = ax + \frac{bA}{x}$$

(a) Find the order size that will minimize T if $A = 1000$, $a = 3$, and $b = 120$.

(b) What is the optimal period in months between orders?

19. A firm is engaged in refining a certain manufactured product. It produces A units a year. Let C_1 denote the unit cost of the crude product, C_2 the cost of placing a purchase order, and C_3 the cost of keeping one unit of the crude product in inventory for one year. Let x denote the number of units ordered when placing an order. Then $\dfrac{A}{x}$ gives the

number of orders placed during the year and $C_2 \dfrac{A}{x}$ gives the total or-

dering cost for the year. Since the number of units in inventory will

vary from x to 0 and will average $\dfrac{x}{2}$, the total inventory cost for the year

may be assumed to be $C_3 \dfrac{x}{2}$. The total cost of producing A units during

the year is therefore given by

$$C(x) = C_1 A + \frac{C_2 A}{x} + \frac{C_3}{2} x$$

Derive a formula that will give (a) the order size that will minimize total costs, and (b) the time interval, in months, between orders.

Section 3

1. If the motion of a particle along the x axis is given by the formula $x = t^3 - t$, where x is measured in feet and t is measured in minutes, what is the velocity of the particle at time (a) $t = 1$, and (b) $t = 0$?

2. A rock falling from the top of a vertical cliff drops a distance of $s = 16t^2$ feet in t seconds.
   (a) What is its speed at the end of t seconds?
   (b) What is its speed when it has fallen 144 feet?

3. A car is traveling east along a straight road according to the formula

$$s = \frac{1}{4} t^4 + \frac{1}{3} t^3 - 3t^2$$

   where s is measured in miles and t is measured in hours. This formula is assumed to hold true for the time interval $[0,4]$.
   (a) What is its average velocity between $t = 1$ and $t = 2$?
   (b) What is its instantaneous velocity at $t = 1$?
   (c) At what time does the car first stop?
   (d) When is the car traveling east?

4. If a ball is thrown vertically up from the ground with an initial velocity of 200 feet per second, the height of the ball at the end of t seconds will be given by

$$y = 200t - 16t^2$$

(a) Find its velocity at the end of 3 seconds.
(b) Find when its velocity is zero.
(c) What is the maximum height reached by the ball?
(d) What is the velocity of the ball when it just hits the ground?

5. The distance in feet that a particle has traveled along the x axis in t minutes is given by the formula

$$s(t) = t + \frac{1}{t}$$

(a) Calculate its average velocity over the interval [2,3].
(b) Calculate its instantaneous velocity at time $t = 2$.
(c) Calculate its instantaneous velocity at time $t = 2.5$ and compare it with the answers to (a) and (b).

6. If the distance function for a particle traveling along the x axis is given by $s(t) = at + b$, where a and b are constants, show that the average velocity in any interval $[t_1, t_2]$ is equal to the instantaneous velocity at time $t = t_1$.

7. If the motion of a particle along the x axis is given by the formula $x = t^2 - t$, where x is measured in feet and t is measured in minutes, describe the motion of the object during the time interval [0,4].

8. In the following problems a particle is moving along the x axis according to the equation of motion given, where s denotes the number of feet the particle is from the origin at the end of t minutes. Determine the intervals of time when the particle is moving to the right and when to the left. Sketch its behavior by means of a line graph above the x axis as shown in Fig. 12.
(a) $s(t) = 2t^3 - 3t^2 - 12t + 8$

(b) $s(t) = \dfrac{t}{t^2 + 1}$

9. If $v(t)$ is the velocity of a particle traveling along the x axis at time t, the acceleration of the particle at time t is defined to be $a(t) = v'(t)$, that is the derivative of the velocity function with respect to time. This implies that the acceleration at time t is obtained by calculating the second derivative of the distance function $s(t)$, written $s''(t)$. Using this definition, find the acceleration functions for the following distance functions.

(a) $s(t) = t - t^2$ (b) $s(t) = t + 4t^2$

(c) $s(t) = t + \dfrac{1}{t}, t > 0$ (d) $s(t) = \dfrac{1}{t+1}$

10. If a projectile is fired upward, its height above the surface of the earth at time t is given by the quadratic function $s(t) = at^2 + bt + c$, where a, b, and c are constants determined by the conditions of the problem. Suppose a projectile is fired straight up from an airplane that is at a height of 2000 feet, with the projectile having an initial velocity of 1000 feet per second. If the acceleration of the projectile is -32 feet per second-per second due to gravity (see Problem 9)
    (a) determine the values of a, b, and c.
    (b) determine how high the projectile travels.
    (c) determine how long it takes for the projectile to hit the ground.
    (d) determine the velocity of the projectile when it just hits the ground.

11. The weight, W, of a limb of a plant or animal as a function of its age, t, can be approximated by the relation

$$W(t) = at^b$$

where a and b are constants that depend on the particular type of plant or animal and the environment. In studying the rates of growth, it is useful to determine the rate of growth per unit weight. This is called the specific rate of growth and is given by

$$S = \frac{W'(t)}{W}$$

How does the specific rate of growth of a limb vary with its age?

12. It has been determined experimentally that the specific heat of ethyl alcohol can be approximated over the temperature range $[0°,60°]$ centigrade by the formula

$$S = 5.068 \times 10^{-1} + 2.86 \times 10^{-3}t + 5.4 \times 10^{-6}t^2$$

    (a) What is the rate of change of the specific heat at $0°$ and at $10°$?
    (b) At what temperature in the interval $[0°,60°]$ is the rate of change a maximum?

13. If the cost function for producing a product is given by

$$C(x) = 40 + 4x + \frac{\sqrt{x^2 + 1}}{10}$$

where x is the number of units produced, calculate the marginal cost when 20 units are produced.

14. If the price-demand function for apartments in a development is given by

$$p(x) = 5\sqrt{400 - 3x}$$

where p is the price in dollars per month and x is the number of apartments rented, for what number of rentals will the marginal revenue be zero?

15. If the price-demand function is defined implicitly by the equation $xp + p = 20$, find the revenue function and the marginal revenue.

16. Suppose that the cost of producing x units of a product is given by

$$C(x) = \frac{x^2}{2} - 2x + 5$$

(a) What is the average cost function? Graph it.
(b) What is the marginal cost function? Graph it.
(c) Where is the point of intersection of the graphs in (a) and (b)?
(d) What value of x minimizes the average cost?
(e) What is the relationship between (c) and (d)?

17. An economist studied models to represent the relation between the price, p, and the demand, x, for various commodities. Among others, he obtained the following implicit functional relationships.

Cotton in the United States (1915–1929) $px^{1.4} = .11$
Butter in Stockholm (1925–1937) $xp^{1.2} = 38$

Using the total revenue formula $R(x) = xp(x)$, calculate the marginal revenue for (a) cotton, and (b) butter.

18. Suppose that the total revenue function in producing x units of a product is given by

$$R(x) = 3 + 5\sqrt{x - 1}, \qquad 2 \leqslant x \leqslant 26$$

(a) What is the price-demand function?
(b) What is the marginal revenue function?
(c) What is the value of x that maximizes the total revenue?

19. The price-demand relationship for a commodity is given by the equation $x^2 + p^2 = 100$, where $100x$ units are demanded when the price is p dollars per unit. (a) Find the total revenue function. (b) Find the marginal revenue function. (c) Find the price that will maximize the total revenue.

20. Work Problem 19 if the price-demand relationship is given by $(p + 3)(x + 4) = 96$.

21. The average cost function is given by $A(x) = \dfrac{C(x)}{x}$, where x denotes the amount produced. Show that the intersection point of the average cost curve and the marginal cost curve gives the most efficient operating volume of production, that is the volume at which the average cost per unit of production is a minimum.

22. The reaction of the body to a dose of a drug can be represented by the function

$$R(x) = x^2 \left(a - \frac{x}{3} \right)$$

where R is the strength of the reaction, x is the amount of the drug that is administered, and a is a positive constant. The sensitivity of a drug is defined to be the rate of change of the strength, R, with respect to the size of the dose, x, that is by $R'(x)$. Find the dosage that will maximize the sensitivity.

23. A trucking company has an average overhaul cost of $1000 per truck and a routine maintenance cost of

$$M(x) = \frac{x^2}{10^6} + .02x$$

where x is the interval in miles between engine overhauls. The total engine maintenance cost in dollars per mile is therefore given by

$$C(x) = \frac{1000}{x} + .02 + \frac{x}{10^6}$$

(a) Find the rate of change of total engine maintenance costs with respect to the overhaul interval.
(b) Compare the rates found in (a) for $x = 50{,}000$ and $x = 100{,}000$.
(c) What interval minimizes the total engine maintenance cost per mile?

24. Calculate the elasticity of demand for the price-demand relationship given in Problem 15.

25. Calculate the elasticity of demand for the price-demand relationship given in Problem 14.

26. Calculate the elasticity of revenue for the revenue function given in Problem 15.

27. If the total revenue function is denoted by $R(x) = xp(x)$, where $p(x)$ is the price-demand function and x is the number of units produced, show that R will be maximized when the elasticity of demand has the value -1.

10 | Exponential and Logarithmic Functions

Problems dealing with growth phenomena give rise to mathematical functions that we have not considered thus far. The only functions that we have treated in detail are polynomials or the ratio of two polynomials. For growth problems these functions are not very satisfactory as models. A function that arises naturally in these problems is the exponential function. It is closely related to the logarithmic function; therefore, we study them together in this chapter. To do so, we need a particular limit, which we now discuss.

1. DEFINITION OF e

Consider the function $f(x) = (1 + x)^{1/x}$, $x > 0$. We wish to study this function as $x \to 0$. To do so, we consider a special case of it by letting $x = \dfrac{1}{n}$, where n is a positive integer, and then allow $n \to \infty$. That is, we study the function $g(n) = \left(1 + \dfrac{1}{n}\right)^n$ as $n \to \infty$. We first calculate a few values of this function for increasingly large values of n to observe whether this function appears to be approaching a limit. For $n = 1,2,3,4,$ and 5, calculations give the values $g(1) = 2$, $g(2) = 2.25$, $g(3) = 2.37$, $g(4) = 2.44$, and $g(5) = 2.49$. It is not obvious from these numbers that $g(n)$ is approaching some value because $g(n)$ is growing, although not rapidly. A few additional calculations might be in order, say, for $n = 10, n = 20,$ and $n = 50$. Such calculations can be made without much difficulty if the proper calculating tools are available. They yield the values $g(10) = 2.59$, $g(20) = 2.65$, and $g(50) = 2.69$. The growth of $g(n)$ is now very slow and it should come as no surprise to learn that $g(n)$ does approach a limit as $n \to \infty$. This limit is a number denoted by the letter e. If the function $f(x) = (1+x)^{1/x}$ has a limit as $x \to 0$, that limit must be the same as the limit of $g(n) = \left(1 + \dfrac{1}{n}\right)^n$ as $n \to \infty$. Therefore, assuming the existence of such a limit, we have by definition

$$e = \lim_{x \to 0} (1+x)^{1/x}$$

This limit has been evaluated to a large number of decimal places. Its value to five decimal places is

$$e = 2.71828$$

From our earlier calculations we might have guessed that the value is close to 2.7.

2. THE EXPONENTIAL FUNCTION AND THE LOGARITHMIC FUNCTION

Consider the two functions $g(x) = 2^x$ and $h(x) = 3^x$. By calculating the values of these two functions for, say, $x = -2, -1, 0, 1, 2$, we should be able to obtain rough sketches of their graphs. This is done in Fig. 1, based on the following table of calculated values:

x	-2	-1	0	1	2
$g(x)$	$\dfrac{1}{4}$	$\dfrac{1}{2}$	1	2	3
$h(x)$	$\dfrac{1}{9}$	$\dfrac{1}{3}$	1	3	9

Now we introduce a third function:

$$f(x) = e^x$$

Since the number e lies between 2 and 3, the graph of this function will lie between the two graphs just constructed in Fig. 1 and will resemble them

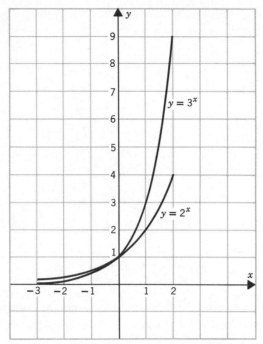

FIGURE 1   Graphs of $g(x) = 2^x$ and $h(x) = 3^x$.

in shape. This new function is called the *exponential function*. It plays a very important role in studying growth problems by means of calculus.

It is clear that e^x increases as x increases; therefore, the exponential function is a strictly increasing function. As a consequence, to each value of y there exists a unique value of x. As we observed in Section 7, Chapter 1, this implies that if we let $y = e^x$, we may solve for x in terms of y to obtain an inverse function. This inverse function is called the *logarithmic function*. The inverse relationship is written in the form $x = \log y$. The graphs of $y = e^x$ and its inverse $x = \log y$ are shown in Figs. 2 and 3.

Those students who have studied logarithms previously for their computational advantages undoubtedly have studied a somewhat different function. If we let $y = 10^x$ and solve for x in terms of y to obtain the inverse relationship, we obtain what is known as the logarithm of y to the base 10. This inverse relationship is written in the form

$$x = \log_{10} y$$

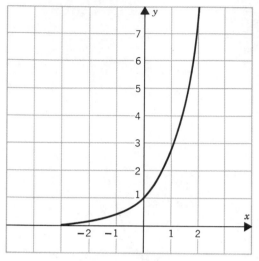

FIGURE 2   A graph of $y = e^x$.

This is the logarithmic function commonly used for computing. When we speak of *the exponential function* or *the logarithmic function* we mean the functions that use the base e. Any other exponential type of function, such as $y = a^x$, where a is some positive constant, is called the exponential function with base a and its inverse $x = \log_a y$ is called the logarithmic function with base a. When no base is indicated on log y, it is understood to be the

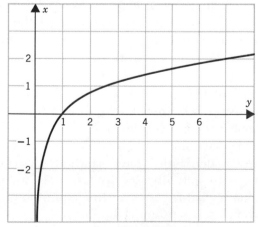

FIGURE 3   A graph of $x = \log y$.

base e. Some books, however, use the special symbol ln y to represent a logarithm with base e. Presently we shall observe why the base e is so important and useful.

3. THE DERIVATIVE OF THE LOGARITHMIC FUNCTION

Since it is customary to denote the variable of a function by x rather than by y, as in Fig. 3, we express the logarithmic function as a function of x and write

$$f(x) = \log x$$

Our objective in this section is to calculate the derivative of this function. In this derivation it is assumed that students are familiar with the basic properties of logarithms (these may be found in the appendix). Those properties do not depend on what base is being used. As usual, we hold x fixed and calculate the quotient

$$\frac{f(x+h) - f(x)}{h} = \frac{\log(x+h) - \log x}{h}$$

Using properties of the logarithmic function, we perform these algebraic operations:

$$\frac{f(x+h) - f(x)}{h} = \frac{1}{h} \log \frac{x+h}{x}$$

$$= \frac{1}{h} \log \left(1 + \frac{h}{x}\right)$$

$$= \frac{1}{x} \frac{x}{h} \log \left(1 + \frac{h}{x}\right)$$

$$= \frac{1}{x} \log \left(1 + \frac{h}{x}\right)^{x/h}$$

The expression $\left(1 + \dfrac{h}{x}\right)^{x/h}$ is very similar to the one used in defining e in

Section 1. To express it in that form, we let $z = \dfrac{h}{x}$, in which case

$$\left(1 + \frac{h}{x}\right)^{x/h} = (1 + z)^{1/z}$$

Since x is any positive number and h is approaching 0, the variable $z = \dfrac{h}{x}$ is approaching 0. As a result, it follows from the definition of e that

$$\lim_{h \to 0} \left(1 + \frac{h}{x}\right)^{x/h} = \lim_{z \to 0} (1 + z)^{1/z} = e$$

With this result available, we are now in a position to calculate the limit of the preceding quotient and obtain a formula for $f'(x)$. Thus

$$f'(x) = \lim_{h \to 0} \frac{1}{x} \log \left(1 + \frac{h}{x}\right)^{x/h}$$

$$= \lim_{z \to 0} \frac{1}{x} \log (1 + z)^{1/z}$$

Although we did not demonstrate this property, $\log x$ is known to be a continuous function of x; therefore, its limiting value is the value of the function at the limiting point. Hence, it follows that

$$f'(x) = \frac{1}{x} \log \left[\lim_{z \to 0} (1 + z)^{1/z} \right]$$

$$= \frac{1}{x} \log e$$

If we choose $y = e$ in the dual relationships $y = e^x$ and $x = \log y$, we obtain $e = e^x$ and $x = \log e$. From the first of these equations it follows that $x = 1$ and, therefore, from the second equation that $1 = \log e$. Applying this result to the preceding expression for $f'(x)$, we obtain

$$f'(x) = \frac{1}{x}$$

Accepting the previous assumptions made concerning the logarithmic function, this demonstrates the validity of the following theorem.

Theorem 1. *The logarithmic function $f(x) = \log x$, $x > 0$, is differentiable and possesses a derivative given by the formula*

(1) $$f'(x) = \frac{1}{x}$$

The importance of the number e in calculus should now be apparent. It arises in finding the derivative of the logarithmic function. It should also be apparent why in calculus we prefer to work with the logarithmic function to the base e rather than to some other base, such as 10. It is only when we choose the base e that we can replace the factor log e in the preceding derivation by 1, thereby obtaining a very neat formula. If we had used some other base, say a, we would have finished with the formula

(2) $$f'(x) = \frac{1}{x}\log_a e$$

This, incidentally, is a formula that may be used to calculate the derivative of a logarithmic function when the base is something other than e. In doing so, however, it is necessary to know the value of the constant $\log_a e$. For the base 10, for example, it is known that $\log_{10} e = .434$.

Formulas 1 and 2 may be extended to more general logarithmic functions by means of the chain rule. Thus, let us suppose that we have the function

$$r(x) = \log(x^2 + x)$$

We may write this in the form

$$r(x) = f(g(x))$$

where $f(u) = \log u$ and $u = g(x) = x^2 + x$. Then, applying the chain rule given by Theorem 7, Chapter 8, we obtain

$$r'(x) = f'(u)g'(x)$$
$$= \frac{1}{u}(2x + 1)$$
$$= \frac{2x + 1}{x^2 + x}$$

More generally, if we have

$$r(x) = \log[g(x)]$$

we let $u = g(x)$ and $f(u) = \log u$. The chain rule then gives

(3) $$r'(x) = \frac{1}{u}g'(x) = \frac{g'(x)}{g(x)}$$

In words, this formula states that the derivative of the logarithm of a function of x is equal to the derivative of the function divided by the function.

In calculating the derivative of logarithmic functions we can often simplify the calculations by using properties of logarithms. For example, consider the problem of calculating $f(x)$ for

$$f(x) = \log \sqrt{\frac{x+2}{x+4}}$$

By replacing the square root sign by the exponent $\frac{1}{2}$ and using properties of logarithms, we may write this in the form

$$f(x) = \frac{1}{2}\left[\log(x+2) - \log(x+4)\right]$$

Then

$$f'(x) = \frac{1}{2}\left[\frac{1}{x+2} - \frac{1}{x+4}\right] = \frac{1}{(x+2)(x+4)}$$

If we had calculated the derivative directly we would have applied formula (3) to obtain

$$f'(x) = \frac{\dfrac{1}{2\sqrt{\dfrac{x+2}{x+4}}}\left[\dfrac{x+4-(x+2)}{(x+4)^2}\right]}{\sqrt{\dfrac{x+2}{x+4}}}$$

Although this result simplifies to the answer obtained previously, the entire procedure is considerably more difficult than the first one.

The technique of applying properties of logarithms is also quite useful in calculating the derivative of a function that is made up of the product or quotient of several other functions. For example, suppose that we wish to calculate $f'(x)$ for

$$f(x) = \frac{(x^2+x)(x-1)^2}{x^2+1}$$

By taking logarithms of both sides, we obtain

$$\log f(x) = \log(x^2+x) + 2\log(x-1) - \log(x^2+1)$$

We now differentiate both sides with respect to x by means of formula (3) to obtain

$$\frac{f'(x)}{f(x)} = \frac{2x+1}{x^2+x} + \frac{2}{x-1} - \frac{2x}{x^2+1}$$

Hence, multiplying by $f(x)$,

$$f'(x) = \left(\frac{2x+1}{x^2+x} + \frac{2}{x-1} - \frac{2x}{x^2+1}\right) \frac{(x^2+x)(x-1)^2}{x^2+1}$$

This is a simpler procedure than differentiating $f(x)$ directly.

4. THE DERIVATIVE OF THE EXPONENTIAL FUNCTION

We can obtain the derivative of the exponential function by using implicit differentiation on the inverse relationship between the exponential and logarithmic functions. We first write down the exponential function, which is denoted by y, and its inverse. Thus

$$y = e^x \quad \text{and} \quad x = \log y$$

Next, we differentiate the inverse relationship with respect to x, treating this relationship as defining y implicitly as a function of x. In view of the chain rule result given by formula (3) with $g(x) = y$, we obtain

$$1 = \frac{y'}{y}$$

Solving for y' and realizing that $y = e^x$, we obtain

$$y' = e^x$$

This result may be summarized as follows:

Theorem 2. *The derivative of the exponential function* $f(x) = e^x$ *is given by the formula*

$$f'(x) = e^x$$

This formula expresses a very interesting geometrical property of the exponential function because it states that the slope of its graph at any point is equal to the y coordinate at that point.

Just as is the case for the logarithmic function, this formula is neat because the base e is used. If we had used an exponential function to some other base, say $f(x) = a^x$, we would have arrived at the more complicated formula

$$f'(x) = a^x \log a$$

Our theorem formula may be extended to more general functions of the exponential type by means of the chain rule. For example, suppose that we have the function

$$r(x) = e^{x^2 + x}$$

We can write this in the form

$$r(x) = f(g(x))$$

where $f(u) = e^u$ and $u = g(x) = x^2 + x$. Then, applying the chain rule, we obtain

$$\begin{aligned} r'(x) &= f'(u)g'(x) \\ &= e^u(2x + 1) \\ &= e^{x^2 + x}(2x + 1) \end{aligned}$$

More generally, if we have

$$r(x) = e^{g(x)}$$

we let $f(u) = e^u$ and $u = g(x)$ and apply the chain rule to obtain

(4)
$$\begin{aligned} r'(x) &= f'(u)g'(x) \\ &= e^u g'(x) \\ &= e^{g(x)}g'(x) \end{aligned}$$

In words, this states that the derivative of an exponential function is equal to the exponential function multiplied by the derivative of the function in the exponent.

Here are a few examples of how formula (4) is used to calculate the derivative of complicated exponential functions.

Example 1.

$$f(x) = x\, e^{x^2}.$$

Application of the product rule gives

$$f'(x) = x\, \frac{d}{dx}(e^{x^2}) + e^{x^2}$$

Then formula (4) gives

$$f'(x) = x\, e^{x^2}\, 2x + e^{x^2} = e^{x^2}(2x^2 + 1)$$

Example 2.

$$f(x) = e^{x \log x}.$$

Formula (4) gives

$$f'(x) = e^{x \log x}\left[x \cdot \frac{1}{x} + \log x\right]$$

$$= e^{x \log x}[1 + \log x]$$

Example 3.

$$f(x) = 10^{x + \sqrt{x}}.$$

First take the logarithm of both sides to obtain

$$\log f(x) = (x + \sqrt{x}) \log 10$$

Then differentiate both sides with respect to x to obtain

$$\frac{f'(x)}{f(x)} = \left(1 + \frac{1}{2\sqrt{x}}\right) \log 10$$

Hence, $f'(x) = f(x)\left(1 + \frac{1}{2\sqrt{x}}\right) \log 10$

$$= 10^{x + \sqrt{x}}\left(1 + \frac{1}{2\sqrt{x}}\right) \log 10$$

5. APPLICATIONS

(a) Continuous Interest

Suppose that an investor receives interest at the rate of $100\,r$ percent a year. If he invests P dollars at the beginning of a year he will have $P(1 + r)$ dollars at the end of the year. However, if interest were compounded twice a year he would have $P\left(1 + \frac{r}{2}\right)$ dollars at the end of 6 months and $P\left(1 + \frac{r}{2}\right)\left(1 + \frac{r}{2}\right)$ dollars at the end of 12 months. More generally, suppose that interest is compounded n times per year. Then the investor will have $P\left(1 + \frac{r}{n}\right)^n$ dollars at the end of the year. If we allow n to become infinite, the investor at the end of one year should receive the amount in dollars given by

$$\lim_{n \to \infty} P \left(1 + \frac{r}{n}\right)^n = P \lim_{n \to \infty} \left(1 + \frac{r}{n}\right)^n$$

$$= P \lim_{n \to \infty} \left[\left(1 + \frac{r}{n}\right)^{n/r}\right]^r$$

$$= Pe^r$$

This limiting value is what is known as the amount that will be received at the end of one year if the interest rate is $100\,r$ percent a year and interest is *compounded continuously.* As a numerical illustration, suppose the interest rate is 5 percent per year. Tables of the exponential function (see the Appendix) show that $e^{.05} = 1.0513$; consequently, continuous compounding increases the nominal interest rate from 5 percent to approximately $5\frac{1}{8}$ percent per year. This increase is undoubtedly considerably less than the ordinary investor assumes it to be when bombarded by the sales propaganda of savings institutions. The continuous compounding formula is, however, much more convenient to work with mathematically when calculating the present or future value of money than the cumbersome formula based on n compoundings a year.

The formula Pe^r determines what P dollars today is worth one year from now. If, however, we are promised A dollars at the end of one year and wish to know what that promise is worth today, we must use the formula Ae^{-r}, because when this is multiplied by the factor e^r we obtain A.

Both of these formulas can be generalized to determine what P dollars today will be worth t years from now, and what A dollars promised t years from today are worth today. The letters P and A are customarily used in problems of this type because they correspond to the *present value, P,* of money and the future *amount, A,* that will be received. Using these symbols, the general formula for the amount is

$$A = Pe^{rt}$$

and for the present value is

$$P = Ae^{-rt}$$

As a numerical illustration of how such a formula is used, let us calculate how many years are required for $1 invested at 5 percent interest compounded continuously to be worth \$2. We solve for t in the equation

$$2 = e^{.05t}$$

Tables of exponentials give $t = 13.8$; hence, it takes 14 years for money to double its value at this interest rate.

The present value factor e^{-rt} is very useful in determining the best time to sell a piece of property whose value increases with time. For example, suppose that experience with land prices in a city shows that a parcel of land worth \$70,000 today will increase in value \$6000 a year. If money is worth 6 percent compounded continuously to the owner, when should he plan on selling this parcel?

The value of this property in dollars t years from now is $70,000 + 6000t$. If the property is sold at that time, the present value of the sale price is

$$f(t) = (70,000 + 6000t)e^{-.06t}$$

This function will be maximized by the value of t that satisfies the equation

$$f'(t) = (70,000 + 6000t)e^{-.06t}(-.06) + e^{-.06t} \cdot 6000 = 0$$

This equation is equivalent to the equation

$$(70,000 + 6000t)(.06) = 6000$$

Its solution is given by $t = 5$; hence, the owner should wait 5 years before selling. The sale price at that time will be \$100,000, which has a present value of \$74,000, as compared to today's market value of \$70,000.

(b) Bacterial Growth

Suppose that a colony of bacteria is grown in a medium that is conducive to growth. Let y denote the number of bacteria in the colony at time t. Under ideal growing conditions the rate at which the bacteria increase in number has been found to be proportional to the number of bacteria in the colony. For example, if one colony has twice as many bacteria as another colony, it is expected to grow twice as fast as the second colony. This is true for other types of biological groups as well. Twice as many adult flies, for example, are likely to produce twice as many offspring, if crowding and lack of food do not interfere. Although the number of bacteria in a colony is an integer, the number is so large that as a function of time it may be treated as a continuous function. Hence, the rate at which the colony size, denoted by y, increases with time t may be treated as the derivative of y with respect to t. Our assumption concerning the nature of the rate of growth then assumes the form

(5)
$$\frac{dy}{dt} = ky$$

where k is the proportionality factor. It depends on the nature of the bacteria and the growing conditions. Since we know that the exponential function $y = e^x$ possesses the property that its derivative is equal to its functional value, it would appear from (5) that y ought to be some type of exponential function. By differentiation it is easily seen that

(6)
$$y = ce^{kt}$$

where c is some constant, is a solution of (5). Furthermore, it can be shown that any solution of (5) must be of the form of (6). Formula (6) defines what is known as exponential growth. This formula has been shown to be a satisfactory model, under ideal growth conditions and over a limited period of time, of things such as bacteria, fruit flies, and the length of certain biological organs.

As an illustration, suppose that we have a colony of 10,000 bacteria and that this colony is expected to double its size every hour. How many bacteria will be present at the end of 5 hours? This information enables us to evaluate c and k in formula (6). When $t = 0$, $y = 10,000$; therefore

$$10,000 = ce^0 = c$$

Next, when $t = 1$, $y = 20,000$; therefore

$$20,000 = 10,000\ e^k$$

This equation is equivalent to $e^k = 2$. From tables of the exponential function it will be found that $k = .69$; consequently, formula (6) assumes the form

$$y = 10,000\ e^{.69t}$$

Substituting the value $t = 5$ and consulting the exponential function tables for $e^{3.45}$, we find that $y = 315,000$.

A more sophisticated model for the growth of a biological population is one that also takes into account the deaths occurring in the population. For such a model it is assumed that the rate of growth with respect to time is given by the formula

(7)
$$\frac{dy}{dt} = \alpha y - \beta y$$

Here α is the proportionality factor associated with births and β is the factor associated with deaths. This model assumes that the rate at which the population is increasing as a result of births is proportional to the size of the population and also that the rate at which it is decreasing as a result of deaths is proportional to its size. The constants α and β are called the birth and death rates, respectively. Since this equation is of the same form as (5) with $k = \alpha - \beta$, the solution is obtained by replacing k by $\alpha - \beta$ in (6). Hence, the solution of (7) is

$$(8) \qquad\qquad\qquad y = ce^{(\alpha-\beta)t}$$

An interesting feature of this formula is that when the death rate β exceeds the birth rate α, the population size will approach 0 as time goes on. That is,

$$\lim_{t\to\infty} y = \lim_{t\to\infty} ce^{(\alpha-\beta)t} = 0$$

If, however, $\alpha > \beta$, the population will become increasingly large as time goes on, but not as rapidly as when $\beta = 0$. The formulation given by (6) assumes that no deaths occur and is used as a model over short intervals of time only.

Although the model given by (7) is more general than that given by (5), it may be treated as a version of (5) by allowing k to represent a proportionality factor that includes the effect of deaths as well as births. Thus the model given by (5) is more general than appearances would indicate, particularly if the constant k represents other growth factors in addition to birth and death rates.

(c) Memory Decay Curves

Consider a memory experiment in which an item is to be learned and retained in an individual's short-term memory. Suppose that new items are presented at a regular rate to the individual for study and memorization, say, r items per unit of time. Let t be the time elapsed between when an item is first presented to the individual and when he is tested on it. Finally, let $s(t,r)$ denote the strength in short-term memory of an item that was presented t units of time earlier if the rate of presenting items is r per unit of time. Then it is often assumed in studies of this kind that the rate with which the strength of short-term memory decreases is proportional to the strength at that time. This assumption may be expressed by the formula

$$(9) \qquad\qquad\qquad \frac{ds}{dt} = -\beta s$$

where β is a constant that depends on the various factors involved in the experiment such as the individual, the difficulty of the items, and the value of r.

Given this relationship, the problem is to find an expression for the function $s(t,r)$. Since r is a constant here, $s(t,r)$ is a function of t only; consequently, the solution of equation (9) is the same as that of equation (5) with $k = -\beta$. Thus, we find that the strength of short-term memory is given by the familiar function

$$s = ce^{-\beta t}$$

(d) Radioactive Disintegration

The exponential function arises naturally as a model for radioactive decay of a radioactive material. Thus suppose that we have a large number of atoms of a radioactive material, such as radium. Then measurements will show that in a short time interval the number of atoms that will disintegrate in a given quantity of the material will be about twice as large as it will be for one half as much material. Therefore, the rate of decrease with respect to time may be assumed to be proportional to the amount of material present. If we let y denote the amount of a radioactive material we have at time t, then the rate at which y decreases is given by the formula

$$\frac{dy}{dt} = -ky$$

where k is a constant of proportionality that depends on the type of material that is disintegrating.

In view of (5) and (6), the solution of this equation is

$$y = ce^{-kt}$$

If y_0 denotes the amount of material at time $t = 0$, then by setting $t = 0$ in this equation, we find that $c = y_0$; hence, we may rewrite this equation as

(10) $$y = y_0 e^{-kt}$$

The time t at which the original amount of material will be reduced by one half is called the *half-life* of the material. This value of t can be obtained by replacing y by $\frac{y_0}{2}$ in (10) and solving for t. Hence, we solve for t in the equation

$$\frac{y_0}{2} = y_0 e^{-kt}$$

Canceling y_0 and taking logarithms to the base e on both sides, we obtain

$$\log \frac{1}{2} = -kt$$

Since $\log \frac{1}{2} = \log 1 - \log 2 = -\log 2$, it follows that the half-life of a radioactive material is given by the formula

(11) $$t = \frac{\log 2}{k} = \frac{.693}{k}$$

The half-life for radium, for example, is 1656 years. Some isotopes of radium, however, have a half-life of only a few minutes or seconds.

(e) Atomic Fission

Consider a sphere of radius r of fissionable material, such as Uranium 235. If a neutron of this material strikes the nucleus of an atom, it will break the nucleus into several parts. This breakup is called atomic fission. It produces new neutrons, as well as other particles, which in turn can cause the fission of other atoms. Some of the neutrons, however, will escape from the material. This breakup and the production of neutrons produces large amounts of energy in the form of heat. If the number of neutrons at time t is denoted by y, then for a short period of time the number of new neutrons produced will be proportional to y. During this same period of time a certain proportion of the neutrons will escape. However, only neutrons close to the surface of the sphere of radius r are likely to escape; therefore, the constant of proportionality, which will be denoted by β, will depend on r. Now it can be shown that the ratio of the area of the sphere to the volume of the sphere is a measure of the proportion of the neutrons that are close enough to the surface of the sphere to escape. Since this ratio is $\frac{a}{r}$, where a is a known constant, we may choose β to be of the form $\frac{\beta_0}{r}$, where β_0 is some constant.

If we let α denote the birthrate of neutrons and β the escape rate, our assumptions can be expressed in the form

$$\frac{dy}{dt} = \alpha y - \beta y$$

As previously, the solution is

$$y = ce^{(\alpha - \beta)t}$$

The value of r for which $\alpha = \beta$ is called the *critical radius*. Since α and β_0 are known constants for any given material and $\beta = \dfrac{\beta_0}{r}$, this requires that

$$\alpha = \frac{\beta_0}{r}$$

Or

$$r = \frac{\beta_0}{\alpha}$$

If r exceeds this critical value, α will be larger than β and then y will become increasingly large as time increases. The amount of energy released therefore grows extremely fast when $r > \dfrac{\beta_0}{\alpha}$. An atomic explosion occurs when a sufficient amount of energy is released in this manner.

COMMENTS

The exponential function and its companion the logarithmic function were introduced in this chapter as arising naturally in the study of certain types of growth or decay problems. Although these applications are very important, the exponential function also occurs in many other situations. It is one of the basic functions in probability and statistical theory. Thus it is exceedingly important, and our illustrations constitute only a small part of its usage.

EXERCISES

Section 2

1. Use the values listed in Section 1 to graph the function

$$f(x) = (1 + x)^{1/x}$$

2. Experiments have shown that if no promotional efforts are made to sell a product and if market conditions are relatively stable the sales of the

product, S, tend to decrease at a constant rate relative to the amount of sales; hence, it may be assumed that

$$S(t) = S(0)e^{-kt}$$

where $S(t)$ denotes the sales at time t, $S(0)$ denotes the sales at the start of the experiment, that is, when $t = 0$, and k is the sales decay constant. If $k = .2$:

(a) How much would sales have decreased in 5 years?
(b) How many years will elapse before sales will drop to one half their original amount?

3. The concentration of a drug in the body fluids depends on the time elapsed since administering the drug. Let $C(t)$ denote the concentration after t units of time. It is frequently assumed that $C(t)$ can be approximated by the function

$$C(t) = C(0)e^{-t/T}$$

where T, called the elimination time, is the time it takes for the concentration to fall to e^{-1} times its initial value.

(a) Calculate $C(t)$ for $\dfrac{t}{T} = 1$, 2, and 5.

(b) What happens to $C(t)$ as $t \to \infty$?

4. A formula used to predict the population of the world is as follows:

$$P(t) = 3 \cdot 10^9 \cdot e^{.018t}$$

Here $P(t)$ is the population t years after the year 1965, which is therefore the base year for calculations. This formula was obtained by using a rate of increase of 1.8%, which was the estimated rate of increase in the year 1965, and the estimated (1965) world population of three billion persons. Use this formula:

(a) To predict the population of the world in the year 2000.
(b) To determine how long it will take for the world population to double the 1965 population.
(c) To determine when the world will have a population of 30 billion.

5. A function that is frequently used as a model in studying the growth of things such as populations of humans or animals or the sales of a product is the Gompertz function defined by the formula

$$G(t) = ca^{b^t}$$

where $a > 0$, $b > 0$, and $c > 0$ are constants. In this formula a has the exponent b^t, where t is the exponent of b. Suppose an individual purchased a piece of property that is assumed to increase in value according to the Gompertz function with $a = .4$, $b = .8$, and $c = 10,000$. How many years will be required for the property to double in value?

Section 3

1. Calculate $f'(x)$ if
   (a) $f(x) = \log(1 + x^2)$
   (c) $f(x) = \log x^2$

   (e) $f(x) = \log \dfrac{x + 1}{x - 1}$

   (g) $f(x) = \log (\log x)$

   (b) $f(x) = \log \sqrt{1 + x^2}$
   (d) $f(x) = x \log x$

   (f) $f(x) = \log \sqrt{\dfrac{x + 1}{x - 1}}$

   (h) $f(x) = \sqrt{x^2 + 1} \, \log (x^2 + 1)$

2. Calculate $f'(x)$ by first calculating $\dfrac{d}{dx} \log f(x)$ if
   (a) $f(x) = 2^x$

   (c) $f(x) = \dfrac{x}{\sqrt{x^2 + x}}$

   (e) $f(x) = \dfrac{\sqrt{x^2 + 1}}{x + 1}$

   (g) $f(x) = x^x$

   (b) $f(x) = (x + 1)(x^2 + 3)(x + x^2)$

   (d) $f(x) = x \cdot 10^x$

   (f) $f(x) = \left(\dfrac{x + 1}{x + 3}\right)^{3/2}$

   (h) $f(x) = \sqrt{x + \sqrt{x}}$

3. Calculate $f'(x)$ if
   (a) $f(x) = \dfrac{\log x}{x}$

   (c) $f(x) = \log \dfrac{x}{x + 1}$

   (e) $f(x) = 3^{2x}$

   (b) $f(x) = \sqrt{\log x}$

   (d) $f(x) = \log [\log (x + 1)]$

   (f) $f(x) = x^{\log x}$

4. Find the equation of the tangent to the curve $y = \log x$ at the point $(2, \log 2)$.

5. Find the equation of the tangent to the curve $y = \log(x^2 + 1)$ at the point $(1, \log 2)$.

6. Find the equation of the tangent to the curve $y = \log\left(\dfrac{1}{x}\right)$ which has the slope -2.

7. Graph the functions $f(x) = \log x$ and $g(x) = \log_{10} x$ on the same graph paper. Use tables of logarithms to the base e and to the base 10 to assist you (see the Appendix).

8. Graph the following functions by first finding critical points and determining where the graph is rising and falling.
   (a) $f(x) = \log(x^2 + 1)$
   (b) $f(x) = x - \log x$

9. Find the local maxima and minima, if any, of the following functions:
   (a) $f(x) = \log(x^2 + 4)$
   (b) $f(x) = \log(x^2 + x + 1)$
   (c) $f(x) = x \log x$
   (d) $f(x) = x^2 - \log x$

10. The weight, y, of a limb of a plant or animal as a function of its age, t, can be approximated by the relation

$$y = at^b$$

   where a and b are constants that depend on the particular type of plant or animal and the environment. In making comparisons between groups of plants or animals, it is useful to determine the rate of growth of a unit weight of the limb. This is called the specific rate of growth.

   (a) Show that the specific growth rate is given by $\dfrac{d}{dt}(\log y)$.

   (b) If one plant is twice as old as another similar type plant, what is its specific growth rate relative to that of the younger plant?

Section 4

1. Calculate $f'(x)$ if
   (a) $f(x) = e^{4x}$
   (b) $f(x) = xe^x$
   (c) $f(x) = e^{x^2}$
   (d) $f(x) = \dfrac{e^{2x}}{x^2}$
   (e) $f(x) = e^{x^2 + 2x}$
   (f) $f(x) = \dfrac{e^x - e^{-x}}{e^x + e^{-x}}$

2. Calculate $\dfrac{dy}{dx}$ if y is defined implicitly as a function of x through the relation
   (a) $e^x + e^y = e^{x+y}$
   (b) $\log(y + x) = e^x$
   (c) $xe^y + ye^x = 1$
   (d) $x \log xy = \log y$

3. Find the equation of the tangent to the curve $y = e^{-x}$ that is parallel to the line $x + y = 2$.

4. Find the equation of the tangent to the curve $y = e^{-x^2/2}$ where the second derivative has a zero value.

5. Graph the following functions using calculus techniques and plotting a few points. (Refer to the Appendix for exponential function values).

(a) $f(x) = e^{-x}$ (b) $f(x) = e^{-x^2}$

(c) $f(x) = xe^{-x}$ (d) $f(x) = \dfrac{e^{-x}}{x}$

(e) $f(x) = xe^{-x^2}$ (f) $f(x) = \dfrac{e^x}{e^x - 1}$

6. By first taking the logarithm, or by using the formula in Section 4 if it is applicable, calculate $f'(x)$ if

(a) $f(x) = 3^{4x}$ (b) $f(x) = 10^{x^2+x}$
(c) $f(x) = 2^{3x} \cdot 4^{2x}$ (d) $f(x) = x^{\log x}$
(e) $f(x) = (x + 2)5^{-x}$ (f) $f(x) = (\log x)^{\log x}$

7. The growth of a population of cells that reproduce by simple division can be expressed by means of the function

$$y(x) = y(0)2^x$$

where $y(0)$ is the number of cells at the time observations began and x denotes the generation. If the generation time, denoted by T, is assumed to be constant, this may be written in the form

$$y(t) = y(0)2^{t/T}$$

where t is the time elapsed since observations began. It is assumed that t and T are measured in the same time units, such as minutes or hours.
(a) Find the rate of increase of this population with respect to time.
(b) If the value of $y(t)$ is known for a particular value of t, say $t = \tau$, find a formula for the generation time T.

8. Suppose that the growth of a bacterial culture is growing in such a manner that there are 10^t milligrams of the culture after t hours of growth.
(a) Find the rate of increase of this culture at the time $t = 2$.
(b) Calculate the average rate of increase from $t = 2$ to $t = 2.1$ and compare it with the answer in (a).

9. Let $f(t)$ be a function that describes the concentration of a certain drug in the blood t units of time after administration. A model that is often used to approximate $f(t)$ is given by the formula

$$f(t) = c \, \frac{e^{-at} - e^{-bt}}{b - a}$$

where $a > 0, b > 0$, and $c > 0$ are constants with $a < b$. Determine the value of t that maximizes the concentration.

10. Consumers who go to sales that are advertised in the newspapers tend to shop in the early stages of the sales promotion. If $f(t)$ denotes the increase in sales over the unadvertised normal amount of sales as a function of the number of days since the sales promotion began, then experience shows that $f(t)$ may be approximated by the function

$$f(t) = A(1.7)^{-t}$$

where A represents the level of sales for a given company at the start of the promotion.
   (a) Calculate the rate at which sales are decreasing five days after the start.
   (b) When will the increased sales be only one fourth of the first day's increase?

Section 5

1. Which of these two offers would you rather have? A sum of $10,000 now or a note that promises to pay you $20,000 ten years from now? Assume that you assign the value of money to be 7 percent compounded continuously.

2. At what rate compounded continuously will $2000 today be worth $3000 seven years from now?

3. How much should be deposited in a savings account today if it is desired to have $5000 at the end of 15 years and if the interest rate is 5 percent
   (a) simple interest?
   (b) compounded continuously?

4. At the end of what year will the principal have doubled itself if it is invested at 4 percent compounded continuously?

5. If one dollar doubles itself in 10 years at an interest rate that is compounded continuously, how long will it take to triple itself?

6. If the purchasing power of the dollar is decreasing at the rate of 4 percent annually because of inflation, how long will it be before the dollar is worth only 50 cents?

7. The number of bacterial colonies in a culture grew from 10 to 40 in 12 hours. What was the size of the population of colonies at the end of 6 hours?

8. Suppose that the cost of planting a piece of land with timber is $1500 and that its value t years after planting is given by

$$f(t) = \$200 \ e^{\sqrt{t/2}}$$

Assume that money is worth 6 percent compounded continuously, which implies that the present value discount factor is $e^{-.06t}$. Find the value of t that will maximize the present value of the timber if it is cut t years after planting.

9. A bacterial colony grows at the rate of e^t members per minute t minutes after first being observed.
   (a) Find an expression that gives the size of the population t minutes after first being observed.
   (b) What is the size of the population at the end of 5 minutes?
   (c) How much did the population increase during the fifth minute?

10. A wine dealer buys new wine and lays it down in his cellar for later sale. Let $s(t)$ be the selling price per case t years after purchase. Assume that money is worth $100r$ percent compounded continuously, which implies that the present value discount factor is e^{-rt}. The present value of a case of wine sold t years later is therefore given by

$$f(t) = s(t)e^{-rt}$$

   (a) Derive a formula for determining when the dealer should sell his wine to maximize its present value.
   (b) Use the result in (a) to calculate t if $r = .05$ and $s(t)$ is given by the formula $s(t) = 50\sqrt{t}$.

11. An investment company is attempting to keep its total investment capital at a fixed amount A. If y denotes the value of its investment at time t and if y falls below A, then additional investments will be made. If,

however, y exceeds A, then capital will be withdrawn from investment. Suppose that this company invests its capital in such a way that the rate of investment is proportional to the difference $A - y$ at any time t. This implies that

$$\frac{dy}{dt} = k(A - y)$$

(a) Show that if α and β are chosen properly, a solution to this differential equation is

$$y = \alpha + \beta e^{-kt}$$

(b) What do the constants α and β represent?
(c) What happens to the investment level as $t \to \infty$?

12. Let $s(t)$ represent the amount of a solute inside a cell at time t. If v is the volume of the cell, then $y(t) = \frac{s(t)}{v}$ is defined to be the concentration of the solute inside the cell. It is known from Frick's law that there exist constants a and b such that

$$\frac{ds}{dt} = a(b - y)$$

Because of the relation between y and s, this is equivalent to

$$v \frac{dy}{dt} = a(b - y)$$

Or

$$\frac{dy}{dt} = \frac{a}{v}(b - y)$$

(a) Show that if α and β are chosen properly, a solution to this differential equation is

$$y = \alpha + \beta e^{-(a/v)t}$$

(b) What do the constants α and β represent?

13. Assume that the birth rate of a population is proportional to the size of the population but that the death rate is proportional to the square of the population size. This yields the model

$$\frac{dy}{dt} = ay - by^2$$

where y denotes the population size at time t and a and b are positive constants.

(a) Show that a solution of this differential equation is

$$y = \frac{ae^{at}}{c + be^{at}}$$

(b) What happens to the population size as $t \to \infty$?

14. Carbon 14 is radioactive and has a half-life of about 5570 years. If the concentration of carbon 14 in the carbon content of a piece of wood, for example, of unknown age were one half that of the carbon 14 concentration in a present-day living specimen of the same type, then we would place its age at about 5570 years. Show, as is done in arriving at formula (11), that if c_0 and c_1 are the radioactivities of samples prepared from present-day and undated materials, respectively, then the age of the undated material is approximately

$$t = 8040 \log \frac{c_0}{c_1}$$

15. A Norwegian archaeologist, Ingstad, excavated some building sites on the northern coast of Newfoundland and found several old Norwegian artifacts. The charcoal from cooking pits at those sites was analyzed by determining the percentage of carbon 14 remaining in the charcoal. It was found to be 88.6 percent. Assuming that the half-life of carbon 14 is 5570 years, use the answer of Problem 14 to determine the date of this Viking settlement in Newfoundland if this dating occurred in 1972.

16. Assume that the rate of growth of a population is not only proportional to the size of the population, y, but also is proportional to the difference $a - y$, where a is a positive constant. This implies that when the population attains a size equal to a, it will stop growing, and will have a negative rate of increase if its size exceeds a. These assumptions give the differential equation

$$\frac{dy}{dt} = ky(a - y)$$

where $k > 0$ is the proportionality factor.

(a) Show that a solution of this differential equation is given by y that satisfies the relation

$$\frac{y}{a-y} = ce^{akt}$$

where c is a positive constant.

(b) What happens to y when $t \to \infty$?

(c) Show that the solution in (a) can be expressed in the form

$$y = \frac{ab}{b + (a-b)e^{-akt}}$$

This equation is called the logistic equation. It is often used to represent the growth of variables such as the number of fruit flies in a container and the size of an epidemic in a population.

17. A basic problem in certain business situations is to determine the best time to replace a capital asset, such as a machine or truck. Let A denote its purchase price, and let B denote its replacement cost. The annual operating costs consist of a fixed cost C, plus an amount that is proportional to the age of the asset, which may be denoted by at. The salvage (sale) value of the asset is assumed to decrease at a constant rate each year; hence, the salvage value may be denoted by Ar^t, where r denotes the fraction that determines the rate of decreased value at the end of each year. If the asset is replaced at the end of n years the total costs over that period of time will consist of the total operating costs, plus the replacement cost, minus the salvage value. This is given by the formula

$$T(n) = \sum_{t=1}^{n} (C + at) + B - Ar^n$$

The average cost per year, which is denoted by $A(n)$, is obtained by dividing $T(n)$ by n; hence

$$A(n) = \frac{1}{n} \sum_{t=1}^{n} (C + at) + \frac{B}{n} - A\frac{r^n}{n}$$

(a) Show that this simplifies to

$$A(n) = C + a\frac{n+1}{2} + \frac{B}{n} - A\frac{r^n}{n}$$

(b) Replace n by x and treat $A(x)$ as a differentiable function of x. Then calculate $A'(x)$.

(c) Suppose $A = 20{,}000$, $B = 24{,}000$, $C = 5000$, $a = 500$, and $r = \frac{2}{3}$. Use the result in (b) to find the approximate value of x that will minimize the average annual costs. Assume that $\dfrac{r^x}{x}$ is so small that it can be ignored in the calculations.

18. A dosage of amount a of a drug is given to a patient at regular time intervals of length τ. Experiments show that the concentration of this drug in the blood t units of time after being administered is given approximately by the formula ae^{-t}. Since a units are being administered every τ units of time the concentration at the end of τ units of time and just before the next dose is $ae^{-\tau}$. The concentration at the end of 2τ units of time and just before the next dose will be $(a + ae^{-\tau})e^{-\tau} = ae^{-\tau} + ae^{-2\tau}$. In general, the concentration at the end of n such time intervals will be given by

$$C(n) = ae^{-\tau} + ae^{-2\tau} + \cdots + ae^{-n\tau}$$

(a) Use the formula for the sum of a geometric progression to show that this formula can be written

$$C(n) = ae^{-\tau} \left(\frac{1 - e^{-n\tau}}{1 - e^{-\tau}} \right)$$

(b) What happens to $C(n)$ as $n \to \infty$?

(c) Suppose that the initial dose is $r + a$ and each succeeding dose is a. Show that if r is chosen to be $r = a \dfrac{e^{-\tau}}{1 - e^{-\tau}}$, the value of $C(n) = r$ and, therefore, since $C(n)$ is increased by the amount a at the beginning of each τ interval, that $C(n)$ will vary between the limits r and $r + a$.

11 | Integration

In this chapter we study the basic problem of how to find the area of a region under a curve and above the x axis between two points on the x axis. The solution of this geometrical problem will be used to solve certain classes of problems in the social sciences. The advantage of examining those problems from a geometrical point of view first is that the explanation is then much simpler. This was also true in the chapter on differentiation where we first studied the derivative from a geometrical point of view before embarking on practical maximum and minimum problems.

1. THE DEFINITE INTEGRAL

Suppose that we wish to find the area of the region under the curve $y = x^2$ and above the x axis between the points $x = 1$ and $x = 2$. A sketch of this curve, together with a representation of the desired area, is shown in Fig. 1. Although we have not defined what we mean by the area of a region under such a curve, we have a good intuitive idea of what it should be. We use our intuition to arrive at a formal definition. Meanwhile we discuss the concept of area as though our intuitive notion of it were well defined.

The first step in attempting to find the desired area is to divide the x axis between $x = 1$ and $x = 2$ into a number, n, of equal length intervals. Suppose that we choose $n = 5$. Then over each such interval we construct a rectangle with that interval as a base and with the height of the rectangle chosen to be the distance up to the curve from the left boundary of the interval. This is shown in Fig. 2. Since we know how to calculate the area of a

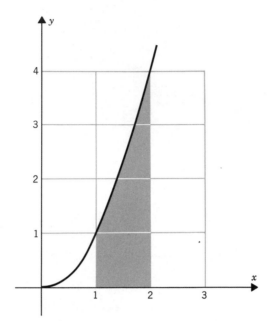

FIGURE 1   An area under $y = x^2$.

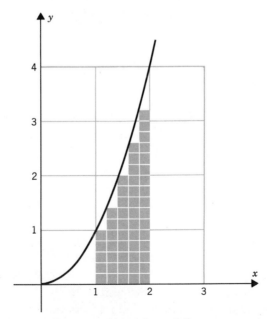

FIGURE 2   A lower bound for an area.

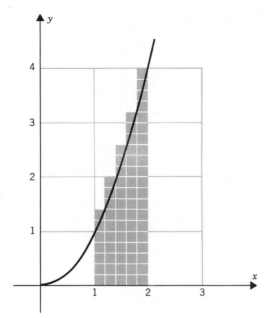

FIGURE 3   An upper bound for an area.

rectangle, we can calculate the sum of the areas of these rectangles. This sum is certainly less than what we would accept as the area of the region under the curve between $x = 1$ and $x = 2$.

Next, we construct rectangles over these same intervals, but this time we choose the height of a rectangle to be the distance up to the curve from the right boundary of the interval. This is shown in Fig. 3. The sum of the areas of these rectangles is certainly greater than what we would accept as the area of the region under the curve between $x = 1$ and $x = 2$. In view of these two results, we write

$$S_L < A < S_U$$

where S_L and S_U denote the lower and upper rectangular area sums and A denotes the area being sought.

Calculations for this problem are easily carried out. First, we calculate the following y values from $y = x^2$.

x	1.0	1.2	1.4	1.6	1.8	2.0
y	1.00	1.44	1.96	2.56	3.24	4.00

Then, since the width of each rectangle is .2, we obtain

$$S_L = .2[1 + 1.44 + 1.96 + 2.56 + 3.24] = 2.04$$

and

$$S_U = .2[1.44 + 1.96 + 2.56 + 3.24 + 4.00] = 2.64$$

Hence, we would agree that the desired area is some number between 2.04 and 2.64.

It should now be clear how to proceed to define the desired area. We divide the x axis between $x = 1$ and $x = 2$ into a very large number, n, of equal length intervals. We calculate the lower and upper sums, S_L and S_U, for this subdivision. We allow the number of intervals, n, to become infinite. The sums S_L and S_U will, as we see presently, approach limits. It seems clear from the geometry of the problem that these limits will be equal. This common limiting value will then be defined to be the desired area A.

We carry out these calculations for this problem to show that our geometrical intuition is justified. Let the division points for the n intervals between $x = 1$ and $x = 2$ be expressed in the form

$$x_i = 1 + \frac{i}{n}, \qquad i = 1, 2, \ldots, n-1$$

The distance up to the curve at the point x_i is given by

$$y_i = \left(1 + \frac{i}{n}\right)^2$$

Then, since the width of each interval is $\frac{1}{n}$, we have

$$S_L = \frac{1}{n}\left[1 + \left(1 + \frac{1}{n}\right)^2 + \left(1 + \frac{2}{n}\right)^2 + \cdots + \left(1 + \frac{n-1}{n}\right)^2\right]$$

Squaring out the parentheses and collecting terms, we find that

$$S_L = \frac{1}{n}\left\{n + \frac{2}{n}(1 + 2 + \cdots + n - 1) + \frac{1}{n^2}[1 + 2^2 + \cdots + (n-1)^2]\right\}$$

To simplify this expression we apply the formula for the sum of the first m positive integers and the formula for the sum of the squares of the first m positive integers. They are

$$1 + 2 + \cdots + m = \frac{m(m+1)}{2}$$

and

$$1^2 + 2^2 + \cdots + m^2 = \frac{m(m+1)(2m+1)}{6}$$

Application of these formulas to the two parentheses in S_L gives

$$S_L = \frac{1}{n} \left[n + \frac{2}{n} \frac{(n-1)n}{2} + \frac{1}{n^2} \frac{(n-1)n(2n-1)}{6} \right]$$

$$= 1 + \frac{n-1}{n} + \frac{2n^2 - 3n + 1}{6n^2}$$

$$= 2\frac{1}{3} - \frac{3}{2n} + \frac{1}{6n^2}$$

Hence

$$\lim_{n \to \infty} S_L = 2\frac{1}{3}$$

Similarly, we find that

$$S_U = \frac{1}{n} \left[\left(1 + \frac{1}{n}\right)^2 + \left(1 + \frac{2}{n}\right)^2 + \cdots + \left(1 + \frac{n}{n}\right)^2 \right]$$

It will be observed that S_U can be obtained from S_L by subtracting the first term inside the brackets of S_L, namely 1, and adding the term $\left(1 + \frac{n}{n}\right)^2$ inside the brackets of S_L. This yields the relation

$$S_U = S_L - \frac{1}{n} + \frac{1}{n}\left(1 + \frac{n}{n}\right)^2$$

$$= S_L + \frac{3}{n}$$

Consequently

$$\lim_{n \to \infty} S_U = \lim_{n \to \infty} S_L + \lim_{n \to \infty} \frac{3}{n} = \lim_{n \to \infty} S_L = 2\frac{1}{3}$$

We have shown that the lower and upper sums possess limits and that those limits are equal. This limiting value, $2\frac{1}{3}$, is therefore by definition the area of the region under the parabola $y = x^2$ and above the x axis between $x = 1$ and $x = 2$.

We were fortunate in this problem to be able to calculate the limiting sums because of the availability of two useful formulas. For more complicated curves, however, we may not be so fortunate. Methods are developed in later sections for solving some of these limit calculations; therefore, we need not be concerned with them here. It will turn out that the calculations are very simple and that for a problem like the present one it is possible to find the answer in less than one minute.

Ignoring computational matters, we now look at the general problem of how to define the area of a region under any curve $y = f(x)$ and above the x axis between two points $x = a$ and $x = b$. To simplify the discussion, we assume that f is a continuous function and that its graph lies entirely above the x axis between the points $x = a$ and $x = b$. We first divide the part of the x axis between $x = a$ and $x = b$ into n intervals. Although we chose the intervals to be of equal length in calculating the area under the parabola, that was done for computational convenience. Upper and lower approximating sums are often more accurate if the intervals are not required to be of equal length; therefore, for the purpose of obtaining a more general theory we drop this restriction and let the n intervals be determined by the points $x_0, x_1, \ldots, x_i, \ldots, x_n$, where $x_0 = a$ and $x_n = b$.

As in the case of the special function $f(x) = x^2$, we construct two sets of rectangles over these intervals in such a way that the sum of the areas of one set will be less than the desired area, and the sum of the areas of the other set will be greater than the desired area.

For the purpose of obtaining S_L we concentrate on the ith interval, which extends from the point x_{i-1} to the point x_i. In that interval we choose a point x at which $f(x)$ assumes its minimum value in the interval. This may occur at one of the end points of the interval or inside it. Since we are assuming that f is continuous, we know that such a point exists. Let this point be denoted by s_i. This choice is shown in Fig. 4. The area of the rectangle that has this interval for its base and whose height is $f(s_i)$ is less than or equal to the area of the region under the curve over that interval. We form the sum of the areas of the rectangles constructed in this manner. If, as before, this sum is denoted by S_L, it follows from our notation that

$$S_L = f(s_1)(x_1 - x_0) + f(s_2)(x_2 - x_1) + \cdots + f(s_n)(x_n - x_{n-1})$$

This can be written in a more compact form by using a summation symbol as follows:

$$S_L = \sum_{i=1}^{n} f(s_i)(x_i - x_{i-1})$$

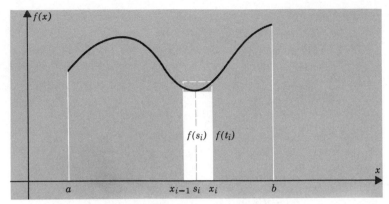

FIGURE 4 The geometry of integration.

To find an upper sum, we choose a point in the ith interval at which the function f assumes its maximum value in that interval. This point is denoted by t_i. For the function graphed in Fig. 4 the point t_i is at x_i; however, for other intervals it might be located somewhere in the interior of the interval or at the left boundary point. The area of the ith rectangle with height $f(t_i)$ is at least as large as the area under the curve over that interval. This is illustrated in Fig. 4 for that function. We form the sum of such rectangular areas to obtain

$$S_U = \sum_{i=1}^{n} f(t_i)(x_i - x_{i-1})$$

The area of the region under the curve and above the x axis between a and b, which is denoted by A, should certainly satisfy the inequalities

$$S_L \leqslant A \leqslant S_U$$

Finally, we calculate the limit of these two sums as $n \to \infty$, with the restriction that the length of the largest of the intervals (x_{i-1}, x_i), $i = 1, \ldots ,n$, must approach zero as $n \to \infty$. With this restriction, let us suppose that

$$\lim_{n \to \infty} S_L = L_1$$

and

$$\lim_{n \to \infty} S_U = L_2$$

If $L_1 = L_2$ we define this common limiting value to be the desired area. Suppose now that we choose any point, denoted by x'_i, in the ith interval and calculate the sum

$$\sum_{i=1}^{n} f(x_i')(x_i - x_{i-1})$$

This sum obviously satisfies the inequalities

$$\sum_{i=1}^{n} f(s_i)(x_i - x_{i-1}) \leqslant \sum_{i=1}^{n} f(x_i')(x_i - x_{i-1}) \leqslant \sum_{i=1}^{n} f(t_i)(x_i - x_{i-1})$$

Therefore, if the preceding limits exist and have a common value, which we now denote by S, then

$$\lim_{n \to \infty} \sum_{i=1}^{n} f(x_i')(x_i - x_{i-1}) = S$$

This last limit is denoted by a special symbol and is called the *integral of f from a to b*. When this limit exists we write

(1) $$\lim_{n \to \infty} \sum_{i=1}^{n} f(x_i')(x_i - x_{i-1}) = \int_a^b f(x)\ dx$$

The integral sign $\int$ is an elongated S to denote that the integral came from a sum. The symbol dx corresponds to the difference $x_i - x_{i-1}$. The letters a and b attached to the integral sign show the interval over which the integral is to be evaluated. The function f that is being *integrated* is called the *integrand* of the integral.

Although we have assumed that f is a continuous function and that its graph lies above the x axis between a and b, these assumptions have been made only for convenience of discussion. The definition given by equation (1) applies to any function for which the limit in that definition exists. It can be shown, however, that this limit always exists for a continuous function and, therefore, that continuous functions can always be integrated.

In terms of the notation introduced in (1), we can represent the area that was calculated previously for the function $f(x) = x^2$ by writing

$$\int_1^2 x^2\ dx = 2\frac{1}{3}$$

2. PROPERTIES OF INTEGRALS

In view of the manner in which the integral is defined, we can write down several of its properties.

(i) If c is any constant

$$\int_a^b cf(x) \; dx = c \int_a^b f(x) \; dx$$

This follows at once by factoring out c in front of the approximating sums and by using the fact that the limit of a constant times a function is equal to the constant times the limit of the function.

(ii) If c is any number satisfying $a < c < b$,

$$\int_a^b f(x) \; dx = \int_a^c f(x) \; dx + \int_c^b f(x) \; dx$$

This follows from considering the interval (a,b) as the sum of two neighboring intervals (a,c) and (c,b) and then breaking up the approximating sums over (a,b) into the sum of the two parts over (a,c) and (c,b). Since the limit of a sum is equal to the sum of the limits, this gives, except for a few minor details, the desired result. This property is also obvious geometrically, since it merely states that the area of the region under f from a to b is equal to the sum of the area from a to c and the area from c to b.

(iii) If f and g are two integrable functions over (a,b),

$$\int_a^b [f(x) + g(x)] \; dx = \int_a^b f(x) \; dx + \int_a^b g(x) \; dx$$

This follows from arguments similar to those used in (ii).

In some problems, manipulations of integrals give rise to an integral for which $a > b$. These manipulations yield valid results provided that we define

$$\int_a^b f(x) \; dx = - \int_b^a f(x) \; dx$$

With this understanding, property (ii) holds whether or not c is a point inside the interval (a,b).

If we choose $b = a$ in this relation, we obtain

$$\int_a^a f(x) \; dx = - \int_a^a f(x) \; dx$$

Solving this equation for the integral, we find that

(2)
$$\int_a^a f(x) \; dx = 0$$

This result is certainly in agreement with geometrical reasoning because the area of a region under the graph of f over an interval of zero length must be zero.

The variable x used in an integral is called a *dummy variable* because the value of the integral does not depend on what symbol is used to represent the variable. For example, we could call the x axis the t axis and then find the area of the region under the graph of $f(t)$ between a and b. It would, of course, be the same as the area using x as our variable. Thus the letter used to represent the variable in integration is irrelevant, and we can write

$$\int_a^b f(x) \, dx = \int_a^b f(t) \, dt$$

The geometrical interpretation of the integral as the area of the region under the graph of $f(x)$ and above the x axis between a and b is valid only if the graph lies above the x axis in that interval. Suppose, however, that the graph of $f(x)$ lies partly above and partly below the x axis in the interval. What is the geometrical interpretation in that case? As an illustration, let the function be $f(x) = x^2 - x$ and the interval be $[0,2]$. The graph of this function is given in Fig. 4, Chapter 9.

From property (ii) we may integrate over $[0,1]$ and over $[1,2]$ and add the results to obtain the integral over $[0,2]$. The second integral obviously gives the usual area because the graph of $f(x)$ lies above the x axis over that interval. The first integral, however, will give the negative of the area of the region between the graph of $f(x)$ and the x axis, because if we had integrated $-f(x)$ over $[0,1]$ we would have integrated a function whose graph lies above the x axis in that interval and which has the same area as the region between it and the x axis. Thus the integral over $[0,2]$ gives the area of the region between the graph of $f(x)$ and the x axis over $[1,2]$ minus the area of the region between the graph of $f(x)$ and the x axis over $[0,1]$. This is true in general. An integral over an interval $[a,b]$ gives the sum of the areas of the regions lying above the x axis minus the sum of the areas of the regions lying below the x axis. If we wish to find the total area of the region between the graph of $f(x)$ and the x axis, we must split up the integral over intervals where the function is only positive and only negative and then change the sign of the integral over the intervals where the function is negative.

As an illustration of how problems of this kind are treated, consider the problem of finding the area of the region between the x axis and the curve

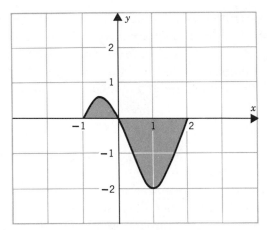

FIGURE 5   A rough graph of $y = x(x + 1)(x - 2)$.

$y = x(x + 1)(x - 2)$ from $x = -1$ to $x = 2$. We first make a rough graph of the function in the interval $[-1, 2]$. Let $f(x) = x(x + 1)(x - 2)$, then the points given by $f(-1) = 0$, $f\left(-\dfrac{1}{2}\right) = \dfrac{5}{8}$, $f(0) = 0$, $f(1) = -2$, and $f(2) = 0$ should suffice for our graphing purposes. A sketch is given in Fig. 5.

The desired area, which is the sum of the areas of the two shaded regions in Fig. 5, is therefore given by

$$\int_{-1}^{0} x(x + 1)(x - 2)\ dx - \int_{0}^{2} x(x + 1)(x - 2)\ dx$$

In the next section we study methods for evaluating integrals of this type.

3. ANTI-DIFFERENTIATION

In this section we develop a technique for evaluating integrals without the necessity of finding the limit of the approximating sum used in its definition. Toward this objective, let f be a continuous function, which is therefore integrable over (a,b), whose graph lies above the x axis in that interval. Let $F(x)$ represent the area of the region under f and above the x axis between the fixed point a and the variable point x, where $a < x < b$. Since this area is given by a corresponding integral, it follows that

$$F(x) = \int_{a}^{x} f(t)\ dt$$

Here we use the variable t as our dummy variable of integration to keep from confusing the variable of integration with the upper limit x. This area is shown in Fig. 6 as the area marked with vertical lines.

We now choose a point $x + h$ to the right of x but such that it lies inside (a,b). The area of the region under f between a and $x + h$ is given by

$$F(x+h) = \int_a^{x+h} f(t)\, dt$$

This area consists of the earlier area $F(x)$ and the area in Fig. 6 marked with horizontal lines. We next take the difference $F(x + h) - F(x)$. This gives the area of the region under f between x and $x + h$ as shown in Fig. 6 and marked with horizontal lines. Since f is assumed to be a continuous function and, hence, possesses an unbroken graph between x and $x + h$, it is clear from the geometry of Fig. 6 that there must exist a point c between x and $x + h$ such that the rectangle whose base is the interval from x to $x + h$ and whose height is $f(c)$ has an area equal to the area $F(x + h) - F(x)$. This geometry is shown in Fig. 7. An algebraic proof of this property of continuous functions, which is called the mean value theorem of integral calculus, is too difficult to present here. The equivalence of these two areas is expressed algebraically by the relation

$$F(x+h) - F(x) = hf(c), \qquad x < c < x + h$$

Dividing both sides of this relation by h, we obtain

(3) $$\frac{F(x+h) - F(x)}{h} = f(c), \qquad x < c < x + h$$

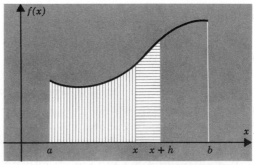

FIGURE 6   The geometry of $F(x)$.

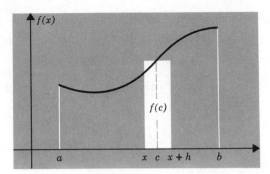

FIGURE 7   A mean value theorem application.

We now take the limit of the left side as $h \to 0$ to obtain the derivative $F'(x)$. But since the point c lies between x and $x + h$, c must approach x when $h \to 0$. In view of the continuity of f, it therefore follows that $f(c)$ must approach $f(x)$ when $h \to 0$; hence, the limit of the right side of equation (3) is $f(x)$. This yields the desired result, which may be expressed in the following manner:

Theorem 1. *If f is a continuous function and if*

$$F(x) = \int_a^x f(t) \, dt$$

then

$$F'(x) = f(x)$$

By means of this theorem we have the desired relationship between differentiation and integration that will enable us to integrate functions without calculating the limit of approximating sums. To see how this is accomplished, consider the problem of evaluating the general integral given by

(4) $$F(b) = \int_a^b f(t) \, dt$$

The notation being used here is that of Theorem 1.

Let $G(x)$ be any function for which $G'(x) = f(x)$. Such a function is called an *antiderivative* of $f(x)$. But by Theorem 1 we know that $F'(x) = f(x)$, and therefore that $F(x)$ is also an antiderivative of $f(x)$. Now by Theorem 2,

Chapter 9, it follows that two functions that possess the same derivative can differ at most by a constant; therefore

(5)
$$F(x) = G(x) + c$$

where c is some constant. From formula (2) and the notation of Theorem 1, we know that

$$F(a) = \int_a^a f(t) \ dt = 0$$

Choosing $x = a$ in (5) and applying this result, we see that c must satisfy the equation

$$G(a) + c = 0$$

Solving for c in this equation and substituting its value into (5), we obtain

$$F(x) = G(x) - G(a)$$

Application of this relationship for $x = b$ to equation (4) gives us the desired formula:

$$\int_a^b f(t) \ dt = F(b) = G(b) - G(a)$$

This result is summarized in the following theorem, which is known as the *fundamental theorem of calculus.*

Theorem 2. *If f is a continuous function and if G is any function satisfying* $G'(x) = f(x)$, *then the integral of f from a to b is given by the formula*

$$\int_a^b f(x) \ dx = G(b) - G(a)$$

Although our derivation has been based on the assumption that f is continuous, the preceding theorems hold true for more general functions, but these functions will not concern us in this book.

The beauty of this formula is that integration is performed by merely finding one antiderivative for the function f and evaluating it at the upper and lower limits of the integral. It makes no difference which antiderivative is found; only one is needed.

As an illustration of how this theorem is applied, consider the problem of evaluating the integral

$$\int_1^2 x^2 \ dx$$

This is the integral that was evaluated in Section 1 by calculating the limiting values of upper and lower approximating sums.

Since $f(x) = x^2$ here, we must find an antiderivative of x^2. That is, we must find a function whose derivative is x^2. One such function is $\dfrac{x^3}{3}$; therefore, we may write

$$G(x) = \frac{x^3}{3}$$

Application of Theorem 2 then gives

$$\int_1^2 x^2 \, dx = G(2) - G(1)$$

$$= \frac{2^3}{3} - \frac{1^3}{3} = 2\frac{1}{3}$$

This, of course, is the result that was obtained in Section 1 by a lengthy limiting process.

As an illustration of the fact that G may be chosen to be any one of the antiderivatives for f, let us assume that we had chosen $G(x) = \dfrac{x^3}{3} + 125$. It satisfies the condition that $G'(x) = x^2$; hence, it may be used in our formula. Then, by Theorem 2, we have

$$\int_1^2 x^2 \, dx = \left(\frac{2^3}{3} + 125\right) - \left(\frac{1^3}{3} + 125\right) = \frac{8}{3} - \frac{1}{3} = 2\frac{1}{3}$$

From this example we observe that the constant that we might add to $\dfrac{x^3}{3}$ to obtain a different antiderivative for x^2 will cancel out in the calculation of $G(b) - G(a)$ and, therefore, in seeking an antiderivative, we should pick the simplest possible one. Here the simplest one is clearly $\dfrac{x^3}{3}$.

As a second illustration of how Theorem 2 is applied, consider the problem of evaluating the integral

$$\int_2^4 (x^2 - x + 2) \, dx$$

From the formula for differentiating x^n we can write down by inspection a function whose derivative is $x^2 - x + 2$. The simplest such function is

$$G(x) = \frac{x^3}{3} - \frac{x^2}{2} + 2x$$

The solution to our problem then is given by

$$G(4) - G(2) = \frac{4^3}{3} - \frac{4^2}{2} + 2(4) - \left[\frac{2^3}{3} - \frac{2^2}{2} + 2(2)\right] = 16\frac{2}{3}$$

Hence

$$\int_2^4 (x^2 - x + 2)\ dx = 16\frac{2}{3}$$

It is not necessary to write out $G(x)$ explicitly in carrying out an integration. It suffices to write down an antiderivative by inspection and indicate the limits at which it is to be evaluated. Thus, in the preceding problem, we could proceed as follows:

$$\int_2^4 (x^2 - x + 2)\ dx = \left[\frac{x^3}{3} - \frac{x^2}{2} + 2x\right]_2^4$$

$$= \left[\frac{64}{3} - 8 + 8\right] - \left[\frac{8}{3} - 2 + 4\right] = 16\frac{2}{3}$$

The problem of finding the area of the closed region between two curves is easily solved by methods similar to those already employed. Thus suppose that we wish to find the area of the region between the two curves $y = x^2$ and $y = 2x - x^2$. Sketches of these two curves are given in Fig. 8.

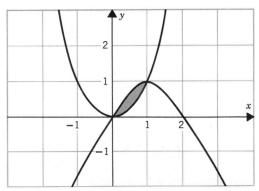

FIGURE 8 Graphs of $y = x^2$ and $y = 2x - x^2$.

First, we must determine where the two curves intersect by solving the two equations

$$y = x^2 \quad \text{and} \quad y = 2x - x^2$$

Equating the two y values, we obtain the equation $x^2 = 2x - x^2$ whose solution is $x = 0$ and $x = 1$. The two curves, therefore, intersect at the points $(0,0)$, and $(1,1)$. The area of the region between the two curves, which has been shaded in Fig. 8, is obtained by finding the area of the region between the x axis and the higher curve from $x = 0$ to $x = 1$ and then subtracting the area of the region between the x axis and the lower curve over this same interval. Since the higher curve over this interval is $y = 2x - x^2$, we need to calculate

$$\int_0^1 (2x - x^2)\ dx - \int_0^1 x^2\ dx$$

But from the properties of integrals, we can write this in the form of a single integral:

$$\int_0^1 (2x - x^2 - x^2)\ dx = \int_0^1 (2x - 2x^2)\ dx$$

The desired area is then given by

$$\left[x^2 - 2\frac{x^3}{3} \right]_0^1 = \frac{1}{3}$$

A special notation is used to distinguish between an antiderivative of a function and the integral of the function between limits a and b. This notation is convenient for constructing a table of antiderivatives, which can then be used to evaluate integrals between limits. This notation consists in writing the integral of a function f without any limits attached. Thus the integral

(6)
$$\int f(x)\ dx$$

represents any antiderivative of $f(x)$. Since any antiderivative of a given function differs from any other by only a constant, we let this symbol denote the family of all such antiderivatives. For example, if we had $f(x) = x^2$, we would write

$$\int x^2\ dx = \frac{x^3}{3} + c$$

The integral in (6) is called the *indefinite integral* of f, as contrasted with the integral with limits a and b which is called the *definite integral* of f. To find the value of a definite integral we therefore first find the indefinite integral and then proceed with our evaluation based on the limits of integration.

From our knowledge of derivative formulas, we can write down the following indefinite integral formulas:

(i) $\displaystyle \int x^n \, dx = \frac{x^{n+1}}{n+1} + c, \qquad n \neq -1$

(ii) $\displaystyle \int e^x \, dx = e^x + c$

(iii) $\displaystyle \int \frac{1}{x} \, dx = \log x + c$

The verification of any indefinite integral formula merely requires us to differentiate the right side and observe whether we obtain the integrand function of the integral on the left side. Incidentally, it is not true for indefinite integrals that the variable of integration is a dummy variable. An indefinite integral is a function of the variable of integration, whereas a definite integral is a number.

4. APPLICATIONS

In this section we study a few problems related to economic theory that involve integration. Some of them involve the area under a curve and others involve anti-differentiation.

(a) In studying the relationship between the price, p, of a commodity and the amount, x, produced, it is important to distinguish between the demand and the supply of the commodity. Thus, if x represents the demand for the product as a function of the price p, it is reasonable to assume that x will increase as p decreases because the demand for a product usually increases with a drop in the price. If we take the inverse of this relationship so that p becomes a function of x, then p will decrease as x increases. This inverse relationship is denoted by $p = f(x)$; hence, we can assume that $f(x)$ is a decreasing function of x. The graph of $p = f(x)$ is called the *price-demand curve*. Such a curve is illustrated in Fig. 9. If, however, x represents the amount that a producer is willing to supply as a function of the price p, it is

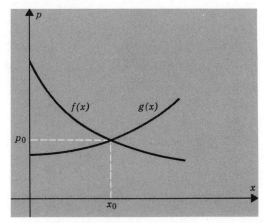

FIGURE 9 Price-demand and price-supply curves.

reasonable to assume that x will increase as p increases because the producer will usually be happy to produce more if the price is increased. If we take the inverse of this relationship so that p becomes a function of x, then p will increase as x increases. This inverse relationship will be denoted by $p = g(x)$; hence, we may assume that $g(x)$ is an increasing function of x. The graph of $p = g(x)$ is called the *price-supply curve*. Such a curve is also shown in Fig. 9. The role of x differs for these two graphs. In one it represents the demand for the product, whereas in the other it represents the amount that is supplied.

When the price of a commodity is such that the demand is equal to the supply, *market equilibrium* is said to occur. The price then is called the *equilibrium price* and the amount produced is called the *equilibrium amount*. These quantities are denoted by p_0 and x_0 in Fig. 9. They can be obtained by solving the pair of equations $p = f(x)$ and $p = g(x)$. These same concepts were defined in Chapter 2 for the case in which both functional relationships are linear; here we define them more generally.

Now suppose that a price-demand curve and a price-supply curve are given and that under a type of monopoly the price and amount produced have been fixed to produce market equilibrium. If a consumer is willing to pay the price given by his price-demand curve, he has gained by having the price fixed at p_0 because for any amount he might desire to buy which is less than the total amount produced, he is paying less than he was willing to pay. For any such x, the saving in price is the difference $f(x) - p_0$. The

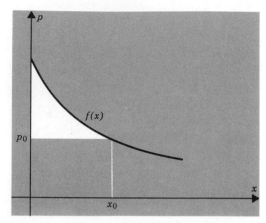

FIGURE 10   The geometry of consumer's surplus.

total consumer gain may be represented geometrically by the area of the region under the graph of $f(x)$ and above the line $p = p_0$ from $x = 0$ to $x = x_0$. This area, which is illustrated in Fig. 10, is called the *consumer's surplus*. Since the evaluation of this area requires integration, we define the consumer's surplus to be the quantity

$$\int_0^{x_0} f(x) \, dx - p_0 x_0$$

As a numerical illustration, suppose that the demand curve is given by $p = f(x) = 20 - \dfrac{x}{5}$ and the price is fixed at $p_0 = 15$. Then x_0 is obtained by solving the equation $15 = 20 - \dfrac{x}{5}$. The solution is $x_0 = 25$; hence

$$\text{Consumer's surplus} = \int_0^{25} \left(20 - \frac{x}{5}\right) dx - 375$$

$$= \int_0^{25} 20 \, dx - \frac{1}{5} \int_0^{25} x \, dx - 375$$

$$= 20(25) - \frac{1}{10} (25)^2 - 375$$

$$= 62.5$$

(b) Now let us concentrate on the producer and his price-supply curve. Assume as before that the price and amount produced have been fixed to realize market equilibrium. If a producer was willing to produce at the price given by $g(x)$, he has gained by having the price set at p_0 because for any amount of production less than the total amount produced he is receiving a higher price than he was willing to accept. For any such x, the gain in price is the difference $p_0 - g(x)$. The total producer gain may be represented geometrically by the area of the region under the line $p = p_0$ and above the graph of $g(x)$ from $x = 0$ to $x = x_0$. This area, which is illustrated in Fig. 11, is called the *producer's surplus*. This area also requires integration; therefore, we define the producer's surplus to be the quantity

$$p_0 x_0 - \int_0^{x_0} g(x)\ dx$$

As a numerical illustration, assume that the price-supply curve is given by $p = g(x) = \dfrac{3}{625} x^2 + 12$. This curve passes through the equilibrium point $(25,15)$. Hence

$$\text{Producer's surplus} = 375 - \int_0^{25} \left(\frac{3}{625} x^2 + 12 \right) dx$$

$$= 375 - \frac{(25)^3}{625} - 12(25)$$

$$= 50$$

(c) The technique of integration is needed when the rate of change of a function is given and the function itself is wanted. For example, suppose we are told that the marginal cost for a given product, which is the rate of

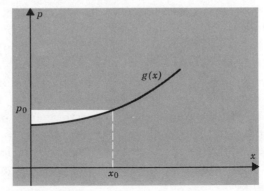

FIGURE 11. The geometry of producer's surplus.

change of the cost with respect to the amount of production, is given by

$$\frac{dC}{dx} = \frac{x}{2000} - \frac{1}{10}$$

Suppose we are also told that it costs \$20 to produce one unit of the product. The problem is to use this information to determine the cost function $C(x)$.

Since integration and differentiation are inverse operations, we can obtain $C(x)$ by integrating $C'(x)$, that is,

$$C(x) = \int C'(x) \, dx$$

Hence, for our problem,

$$C(x) = \int \left(\frac{x}{2000} - \frac{1}{10} \right) dx$$
$$= \frac{x^2}{4000} - \frac{x}{10} + c$$

where c is some constant. But we are given that $C(1) = 20$; hence

$$20 = \frac{1}{4000} - \frac{1}{10} + c$$

Solving for c we obtain $c = 20.10$, correct to two decimals. The desired cost function is therefore given by the formula

$$C(x) = \frac{x^2}{4000} - \frac{x}{10} + 20.10$$

(d) An industrial machine is purchased for A dollars. It produces revenue for the firm, but the amount of revenue varies with its age. Let $f(t)$ denote the revenue in dollars per year that can be expected to be earned by this machine when it is t years old. Although t is normally thought of as assuming only positive integer values, we may consider continuous time and treat $f(t)$ as a continuous function of t. Let $S(t)$ denote the scrap value of the machine if it is sold after being used for t years. Finally, let the value of money be calculated at the nominal interest rate of $100r$ percent per year but let it be compounded continuously. The problem is to determine how long the machine should be operated before being sold for scrap so as to maximize the present value of the revenue and scrap value to be realized minus the original cost.

Assume that the machine is to be operated for T years. From Section 5,

Chapter 10, we know that the present value of c dollars obtained T years from now is given by ce^{-rT}; hence, the present value of the scrap value of the machine is given by

$$S(T)e^{-rT}$$

If we were to divide the interval $[0,T]$ into a large number, n, of subintervals and if we were to calculate the revenue earned during the ith subinterval, we would arrive at the approximate quantity $f(t_i)(t_i - t_{i-1})$ because $f(t)$ represents the revenue earned in a year and, therefore, we must take a fractional part of that revenue corresponding to the length of the subinterval, which is $(t_i - t_{i-1})$. Thus, if each year were divided into 10 subintervals, and t_i were a point in the fifth year, we might have $t_i - t_{i-1} = 5\frac{3}{10} - 5\frac{2}{10} = \frac{1}{10}$. This is only an approximation, but a good one when the intervals are very small, because we are using $f(t_i)$ which gives the revenue rate at the end of each subinterval. The present value of this revenue is therefore approximated by

$$e^{-rt_i}f(t_i)(t_i - t_{i-1})$$

We now sum these present values for the entire interval $[0,T]$ to obtain

$$\sum_{i=1}^{n} e^{-rt_i}f(t_i)(t_i - t_{i-1})$$

We then take the limit of this sum as $n \to \infty$ and apply definition (1) to arrive at the integral

$$\int_0^T e^{-rt}f(t)\ dt$$

This gives the present value of all future revenues calculated on a continuous revenue earning basis and with interest on money being compounded continuously.

The present value of future revenues and scrap value minus the original cost is therefore given by

$$F(T) = \int_0^T e^{-rt}f(t)\ dt + S(T)e^{-rT} - A$$

To find the value of T that maximizes $F(T)$, we calculate $F'(T)$ and set it equal to zero to find the critical points. In doing so, we apply Theorem 1 to differentiate the integral with respect to its upper limit. This gives

$$F'(T) = e^{-rT}f(T) - rS(T)e^{-rT} + S'(T)e^{-rT} = 0$$

Canceling the factor e^{-rT}, we obtain

$$f(T) - rS(T) + S'(T) = 0$$

In any given problem the functions $f(t)$ and $S(t)$ will be available, in which case this equation can be solved for T. Arguments based on practical considerations should suffice to determine which, if any, roots of this equation maximize $F(T)$.

5. INTEGRATION BY SUBSTITUTION

The integration formulas of the preceding sections are very limited in their coverage of functions that can arise in applied problems. The properties of integrals given in Section 2 are helpful because they enable us to integrate a sum of functions by integrating each function separately, and to eliminate constants; however, our coverage is still highly limited. We can extend these formulas considerably by a device called *substitution*. Consider in this connection the problem of finding the indefinite integral of the function $(x + 1)^3$. We could solve this problem by expanding this expression to obtain $x^3 + 3x^2 + 3x + 1$ and then integrate term by term; however, let us try to find a solution by guessing. All we really need to do is to guess a function whose derivative is $(x + 1)^3$. If our function had been x^3, we would have immediately written down $\frac{x^4}{4}$. This suggests that we let $u = x + 1$ and treat our function as though it were u^3 and, as a result, obtain the solution $\frac{u^4}{4} = \frac{(x + 1)^4}{4}$. Now let us apply the chain rule for differentiation to determine whether we have a valid solution. If we let $F(x) = \frac{u^4}{4}$ where $u = x + 1$, then the chain rule gives

$$\frac{dF}{dx} = \frac{dF}{du}\frac{du}{dx} = u^3 = (x + 1)^3$$

Since we have found a function, $F(x) = \frac{(x + 1)^4}{4}$, whose derivative is our integrand, we have solved the anti-differentiation problem; therefore

$$\int (x+1)^3\, dx = \frac{(x+1)^4}{4} + c$$

If our integrand had been $(x^2+1)^3$, this substitution and guessing would not have worked; therefore, we need a systematic way of treating such problems. This is accomplished by using the chain rule for differentiation to obtain a general integration formula.

We need a special version of the chain rule before we can apply it to our general integration problem. Suppose that we introduce a new variable u by means of the relation $x = g(u)$. Assume that we can solve this relationship to obtain the inverse relationship, that is to obtain u as a function of x, say, $u = h(x)$ and that this holds true for some interval on the x axis. If we substitute this expression for u in $x = g(u)$ we obtain an identity in x. Differentiating both sides of this identity with respect to x by means of the chain rule then yields the desired preliminary formula:

(7)
$$1 = \frac{dg}{du}\frac{du}{dx} = g'(u)\frac{du}{dx}$$

The technique for performing an integration by means of the substitution $x = g(u)$, when such a substitution works, is given by means of the following theorem. The formula of the theorem is stated in two equivalent ways because one version is usually easier to apply than the other. In the second version it is understood that x is to be replaced by $g(u)$ wherever it occurs in the integral on the right.

Theorem 3. *If $x = g(u)$ is differentiable and has an inverse, then*

$$\int f(x)\, dx = \int f(g(u))g'(u)\, du$$

or, equivalently,

(8)
$$\int f(x)\, dx = \int \frac{f(x)}{\dfrac{du}{dx}}\, du$$

In both of these formulas the integration on the right produces a function of u, which in turn must be expressed as a function of x by solving for u in terms of x in the relationship $x = g(u)$ and then substituting it in the answer.

Since any indefinite integral formula can be verified by differentiating the right side with respect to x and showing that the result is the integrand function on the left, we can prove the validity of this theorem by differentiating both sides of the preceding formula with respect to x and showing that we obtain the same function of x on both sides. The integral on the right is a function of u; hence, we must apply the chain rule to differentiate it with respect to x. Differentiating both sides of the first formula, we obtain

$$\frac{d}{dx} \int f(x) \, dx = \frac{d}{dx} \int f(g(u))g'(u) \, du$$

$$= \frac{d}{du} \int f(g(u))g'(u) \, du \cdot \frac{du}{dx}$$

But, as was noted in the discussion of (6), the derivative of an indefinite integral with respect to the variable of integration is the integrand function; hence

$$\frac{d}{dx} \int f(x) \, dx = f(x)$$

and

$$\frac{d}{du} \int f(g(u))g'(u) \, du = f(g(u))g'(u)$$

Substituting these results in the preceding equation, we have

$$f(x) = f(g(u))g'(u) \cdot \frac{du}{dx}$$

Application of formula (7) reduces this to

$$f(x) = f(g(u))$$

Since $x = g(u)$, this demonstrates the correctness of the first formula of Theorem 3.

The second version of the formula, namely formula (8), is obtained by using formula (7) to replace $g'(u)$ by $\dfrac{1}{\dfrac{du}{dx}}$ in the first version and postponing the substitution $x = g(u)$ until it is needed for the integration with respect to the variable u. Formula (8) is the formula that is almost always used to perform the integration.

As an illustration of how Theorem 3 is applied, let us return to the problem of integrating $(x + 1)^3$. Here the obvious substitution is $u = x + 1$.

Since $f(x) = (x+1)^3$ and $\dfrac{du}{dx} = 1$, application of formula (8) gives

$$\int (x+1)^3\, dx = \int (x+1)^3\, du$$

Now substituting $x+1 = u$ on the right, we obtain

$$\int (x+1)^3\, dx = \int u^3\, du$$
$$= \frac{u^4}{4} + c$$
$$= \frac{(x+1)^4}{4} + c$$

As another illustration, suppose that we wish to integrate

$$\int (x^2+1)^4 x\, dx$$

The natural substitution to simplify this problem is $u = x^2 + 1$. Then $\dfrac{du}{dx} = 2x$ and formula (8) gives

$$\int (x^2+1)^4 x\, dx = \int \frac{(x^2+1)^4 x}{2x}\, du = \frac{1}{2}\int (x^2+1)^4\, du$$
$$= \frac{1}{2}\int u^4\, du = \frac{u^5}{10} + c = \frac{(x^2+1)^5}{10} + c$$

Suppose that the preceding problem had been changed slightly to the following one:

$$\int (x^2+1)^4 x^2\, dx$$

The same natural substitution of $u = x^2 + 1$ will now fail because our formula gives

$$\int (x^2+1)^4 x^2\, dx = \int \frac{(x^2+1)^4 x^2}{2x}\, du$$
$$= \frac{1}{2}\int (x^2+1)^4 x\, du$$
$$= \frac{1}{2}\int u^4 \sqrt{u-1}\, du$$

This last integral is more complicated than the original integral; therefore, no simplification resulted from the substitution.

As a final illustration, let us integrate

$$\int e^{x^2} x \, dx$$

The problem is difficult because e has the exponent x^2 rather than x; therefore, to simplify it we let $u = x^2$. Since $\dfrac{du}{dx} = 2x$, formula (8) gives

$$\int e^{x^2} x \, dx = \int \frac{e^{x^2} x}{2x} \, du = \frac{1}{2} \int e^{x^2} \, du$$

$$= \frac{1}{2} \int e^u \, du = \frac{e^u}{2} + c = \frac{e^{x^2}}{2} + c$$

Although we have a formula that states how to carry out an integration by substitution, it does not suggest what substitution to make. The natural method to determine whether a substitution will work is to choose u as a function of x that will simplify the integrand considerably. Thus, if the integrand had a function of the form $(a_0 + a_1 x + \cdots + a_k x^k)^n$ in it, the natural substitution to try would be $u = a_0 + a_1 x + \cdots + a_k x^k$. Experience with the technique will often suggest how u should be chosen. As observed in the preceding illustrations, the substitution technique will work only if the function to be integrated is of the right form. If various obvious choices for u do not yield a solution, the chances are that the substitution method will not work and some other technique is needed.

6. INTEGRATION BY PARTS

A second useful technique for calculating complicated integrals is one based on the product rule for differentiation and is called *integration by parts*. Recalling the product rule, if $r = fg$, then

$$r'(x) = f(x)g'(x) + g(x)f'(x)$$

We first rearrange this relationship as follows:

$$f(x)g'(x) = r'(x) - g(x)f'(x)$$

Now suppose that we wish to integrate a function that can be expressed in the form of the left side of this equation. It follows then that

(9) $$\int f(x)g'(x)\ dx = \int r'(x)\ dx - \int g(x)f'(x)\ dx$$

The first integral on the right has the value $r(x) + c = f(x)g(x) + c$ because differentiation gives $r'(x)$. Now it may be easier to evaluate the second integral on the right than it is to evaluate the integral on the left, in which case we have simplified the problem. Summarizing this result, we have

Theorem 4. *If f and g are differentiable functions,*

$$\int f(x)g'(x)\ dx = f(x)g(x) - \int g(x)f'(x)\ dx$$

It is unnecessary to add the constant c in writing down the antiderivative of the first integral on the right in equation (9) because the second indefinite integral on the right will include such an arbitrary constant.

As an illustration of how this formula is used, consider the evaluation of the following integral

$$\int xe^x\ dx$$

Suppose that we choose $f(x) = x$ and $g'(x) = e^x$. Then $f'(x) = 1$ and $g(x) = e^x$, and application of the formula of Theorem 4 gives

$$\int xe^x\ dx = xe^x - \int e^x\ dx$$
$$= xe^x - e^x + c$$

Incidentally, an integration such as this can always be checked for accuracy by differentiating the answer to see whether it yields the integrand on the left. Students should check this one.

Suppose that we had chosen $f(x) = e^x$ and $g'(x) = x$. Then we would have had $f'(x) = e^x$ and $g(x) = \dfrac{x^2}{2}$, in which case the formula would have given

$$\int xe^x\ dx = \frac{x^2}{2}e^x - \int \frac{x^2}{2}e^x\ dx$$

This substitution complicates the problem, instead of simplifying it; therefore, it was a poor choice.

Integration by substitution and by parts are only two of the many techniques available for integrating complicated functions. They are, however, two of the most useful such techniques. Standard mathematical tables usually contain a table of integrals; hence, if an integral arises that cannot be treated by our preceding techniques, a table of this kind should be consulted. However, in doing so, it may be necessary to use the method of substitution to simplify the integrand before it assumes the form of an integrand that can be found in the tables.

COMMENTS

Integral calculus can be a very useful tool for studying problems in which it is necessary to sum a large number of functional values over time. These sums can often be approximated very well by the proper integral. The techniques of both differential and integral calculus can then be applied to solve maximization, or minimization, problems related to such sums. Our illustrations indicated some of the possibilities in this direction.

EXERCISES

Section 1

1. Assume that you do not have a formula for finding the area of a triangle. Calculate the area of the region under the straight line $y = x$ and above the x axis between $x = 0$ and $x = 1$ by dividing that interval into n equal length subintervals, finding expressions for upper and lower approximating sums, and showing that these sums possess a common limit.

2. Estimate the value of the integral

$$\int_1^4 \frac{1}{x}\, dx$$

by choosing intervals of length $\frac{1}{2}$, finding upper and lower approximating sums, and choosing as your estimate a number half way between the values of those two sums.

3. What area does each of the following integrals represent? Illustrate with a sketch.

(a) $\displaystyle\int_0^a x \, dx$ (b) $\displaystyle\int_1^3 (t-1)^2 \, dt$

(c) $\displaystyle\int_0^x t^3 \, dt$ (d) $\displaystyle\int_1^3 \log s \, ds$

4. Estimate the value of the integral

$$\int_0^1 (x - x^2) \, dx$$

by choosing intervals of length $\dfrac{1}{4}$ and then:

(a) Finding upper and lower approximating sums and choosing as your estimate a value half way between the values of those sums.

(b) Using the midpoint of each interval to determine the functional value to use in a single approximating sum. Compare your answer with that obtained in (a). The exact answer here is $\dfrac{1}{6}$.

5. The number of hours, $h(x)$, of labor needed to produce the xth unit of a product, such as an airplane, can often be apppoximated quite well by a function of the form

$$h(x) = ax^b$$

where $a > 0$ and $-1 \leqslant b \leqslant 0$ are constants that depend on the nature of the product and the amount of experience with it or similar products. The graph of this function is called the learning curve for the product. Consider the geometry of the sum $\displaystyle\sum_{x=1}^{n} h(x)$, which gives the total number of man-hours of labor needed to produce n units of the product. This sum can be represented by the sum of the areas of the rectangles shown in the accompanying sketch in which the curve is the graph of $h(x) = ax^b$.

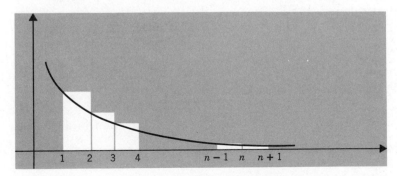

An approximation to this sum is the area under the curve between $x = 1$ and $x = n + 1$.

(a) Write down an integral that may be used to approximate the desired sum.

(b) What limits would you use to find a better approximation to this sum?

6. Assume that the costs of carrying an inventory of size A for the length of time t is proportional to its size and the time interval; hence, these costs are given by

$$I = kAt$$

where k is the proportionality constant. Suppose that the size of the inventory varies over time as given by the function $f(t)$. Suppose also that the inventory is to be carried over the total time interval $[0,T]$. Let that interval be divided into n subintervals by means of the points $t_0 = 0$, $t_1, t_2, \ldots, t_n = T$. If it costs c dollars to carry one unit of inventory over one unit of time, then the cost of carrying $f(t_{i-1})$ units of inventory over the time interval (t_{i-1}, t_i) will be given approximately by

$$ck\, f(t_{i-1})(t_i - t_{i-1})$$

(a) Write down an integral that can be used to determine the total inventory costs.

(b) Why is the preceding expression only an approximation to the inventory costs for the interval (t_{i-1}, t_i)? Could you make it exact by changing $f(t_{i-1})$?

Section 3

1. Find the area of the region between the given curve and the x axis, but only between the indicated points on the x axis. Sketch the curve first.
   (a) $y = 4 - x^2$, $[1,2]$
   (b) $y = \sqrt{x}$, $[0,2]$
   (c) $y = (x + 3)^2$, $[-4,-1]$
   (d) $y = 6x + x^2 - x^3$, $[0,3]$
   (e) $y = x^2 - 1$, $[-1,1]$
   (f) $y = x^2 - 2x$, $[0,3]$

2. Graph the two curves $y = x^2$ and $y = x + 2$ and then find the area of the closed region between them.

3. Graph the two curves $y = \sqrt{x}$ and $y = \frac{x}{2}$ and then find the area of the closed region between them.

4. Find the area of the closed region between the curves $y^2 = x$ and $y = x - 2$.

5. Find the nature of the function $f(x)$ if

(a) $f'(x) = \dfrac{1}{2}$

(b) $f'(x) = \dfrac{x}{2}$

(c) $f'(x) = \sqrt{x}$

(d) $f'(x) = x^3 + \dfrac{1}{x^2}$

6. Determine the indefinite integral for each of the following functions.

(a) $f(x) = x^2 + x^{3/2}$

(b) $f(x) = \dfrac{2}{x^3} - \dfrac{4}{x^5}$

(c) $f(x) = x^{1/3} + 3x^{1/2}$

(d) $f(x) = \dfrac{3}{\sqrt{x}} + \dfrac{4}{\sqrt[3]{x}}$

7. Find the area of the closed region between the curves $y = x^3$ and $y = 4x$.

8. Find the area of the closed region between the curves $y = 3x - 3$ and $y = x^3 + x^2 + x - 3$.

9. Find the area of the region between the curve $y = e^{-x}$ and the x axis from $x = 0$ to $x = b$. What happens to this area as $b \to \infty$?

10. Find the area of the region between the curve $y = \dfrac{1}{\sqrt{x}}$ and the x axis from $x = 1$ to $x = b$. What happens to this area as $b \to \infty$?

11. Let $f(x)$ be a function defined over the interval $[a,b]$. Then the integral

$$\int_a^b x^k f(x) \, dx$$

when it exists is called the kth moment of the function. For each of the following functions calculate the first and second moments.

(a) $f(x) = x^2$, $[a,b] = [0,1]$

(b) $f(x) = \dfrac{1}{x}$, $[a,b] = [1,2]$

12. Write down an integral with the proper limits that will give the area inside the circle whose equation is $x^2 + y^2 = r^2$.

13. The velocity of a car in miles per hour t minutes after passing an observation post is given by the formula $v(t) = 2t + 30$. Find a formula for the distance, s, that it will have traveled in those t minutes.

14. If $F'(x) = x + 1$ and $F(0) = 2$, what is $F(3)$?

15. If $\displaystyle\int_0^x f(t)\, dt = x^3 + x$, what is $f(2)$?

16. Show that if b is any number satisfying $b > 1$, then

$$\int_a^{ab} \frac{1}{x}\, dx = \int_1^b \frac{1}{x}\, dx$$

17. If $F(x) = \displaystyle\int_1^x \sqrt{1 + t^2}\, dt$, what is $F'(2)$?

18. If $F(x) = \displaystyle\int_{-1}^x (1 + t)^{2/3}\, dt$, what is $F'(0)$?

19. If $F(x) = \displaystyle\int_1^x \frac{1}{t+1}\, dt$, what value of x satisfies the relation $F'(x) = 2$?

20. There are certain types of old machines used in industry that are just as reliable as the corresponding new machine if their worn-out parts are replaced by new parts. This can be expressed mathematically by stating that the hazard rate is constant, where the hazard rate is defined as the left side of the equation

$$\frac{f(t)}{1 - \displaystyle\int_0^t f(x)\, dx} = c$$

Here c is the constant hazard rate value and $f(t)$ is a probability function that is related to the expected life of the machine. Show that a solution to this equation is

$$f(t) = ce^{-ct}$$

Section 4

1. Given the following marginal revenue function, find the revenue function such that $R(0) = 0$.

$$R'(x) = -x^3 + 3x^2 - x + 2$$

2. The marginal cost of production for a commodity is given by

$$C'(x) = \frac{3x^2}{10^4} - \frac{18x}{10^2} + 18$$

where x denotes the number of units produced. Determine the nature of the cost function.

3. Given the marginal revenue function

$$R'(x) = 20 - 6x + x^2$$

(a) Find the total revenue function.
(b) Find the price-demand function.

4. Assume that the price-demand curve and the price-supply curve are given by the respective equations

$$p = \frac{24}{x+2} \quad \text{and} \quad p = \frac{x}{2}$$

(a) Find the equilibrium price and quantity.
(b) Calculate the producer's surplus.

5. If the price-demand curve is given by $p = 3 - \sqrt{x}$ and the price is fixed at $p_0 = 2$, find the consumer's surplus.

6. If the price-supply curve is given by $p = (x + 2)^2$ and the price is fixed at $p_0 = 16$, find the producer's surplus.

7. In thermodynamics it is known that the constant k of the rate of a reaction depends on the temperature, t, according to the formula

$$\frac{d(\log k)}{dt} = \frac{c}{t^2}$$

where c is a constant. Use anti-differentiation to find a formula for k as a function of t.

8. Let y denote the amount of invested capital that a company has at time t. Its investment policy is to have its rate of investment proportional to the amount invested at that time; hence

$$\frac{dy}{dt} = ky$$

Let $\frac{dy}{dt} = f(t)$. Show that this policy implies that

$$f(t) = k \left[A + \int_0^t f(x)\, dx \right]$$

where A is the amount invested at time $t = 0$.

9. For certain types of stimulus-response experiments it is assumed that the magnitude of the response, $f(x)$, to a stimulus of magnitude x is such that for a small change in the stimulus the change in the response will

be proportional to the relative change in the stimulus. This implies, for example, that a subject will detect a change in a weight from 4 to 5 pounds as readily as a change from 40 to 50 pounds. Suppose that the stimulus is changed from x to $x + h$. The preceding assumption then states that

$$f(x + h) - f(x) = k\frac{h}{x}$$

where k is the proportionality factor. This is equivalent to

$$\frac{f(x + h) - f(x)}{h} = \frac{k}{x}$$

In the limit as $h \to 0$ this gives

$$f'(x) = \frac{k}{x}$$

Since the derivative is a good approximation to the corresponding quotient when h is small, we can use this as a model for the preceding type of experiments.

(a) Find a formula for the stimulus-response function $f(x)$.

(b) Work the same problem if it is assumed that the relative change in the response is proportional to the relative change in the stimulus. Use the property that

$$\frac{d[\log f(x)]}{dx} = \frac{f'(x)}{f(x)}$$

10. The cost of manufacturing a machine that produces aluminum foil depends on the quality of the machine, which in turn determines how long it will last. Let $C(t)$ denote its cost, where t is the number of years that it will last. Assume that this machine produces a steady stream of income of A dollars a year as long as it operates. If money is worth $100r$ percent interest compounded continuously, the present value of the income to be earned from this machine over a period of t years is given by

$$\int_0^t Ae^{-rx}\, dx$$

Suppose that the buyer of the machine wishes to maximize the ratio of the present value of all future earnings to the cost of the machine. Use Theorem 1 to obtain an equation whose solution will give the value of t that maximizes this ratio.

Section 5

1. Use the method of substitution to evaluate each of the following indefinite integrals.

   (a) $\int \sqrt{2x+3}\ dx$

   (b) $\int \frac{1}{3x+4}\ dx$

   (c) $\int \frac{1}{\sqrt{3x+4}}\ dx$

   (d) $\int x^2 \sqrt{1+x^3}\ dx$

   (e) $\int \frac{x}{\sqrt{x+1}}\ dx$

   (f) $\int \frac{x^4}{x^5+1}\ dx$

   (g) $\int x(x^2+1)^{10}\ dx$

   (h) $\int \frac{x^2}{(x-1)^3}\ dx$

2. Use the method of substitution to evaluate each of the following indefinite integrals.

   (a) $\int e^{4x}\ dx$

   (b) $\int x\,e^{-x^2}\ dx$

   (c) $\int \frac{e^x}{e^x+1}\ dx$

   (d) $\int \frac{1}{x \log x}\ dx$

   (e) $\int \frac{x^2}{\sqrt{2x+1}}$

   (f) $\int \frac{x-3}{(x+3)^{2/3}}\ dx$

   (g) $\int x^3(1+x^2)^{3/2}\ dx$

   (h) $\int \sqrt{x}\sqrt{1+\sqrt{x}}\ dx$

3. Use the method of substitution to evaluate each of the following definite integrals.

   (a) $\int_1^2 \frac{x}{\sqrt{x^2-1}}\ dx$

   (b) $\int_{-1}^0 \frac{1}{2x+3}\ dx$

   (c) $\int_0^1 \frac{1}{\sqrt{1-x}}\ dx$

   (d) $\int_1^2 \frac{\log x}{x}\ dx$

   (e) $\int_0^1 \frac{1}{(x+1)^{2/3}}\ dx$

   (f) $\int_{-2}^{-1} \frac{x^3}{1+x^4}\ dx$

4. The price-demand function for a commodity is given by

   $$p = e^{5-x/10}, \qquad 0 \leqslant x \leqslant 50$$

   Calculate the consumer's surplus if $x_0 = 30$.

5. The price-demand function for a commodity is given by

$$p = 50 \cdot 2^{-x/5}$$

Calculate the consumer's surplus if $x_0 = 10$.

6. For Problem 4, Section 4, calculate the consumer's surplus.

7. Find the area of the region between the curve $y = xe^{-x^2}$ and the x axis from $x = 0$ to $x = 3$.

8. Find the area of the region between the curve $y = \dfrac{1}{x+1}$ and the x axis from $x = 0$ to $x = b$. What happens to this area as $b \to \infty$?

9. Find the area of the region between the curve $y = \dfrac{1}{(x+1)^2}$ and the x axis from $x = 0$ to $x = b$. What happens to this area as $b \to \infty$? Compare this answer with that of Problem 8.

10. Find the area of the region between the following two curves from $x = 0$ to $x = 1$:

$$y = \frac{1}{x+1} \quad \text{and} \quad y = \frac{x^2}{x+1}$$

Section 6

1. Use integration by parts to evaluate each of the following indefinite integrals.

(a) $\displaystyle\int x\, e^{4x}\, dx$

(b) $\displaystyle\int x \log x\, dx$

(c) $\displaystyle\int \log x\, dx$

(d) $\displaystyle\int x^2\, e^x\, dx$

(e) $\displaystyle\int x^3\, e^{x^2}\, dx$

(f) $\displaystyle\int (\log x)^2\, dx$

2. Use the method of substitution and integration by parts to evaluate each of the following definite integrals. Use the answers to Problem 1 to assist you.

(a) $\displaystyle\int_0^1 \log (2x + 3)\, dx$

(b) $\displaystyle\int_{-1}^1 x \log (x + 2)\, dx$

(c) $\displaystyle\int_0^1 x\, e^{3x+4}\, dx$

(d) $\displaystyle\int_{-1}^0 \log^2 (1 - 4x)\, dx$

3. Find the area of the region between the curve $y = \log x$ and the x axis from $x = 1$ to $x = e$.

4. Find the area of the region between the curve $y = x \log x$ and the x axis from $x = 3$ to $x = 5$.

5. Find the consumer's surplus and the producer's surplus if the price-demand and price-supply functions are, respectively,

$$p = \log (16 - x) \quad \text{and} \quad p = \log (2x + 4)$$

Assume that there is market equilibrium.

6. Use integration by parts to show that if k is any positive integer

$$\int x^k e^x \, dx = x^k e^x - k \int x^{k-1} e^x \, dx$$

7. Use the formula of Problem 6 to evaluate the integral

$$\int_0^1 x^3 e^x \, dx$$

8. Use integration by parts to show that if k is any positive integer

$$\int x^k e^{-x^2/2} \, dx = -x^{k-1} e^{-x^2/2} + (k-1) \int x^{k-2} e^{-x^2/2} \, dx$$

9. Given the probability function $f(x) = ae^{-ax}$, $x > 0, a > 0$, calculate its first two moments by first evaluating the following definite integral for $k = 1$ and $k = 2$ and then allowing $b \to \infty$. Assume that $x^r e^{-ax} \to 0$ as $x \to \infty$ for $r \geq 0$.

$$\int_0^b x^k a e^{-ax} \, dx$$

10. Given the probability function $f(x) = ce^{-x^2/2}$, where c is a constant such that $\int_{-\infty}^{\infty} f(x) \, dx = 1$, calculate its first and second moments by first evaluating the following definite integral for $k = 1$ and $k = 2$ and then allowing $b \to \infty$. Use the answer to Problem 8 for $k = 2$.

$$\int_{-b}^b x^k c e^{-x^2/2} \, dx$$

12 | Functions of Several Variables

The problems in calculus that we have encountered thus far have involved functions of a single variable. There were some problems concerned with maxima and minima that involved two variables, but because of restrictions on those variables it was always possible to reduce the problem to a one-variable problem. In this chapter we consider functions of several variables when such a reduction is not possible. Most of our discussion is concerned with functions of two variables because they are easy to visualize geometrically; however, our methods are applicable to functions of more than two variables.

We encountered functions of several variables in Chapter 2 when we studied the pivotal reduction method for solving sets of linear equations; however, that technique involves only linear functions. Here, we are interested mostly in nonlinear functions.

In calculus a function $f(x,y)$ is assumed to have as its domain the entire x,y plane or a subset of it. Hence, a function is a rule that assigns a unique value to each point in a set of points in the x,y plane. As in the case of a function of a single variable, here also it is useful to graph a function when studying its properties. Thus, if we graph $f(x,y)$ in three-dimensional space with an axis perpendicular to the x,y plane as the functional axis, we usually obtain a surface because to each point in the x,y plane where $f(x,y)$ is defined, there is only one value of $f(x,y)$. As an illustration, suppose we are given the function $f(x,y) = x^2 + y^2$. Since all points on the circle $x^2 + y^2 = r^2$ in the x,y plane, where r is any positive number, give the same value of $f(x,y)$, namely r^2, we can find a circle of points lying on the surface by going up the $f(x,y)$ axis a distance of r^2 and then drawing a circle of

374

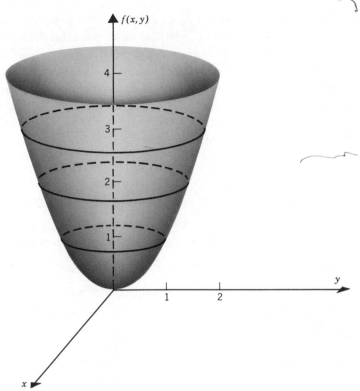

FIGURE 1   A graph of $f(x,y) = x^2 + y^2$.

radius r with its center on that axis. Hence, this surface has only circular cross sections when it is cut by horizontal planes. A sketch of the surface is easily made by drawing a few of those circles for various values of r and then joining them with a smooth surface. This is done in Fig. 1.

1.   PARTIAL DERIVATIVES

Given a function $f(x,y)$ of the two variables x and y, we represent it by the letter u. Hence, we write

$$u = f(x,y)$$

This is similar to what we did previously with a function $f(x)$ of the single variable x. We often denoted it by the letter y and wrote $y = f(x)$.

Suppose that we are interested in knowing how the function u is changing as one of the variables changes but the other variable is held fixed at some value. If one of the variables is held fixed, the function u becomes a function of the other variable only, and we have a one-variable problem. Since the rate of change of a one-variable function is given by the derivative with respect to that variable, we can solve our problem by calculating the derivative of u with respect to that variable. Suppose, for example, that y is held fixed at the value $y = y_0$. Then

$$u = f(x,y_0)$$

is a function of x only and we may calculate $\dfrac{du}{dx}$. When dealing with functions of two variables, however, we use a slightly different symbol to show that the function is a function of two variables but that one of them is being held fixed. Thus, we write $\dfrac{\partial u}{\partial x}$ rather than $\dfrac{du}{dx}$ to represent this derivative. Similarly, we can hold x fixed at some point, say $x = x_0$, and then calculate the rate at which u is changing as y changes by calculating the derivative $\dfrac{\partial u}{\partial y}$. The derivatives $\dfrac{\partial u}{\partial x}$ and $\dfrac{\partial u}{\partial y}$ are called *partial derivatives*. They are calculated in the same manner as ordinary derivatives; only the labels are different. Thus, suppose that $f(x,y) = x^2 + y^2$. Then letting $u = x^2 + y^2$, we obtain

$$\frac{\partial u}{\partial x} = 2x$$

because when y is held fixed at, say, $y = y_0$, the term y^2 becomes the constant y_0^2 and its derivative with respect to x is zero. Similarly

$$\frac{\partial u}{\partial y} = 2y$$

As another illustration, suppose that $f(x,y) = xy + x^2y^2$. Then, writing $u = xy + x^2y^2$ and treating y as a constant, we obtain

$$\frac{\partial u}{\partial x} = y + 2xy^2$$

Similarly, treating x as a constant,

$$\frac{\partial u}{\partial y} = x + 2x^2y$$

Although we are using the new symbols $\dfrac{\partial u}{\partial x}$ and $\dfrac{\partial u}{\partial y}$ to denote partial derivatives, other notations are in common use. We may write $\dfrac{\partial f}{\partial x}$ and $\dfrac{\partial f}{\partial y}$ to represent these derivatives if we do not wish to introduce the letter u to represent the function $f(x,y)$. Another common notation for these same partial derivatives is given by the symbols f_x and f_y. This last notation is particularly useful when we wish to display the point at which the derivatives are to be evaluated.

Partial derivatives have an interesting geometrical interpretation. In this connection, consider the geometrical meaning of $\dfrac{\partial u}{\partial y}$ for the function $u = f(x,y) = x^2 + y^2$ whose graph is sketched in Fig. 1. Suppose that we hold x fixed at the value $x = 1$. The only points lying on the surface that have $x = 1$ are those that lie in a vertical plane that is perpendicular to the x axis and which cuts the x axis at $x = 1$. This is shown in Fig. 2. Hence, the two equations $u = x^2 + y^2$ and $x = 1$ taken together represent the equation of the curve lying on the surface $u = x^2 + y^2$ and in the cutting plane whose equation is $x = 1$. Observe from Fig. 2 that this curve has the appearance of a parabola.

If this curve is projected perpendicularly onto the plane formed by the y axis and the vertical u axis, it would be the curve in that plane with the equation $u = 1 + y^2$, and it would have a graph like that shown in Fig. 3.

Now that we are in two dimensions, the ordinary derivative $\dfrac{du}{dy}$ will represent the slope of the curve $u = 1 + y^2$ at the point where it is calculated. Since this slope is the same as the slope of the corresponding curve in the $x = 1$ plane, the partial derivative $\dfrac{\partial u}{\partial y}$ gives the slope of the curve in the $x = 1$ plane at the point where it is calculated. That is, it gives the slope of the tangent line drawn to the curve of intersection in the $x = 1$ plane at that point. Such a tangent line is shown in Fig. 2 for $y = y_0$.

Similarly, the partial derivative $\dfrac{\partial u}{\partial x}$ gives the slope of the curve that is obtained by intersecting the surface $u = x^2 + y^2$ by the vertical plane that is perpendicular to the y axis at the point $y = y_0$, where y_0 is the fixed value of y.

If we have a function of three variables, say $f(x,y,z)$, we assume that the domain consists of the points in three-dimensional space, or a subset of it.

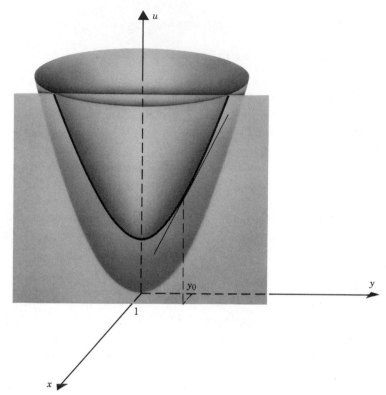

FIGURE 2   The geometry of partial differentiation.

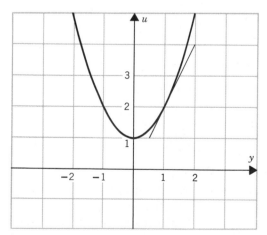

FIGURE 3   A graph of the curve $u = 1 + y^2$.

As previously, we denote our function by u and consider the problem of measuring the rate with which u changes as the variables change. If we fix two of the variables, say, $y = y_0$ and $z = z_0$, then $u = f(x,y_0,z_0)$ becomes a function of the single variable x, and we can calculate its derivative. As in the case for a function of two variables, we call this a partial derivative and denote it by $\dfrac{\partial u}{\partial x}$. If we hold x and z fixed, we can calculate the partial derivative $\dfrac{\partial u}{\partial y}$. Similarly, if we hold x and y fixed, we can calculate $\dfrac{\partial u}{\partial z}$. These partial derivatives are calculated in the same manner as ordinary derivatives, always keeping in mind that the other two variables are treated as constants. For functions of three or more variables it is not possible to give our earlier geometrical interpretation to partial derivatives because that would require us to graph a surface in four or more dimensions.

2. MAXIMA AND MINIMA

Partial derivatives are very helpful in finding maxima and minima of functions of two or more variables. We begin by considering a function of two variables. We treat the problem geometrically and attempt to determine where a function $f(x,y)$ assumes its maximum, or minimum, value in a domain in the x,y plane. For convenience we assume that the domain is a rectangle with its sides parallel to the coordinate axes. Other reasonable domains would also be satisfactory. In this connection, suppose that we have a function whose graph is like that shown in Fig. 4, and we wish to determine where the maximum shown there is located.

Let us pass a vertical plane perpendicular to the y axis through the maximum point. The resulting curve of intersection will necessarily have a maximum point on it at this three-dimensional maximum point. Since we know from earlier work with two-dimensional curves that the derivative must vanish at a local maximizing point if the derivative exists there, it follows that the partial derivative f_x must have a zero value at such a maximum. Similarly, if we cut the surface with a vertical plane perpendicular to the x axis that passes through the maximum point, the resulting curve must have a local maximum at the three-dimensional maximum point. Therefore, f_y must also vanish at that point. Geometrically, since the partial derivatives give the slopes of the tangent lines to the curves lying in

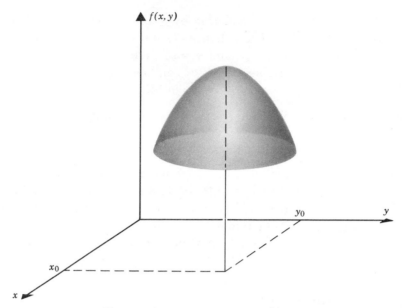

FIGURE 4   A maximizing problem.

the cutting planes, this implies that the tangents to these two curves at the maximum point must be horizontal. The geometry of this discussion is shown in Fig. 5. We can express these facts in the form of a theorem as follows:

Theorem 1. *If a function $f(x,y)$ is defined inside a rectangle in the x,y plane and possess a maximum, or minimum, at a point (x_0,y_0) inside the rectangle and if the partial derivatives f_x and f_y exist at (x_0,y_0), then*

$$f_x(x_0,y_0) = 0 \quad \text{and} \quad f_y(x_0,y_0) = 0$$

Our search for a point where a function $f(x,y)$ takes on its absolute maximum value in a domain is similar to that which we pursued for a function of a single variable. We first find the critical points. Here a critical point is one where the two partial derivatives vanish. Then we calculate the value of $f(x,y)$ at all such critical points. If the maximum value is assumed inside the domain, the largest of those functional values will be the absolute maximum value. It may happen, however, that the function assumes its maximum value on the boundary of the domain. Therefore, before deciding

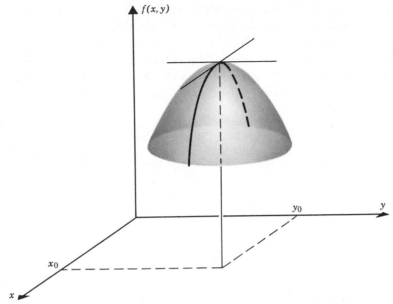

FIGURE 5 Calculus properties at a maximum point.

that we have found the point where the function assumes its absolute maximum value by means of the partial derivative technique, we must give
arguments why the maximum is not attained on the boundary; otherwise it
would be necessary to check the value of $f(x,y)$ on the boundary of the
domain and determine whether the maximum value on the boundary
exceeds the maximum value at critical points. In most applied problems it
is easy to show that the maximum cannot occur on the boundary. We are
assuming in this discussion that the function $f(x,y)$ has a smooth graph so
that we need not be concerned with points where the derivatives do not
exist as we did for functions of one variable.

   Minimizing problems are treated in the same manner as maximizing
problems.

   Just as in the case of a function of one variable, there exists a theorem in
calculus which states that a continuous function in two or more variables
defined over a rectangular domain that includes the boundary will assume
a maximum and a minimum value in the domain. We do not, however, attempt to define continuity for functions of more than one variable. Fortunately, most of the functions arising in actual problems are continuous

functions possessing partial derivatives; therefore, the suggested procedure for finding maxima and minima is a realistic one.

As an illustration, let the function $f(x,y) = x^2 - 12y^2 - 4y^3 + 3y^4$ have as its domain the entire x,y plane and consider the problem of finding where the function assumes its minimum value. We first calculate the partial derivatives and set them equal to zero. Thus

$$f_x = 2x = 0$$
$$f_y = -24y - 12y^2 + 12y^3 = 0$$

From the first equation we obtain $x = 0$ and from the second we have

$$12y(-2 - y + y^2) = 0$$

Or, factoring the quadratic,

$$12y(y - 2)(y + 1) = 0$$

Hence, $y = 0$, $y = 2$, and $y = -1$. Combining these results, it follows that the critical points are $(0,0)$, $(0,2)$, and $(0,-1)$. Next we calculate the value of the function at these three points. This gives

$$f(0,0) = 0, \qquad f(0,2) = -32, \qquad f(0,-1) = -5$$

Since $(0,2)$ produces the smallest value of the function, it is our candidate for the point that minimizes $f(x,y)$. All that remains to be done is to check that the function cannot take on a value smaller than -32 on the boundary of its domain. The domain here is the entire x,y plane; therefore, it suffices to see what happens to the value of $f(x,y)$ as x and y become infinite. Since $f(x,y)$ has x^2 as its only x term, $f(x,y)$ becomes increasingly large as $|x| \to \infty$. Furthermore, since the highest power term in y is $3y^4$, $f(x,y)$ will also become increasingly large as $|y| \to \infty$. Thus, $f(x,y)$ must assume its minimum value inside a sufficiently large rectangle and, hence, by Theorem 1 it must occur at the point $(0,2)$. We do not attempt to look at this problem geometrically by sketching the surface whose equation is $f(x,y) = x^2 - 12y^2 - 4y^3 + 3y^4$ because of the complexity of the surface.

The preceding technique can be extended to finding maxima and minima of functions of any number of variables. We merely calculate all the partial derivatives, set them equal to zero, find the resulting critical points, check the critical points to determine which one yields the largest, or smallest, value, and finally give arguments why the maximum, or minimum, will not occur on the boundary of the domain selected.

3. SECOND DERIVATIVE TEST

Theorem 1 usually suffices for determining where a function assumes its absolute maximum, or minimum, value. It leaves much to be desired, however, if we are also interested in determining local maxima and minima. The conditions of Theorem 1 must be satisfied at a local maximum or minimum inside a domain; however, it does not follow that if those conditions are satisfied at a point (x_0, y_0), then that point will be a local maximizing or minimizing point.

As an illustration of the difficulty that can occur, consider the problem of locating the local maxima and minima of the function $f(x,y) = y^2 - x^2$. Here

$$f_x = -2x \quad \text{and} \quad f_y = 2y$$

The only critical point is $(0,0)$. The value of f at this critical point is $f(0,0) = 0$. This value is certainly not a local maximum because if we choose $y = 2x$ we obtain $f(x,y) = 3x^2$, and this will be positive no matter how small $|x| \neq 0$ becomes. Similarly, $f(0,0)$ cannot be a local minimum because if we choose $y = \dfrac{1}{2}x$ we obtain $f(x,y) = -\dfrac{3}{4}x^2$, and this will be negative no matter how small $|x| \neq 0$ becomes.

A graph of the function $f(x,y) = y^2 - x^2$ is shown in Fig. 6. It should be clear from this graph why the critical point $(0,0)$ produces neither a local

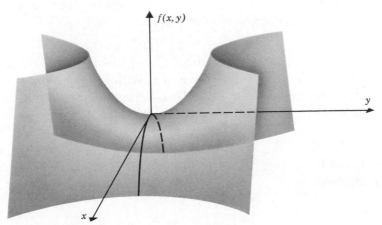

FIGURE 6   A saddle point at $(0,0)$.

maximum nor minimum. A point such as this is called a *saddle point* because the surface on which it is found has the shape of a saddle.

Since determining whether a critical point is a local maximizing or minimizing point and not a saddle point often requires a considerable amount of investigation, it is desirable to have a routine method for making this determination. Fortunately, there exists a test based on second-order partial derivatives of the function f that does just this for most problems. The test is not a universal one because there are situations to which it does not apply; however, these are seldom encountered in realistic problems. The test may be expressed as follows:

Theorem 2. *Let $f(x,y)$ together with its first and second-order partial derivatives be continuous functions over some rectangle in the x,y plane. Suppose that*

$$f_x(x_0,y_0) = 0 \quad \text{and} \quad f_y(x_0,y_0) = 0$$

where (x_0,y_0) is some point inside this rectangle. Then

(i) *if $f_{xx}(x_0,y_0)f_{yy}(x_0,y_0) - f_{xy}^2(x_0,y_0) > 0$, the function f will have a local maximum at (x_0,y_0) provided that $f_{xx}(x_0,y_0) < 0$, and a local minimum at (x_0,y_0) provided that $f_{xx}(x_0,y_0) > 0$.*

(ii) *if $f_{xx}(x_0,y_0)f_{yy}(x_0,y_0) - f_{xy}^2(x_0,y_0) < 0$, the function f will have neither a local maximum nor minimum at (x_0,y_0).*

(iii) *if $f_{xx}(x_0,y_0)f_{yy}(x_0,y_0) - f_{xy}^2(x_0,y_0) = 0$, no conclusion can be drawn concerning the nature of the function at (x_0,y_0).*

The second partial derivative symbol f_{xy} tells us first to calculate f_x and then to calculate the partial derivative of f_x with respect to y.

As an illustration, consider the problem that was used as an illustration in the preceding section. There we obtained the partial derivatives

$$f_x = 2x \quad \text{and} \quad f_y = -24y - 12y^2 + 12y^3$$

We also found the critical points to be $(0,0)$, $(0,2)$, and $(0,-1)$. Calculations give

$$f_{xx} = 2$$
$$f_{yy} = -24 - 24y + 36y^2$$
$$f_{xy} = 0$$

Hence

$$f_{xx}f_{yy} - f_{xy}^2 = 2(-24 - 24y + 36y^2) - 0$$
$$= 24(3y^2 - 2y - 2)$$

This test function has the value -48 at $(0,0)$, 144 at $(0,2)$, and 72 at $(0,-1)$. According to Theorem 2, the function f therefore has neither a local maximum nor a local minimum at $(0,0)$. Since $f_{xx} > 0$ at both of the other critical points, it follows that f has a local minimum at both $(0,2)$ and $(0,-1)$.

In Section 2 we were looking only for absolute maxima and minima and in the process calculated $f(0,0) = 0$, $f(0,2) = -32$, and $f(0,-1) = -5$. From our latest calculations we now know that $(0,0)$ is a saddle point and that $(0,2)$ and $(0,-1)$ are local minimizing points. Thus Theorem 2 has given us additional information concerning the nature of the function at its critical points. Although this theorem is useful for determining where local maxima and minima occur, it does not solve the problem of where absolute maxima and minima occur. It is still necessary in such problems, as was done in Section 2, to evaluate the function at the critical points and, if necessary, to compare those values with the boundary values of the function.

4. APPLICATIONS

In this section we work a few problems that illustrate how Theorem 1, or a generalization of it, can be used to solve maxima and minima problems in the social sciences.

(a) A firm has three separate plants producing the same product. It has an order to produce 100 units. The cost functions for the three plants are given by

$$C_1(x) = 200 + \frac{1}{10}x^2$$

$$C_2(y) = 200 + y + \frac{y^3}{300}$$

$$C_3(z) = 200 + 10z$$

Here x, y, and z denote the number of units that the three plants are to produce. The total cost is then $u = C_1(x) + C_1(y) + C_3(z)$. The problem is to find the values of x, y, and z that minimize u subject to the restriction that $x + y + z = 100$.

We first reduce the problem to one involving only two variables by using the restriction equation to eliminate z from u. Since $z = 100 - x - y$, it follows that

$$u = 600 + \frac{x^2}{10} + y + \frac{y^3}{300} + 10(100 - x - y)$$

$$= 1600 + \frac{x^2}{10} - 10x - 9y + \frac{y^3}{300}$$

Calculating the partial derivatives, we obtain

$$\frac{\partial u}{\partial x} = \frac{x}{5} - 10$$

$$\frac{\partial u}{\partial y} = -9 + \frac{y^2}{100}$$

Setting the partial derivatives equal to zero and solving the resulting equations gives $x = 50$ and $y = 30$. Hence, $z = 100 - 50 - 30 = 20$. These values give $u = 1170$.

The critical point $x = 50$, $y = 30$ produces an absolute minimum because it can be shown that the value of u will exceed 1170 for any point on the boundary of the domain, which is determined by the x and y axes and the line $x + y = 100$. Therefore, by Theorem 1 the minimum must be assumed at an interior point where the partial derivatives vanish.

(b) If a function of a single variable x is believed to be a linear function but the values of the coefficients are not known, it is often possible to estimate those coefficients quite accurately by using experimental data relating the variable and the function. If the function is denoted by y, the problem is to estimate the coefficients a and b in the linear relation $y = ax + b$.

Suppose a number, n, of experiments are conducted in which various values of x are chosen and the corresponding experimental values of y are obtained. These pairs of values may be designated (x_i, y_i), $i = 1, \ldots, n$. Since experimental errors of various kinds will prevent the observed value y_i from being exactly equal to its value $ax_i + b$ if x and y are exactly linearly related, we may use $y_i - ax_i - b$ as a measure of the experimental error in predicting the exact value $ax_i + b$ corresponding to this value of x. We now form the expression

$$G(a,b) = \sum_{i=1}^{n} [y_i - ax_i - b]^2$$

This represents the sum of the squares of all the errors of this type. In this formulation we assume that the values of a and b are known and that this sum of squares can be evaluated. Since they are unknown and the problem is to estimate them, we do so by choosing them to have the values that minimize the preceding sum of squares of errors, $G(a,b)$.

To minimize $G(a,b)$, we calculate the two partial derivatives by using the chain rule of differentiation and the property that the derivative of a sum is equal to the sum of the derivatives. This gives

$$\frac{\partial G}{\partial a} = \sum_{i=1}^{n} 2[y_i - ax_i - b][-x_i]$$

$$\frac{\partial G}{\partial b} = \sum_{i=1}^{n} 2[y_i - ax_i - b][-1]$$

Equating these to zero and performing some algebra, we obtain

$$\sum_{i=1}^{n} [-x_i y_i + ax_i^2 + bx_i] = 0$$

$$\sum_{i=1}^{n} [-y_i + ax_i + b] = 0$$

We can sum these expressions term by term to obtain

$$a \sum x_i^2 + b \sum x_i = \sum x_i y_i$$

(1)

$$a \sum x_i + bn = \sum y_i$$

Since the x_i and y_i are known numbers obtained from experimental data, the sums are known numbers; hence, we have two linear equations in the two unknowns a and b to solve. It is not difficult to verify that G becomes increasingly large if a and b move off to infinity and, therefore, that an absolute minimum is attained for the values of a and b given by the solution of equations (1). The straight line that has these values of a and b is called the best fitting straight line in the sense of least squares, and the values of a and b given by equations (1) are called the *least squares estimates* of the unknown coefficients.

The preceding technique for estimating the coefficients in a linear model can be applied to other models as well. By means of it an experimenter can determine the coefficients of a model that he believes is appropriate to his problem. Once he has the function explicitly determined, he is in a position to apply calculus techniques to study it further.

5. LAGRANGE MULTIPLIERS

Problems of maximizing or minimizing a function of several variables when those variables satisfy some restriction equations can become quite difficult to solve. There is a technique, called the method of Lagrange multipliers, that often simplifies the calculations. It is named after the eighteenth-century French mathematician who introduced it.

For the purpose of both illustrating the technique and comparing it with the earlier standard method of solution, we solve a simple problem involving only two variables. We wish to find where the function $f(x,y) = x^2 + 2xy$ assumes its minimum value if the variables are subject to the restriction equation $x^2y = 27$. This problem was solved in Section 2, Chapter 9, by solving for y in terms of x in the restriction equation and substituting it into $f(x,y)$ to reduce f to a function of a single variable.

In the Lagrange multiplier method a new function is introduced in the following manner. First, we rewrite the restriction equation so that it assumes the form $g(x,y) = 0$. Hence, we write

$$g(x,y) = x^2y - 27 = 0$$

The new function then is the function defined by

$$F(x,y) = f(x,y) - \lambda g(x,y)$$
$$= x^2 + 2xy - \lambda(x^2y - 27)$$

The parameter λ is the Lagrange multiplier here. It always multiplies the restriction function after the restriction equation has been expressed in the form $g(x,y) = 0$. We now treat $F(x,y)$ as though it were a function of x and y without any restriction on those variables. We therefore proceed to find where the function $F(x,y)$ assumes its minimum value in the manner of the preceding sections. That is, we calculate F_x and F_y, set them equal to zero, find the critical points, and check to see which of those points, if any, yields the minimum. In this problem, we obtain

$$F_x = 2x + 2y - 2\lambda xy = 0$$
$$F_y = 2x - \lambda x^2 = 0$$

Since we have the restriction equation

$$x^2y - 27 = 0$$

in addition to the preceding two equations, we have three equations in the three unknowns x, y, and λ. We proceed to solve them. Since we are not interested in the value of λ we try to eliminate it first. Solving for λ in the second of the two partial derivative equations, we obtain

$$\lambda = \frac{2}{x}$$

This is substituted into the first of those two equations to give

$$2x + 2y - 4y = 0$$

which reduces to

$$x = y$$

The preceding three equations in three unknowns are now reduced to the following two equations in two unknowns:

$$x = y$$
$$x^2y - 27 = 0$$

The solution of this pair of equations is $x = 3$ and $y = 3$. As previously, it is easily shown that this pair of numbers minimizes $f(x,y) = x^2 + 2xy$, subject to the restriction $x^2y = 27$.

The reasoning behind the Lagrange multiplier technique is the following one. Suppose that we have found a point (x_0,y_0) that minimizes $f(x,y)$ subject to the restriction $g(x,y) = 0$. Then this point must also minimize the function $F(x,y)$ regardless of the value of λ because as long as the restriction $g(x,y) = 0$ is satisfied the term $\lambda g(x,y)$ has the value zero and, hence, minimizing $f(x,y)$ minimizes $F(x,y)$. Conversely, any point (x_0,y_0) that minimizes $F(x,y)$ and which also satisfies the restriction equation must minimize $f(x,y)$ subject to this restriction because once more the term $\lambda g(x,y)$ has the value zero for any such point and therefore if $F(x_0,y_0)$ is a minimum, $f(x_0,y_0)$ must be a minimum. Although this shows that minimizing F is equivalent to minimizing f, it does not prove that we may treat F as a function of x and y without any restrictions on those variables. A proof of this fact is rather involved and, therefore, will not be given here.

The same type of reasoning applies to problems involving functions of several variables and with several restrictions. We demonstrate the technique on the following problem.

Given the function $f(x,y,z) = xyz$, subject to the restriction $xy + 2xz + 2yz = 24$, find where it is maximized. We first write the restriction in the form of $g(x,y,z) = xy + 2xz + 2yz - 24 = 0$. Then we write

$$F(x,y,z) = f(x,y,z) - \lambda g(x,y,z)$$
$$= xyz - \lambda[xy + 2xz + 2yz - 24]$$

Next, we calculate the three partial derivatives and set them equal to zero. Thus

$$F_x = yz - \lambda[y + 2z] = 0$$
$$F_y = xz - \lambda[x + 2z] = 0$$
$$F_z = xy - \lambda[2x + 2y] = 0$$

To solve these equations we solve for λ in the last equation and substitute it into the first two equations to obtain

$$yz = [y + 2z]\,\frac{xy}{2x + 2y}$$

and

$$xz = [x + 2z]\,\frac{xy}{2x + 2y}$$

Multiplying through by $2x + 2y$ and collecting terms, we obtain

$$2y^2z = xy^2$$

and

$$2x^2z = x^2y$$

These equations can be written in factored form as

$$y^2(x - 2z) = 0$$

and

$$x^2(y - 2z) = 0$$

From the first equation we obtain $y = 0$ and $x = 2z$. From the second equation we obtain $x = 0$ and $y = 2z$. The possible solutions then consist of the pairs

$$(y = 0, x = 0), \quad (y = 0, y = 2z), \quad (x = 2z, x = 0), \quad (x = 2z, y = 2z)$$

These are equivalent to

$$(x = 0, y = 0), \quad (y = 0, z = 0), \quad (x = 0, z = 0), \quad (x = 2z, y = 2z)$$

We still have the restriction equation to be satisfied, which is

$$xy + 2xz + 2yz = 24$$

The first of the four pairs of partial solutions does not satisfy the restriction equation; hence, it may be discarded. The second and third pairs also fail to satisfy the restriction equation and they may also be discarded. Hence, we are left with the fourth pair only. Substituting those values into the restriction equation, we obtain

$$4z^2 + 4z^2 + 4z^2 = 24$$

This gives $z^2 = 2$ and $z = \pm\sqrt{2}$. There are therefore two legitimate critical points:

$$(2\sqrt{2},\ 2\sqrt{2},\ \sqrt{2}) \qquad \text{and} \qquad (-2\sqrt{2}, -2\sqrt{2}, -\sqrt{2})$$

The values of $f(x,y,z) = xyz$ at these two points are $8\sqrt{2}$ and $-8\sqrt{2}$, respectively. Since we are interested in maximizing f we are left with the point $x = 2\sqrt{2}$, $y = 2\sqrt{2}$, and $z = \sqrt{2}$.

This problem grew out of maximizing the volume of a box subject to the restriction that it could not use more than 24 square feet of material. We know from practical considerations that a finite maximum exists; therefore, this solution must produce an absolute maximum.

COMMENTS

The problem of determining where a function of several variables assumes its maximum or minimum value is obviously an important one in all branches of science. The problems that we solved in this chapter are the basic kind of such problems. From these it is possible to branch out in many directions. There are many theoretical problems, for example, in economic theory, in which there are several functions of several variables with functional relations and restrictions imposed on them, and the problem is to maximize one of those functions subject to the restrictions. This is similar to the problem of linear programming, only here the functions are not linear.

A student who has studied this book should be aware that a large share of the applications are concerned with how to maximize or minimize some function. This obviously is a motivating force in calculus, but it is also the rationale for linear programming. Mathematical methods are certainly a powerful tool for solving this type of problem and would merit study for this reason only.

EXERCISES

Section 1

1. For each of the following functions calculate the two partial derivatives.

   (a) $f(x,y) = 3xy + 6x - y^2$
   (b) $f(x,y) = \sqrt{x^2 + y^2}$

   (c) $f(x,y) = xy^2 - 5y + 4$
   (d) $f(x,y) = \dfrac{x^2 - y}{x + y}$

   (e) $f(x,y) = xy\, e^{xy}$
   (f) $f(x,y) = x \log(x + y)$

2. For each of the following functions calculate the three partial derivatives.

   (a) $f(x,y,z) = xy + xy^2 + yz$
   (b) $f(x,y,z) = x^2 + y^2 - xz^2$

   (c) $f(x,y,z) = \sqrt{x^2 + y^2 + z^2}$
   (d) $f(x,y,z) = \dfrac{x^2 + y^2}{z^2}$

   (e) $f(x,y,z) = xyz + \log xyz$
   (f) $f(x,y,z) = e^{xyz} + \dfrac{xy}{z}$

3. For each of the following functions calculate all partial derivatives.

   (a) $f(x,y) = \dfrac{x - y}{x + y}$
   (b) $f(x,y) = \dfrac{xy}{x^2 + y^2}$

   (c) $f(x,y) = \log(x^2 + y^2)$
   (d) $f(x,y) = xe^y + ye^x$

   (e) $f(x,y) = e^{-xy} + \log xy$
   (f) $f(x,y) = \dfrac{1}{xy} - \dfrac{4}{x^2 y} - \dfrac{4}{xy^2}$

   (g) $f(x,y) = x \log y + y \log x$
   (h) $f(x,y,z) = \log(x^2 + y^2 + z^2)$

   (i) $f(x,y,z) = \dfrac{1}{\sqrt{x^2 + y^2 + z^2}}$
   (j) $f(x,y,z) = ye^x + ze^y + e^z$

Section 2

1. Given $f(x,y) = xy$:

   (a) Find any critical points.

   (b) Determine whether f possesses a maximum or minimum at any critical point. Explain.

2. Let the domain of $f(x,y) = xy$ be the triangular region in the x, y plane determined by the points $(0,0)$, $(1,0)$, and $(0,1)$. Find the maximum value of f over this domain.

3. Given $f(x,y) = x^2 - y^2 + 4xy + 3$, explain why f does not possess a maximum nor a minimum if the entire x, y plane is its domain.

4. Find the maximum value of the function

$$f(x,y) = \frac{1}{xy} - \frac{4}{x^2 y} - \frac{4}{xy^2}$$

Section 3

In the following problems use the test function of Theorem 2 to determine the nature of the function at its critical points.

1. $f(x,y) = xy + \frac{1}{x} - \frac{64}{y}$

2. $f(x,y) = x^3 + y^3 - 18xy$

3. $f(x,y) = 4xy^2 - 2x^2 y - x$

4. $f(x,y) = 2x^4 + y^2 - x^2 - 2y + 3$

5. $f(x,y) = x^4 - y^4$

6. $f(x,y) = y^2 + 3x^4 - 4x^3 - 12x^2 + 24$

7. $f(x,y) = xy(12 - 3x - 4y)$

Section 4

1. A rectangular box without a top is to be made that contains 32 cubic feet. What dimensions will require the least area of material?

2. Show that among all rectangular boxes having a top and having a volume of 1 cubic foot, the cubical box has the least surface area.

3. Rectangular boxes mailed by parcel post cannot have the sum of the length and the girth (shortest distance around the four sides) exceed 100 inches. Find the dimensions of a rectangular box which can be mailed that has maximum volume.

4. A rectangular box is to have a volume of 20 cubic feet. The material for the top and bottom costs 10 cents per square foot. The material for the front and back costs 8 cents a square foot and that for the two sides costs 6 cents a square foot. Find the dimensions of the box that will cost the least.

5. A company has three plants producing the same kind of product. Let x, y, and z, respectively, denote the number of units that each plant is scheduled to produce to meet an order for 2000 units; hence,

$x + y + z = 2000$. The cost functions for these three plants are given by the formulas

$$f(x) = 200 + \frac{x^2}{100}$$

$$g(y) = 200 + y + \frac{y^3}{4000}$$

$$h(z) = 200 + 10z$$

The total cost of filling this order is therefore given by

$$C = f(x) + g(y) + h(z)$$

Find the values of x, y, and z that will minimize C.

6. A manufacturing firm has a total yearly production capacity of, at most, 1000 units of its product. Its product is sold exclusively in two foreign markets. Let x and y denote the number of units that the firm hopes to sell in those two markets. Experience has shown that the price-demand functions for the two markets are given by

$$p_1 = 4000 - 3x, \qquad x \leqslant 1000$$

and

$$p_2 = 3000 - 2y, \qquad y \leqslant 1000$$

The cost to the company of manufacturing z units of its product is given by the function

$$C(z) = 100{,}000 + 2350z - \frac{z^2}{2}, \qquad z \leqslant 1000$$

The profit function is therefore given by

$$P = xp_1 + yp_2 - C(x + y)$$

Find the values of x and y that will maximize P.

7. A manufacturing company sells its product exclusively at two foreign markets. Let x and y denote the number of units that it hopes to sell at those two markets. Let $R_1(x)$ and $R_2(y)$ denote the revenues that will be obtained from those sales and let $C(x + y)$ denote the total cost of production. The profit function is therefore given by

$$P = R_1(x) + R_2(y) - C(z)$$

where $z = x + y$. Calculate the two partial derivatives of P by means of the chain rule and show that if P is to be maximized it is necessary that

$$\frac{dR_1}{dx} = \frac{dR_2}{dy}$$

Hence, a necessary condition for maximum profit is that the marginal revenues shall be equal.

8. Suppose that a company produces two similar commodities and has a monopoly of the market for this commodity. Suppose further that the demands for those two commodities, denoted by x and y, respectively, as functions of their prices, p and q, are given by

$$x = f(p,q) \qquad \text{and} \qquad y = g(p,q)$$

Let $C(x,y)$ denote the cost to the company to produce those quantities of the two commodities. Then the profit function is given by

$$P = px + qy - C(x,y)$$

If the production costs are \$300 per unit for the first commodity and \$200 per unit for the second commodity and if the price-demand equations are

$$x = 600 - 2p + q \qquad \text{and} \qquad y = 800 + p - 3q$$

find the prices that will maximize the profit.

Section 5

The following problems are to be solved by employing Lagrange multiplier techniques. In each case, assume that the maximum, or minimum, value of the function is taken on at a critical point.

1. Use Lagrange multiplier techniques to find the values of x and y that minimize
$$f(x,y) = x^2 + xy + 2y^2 - 2x \qquad \text{if} \qquad x - 2y + 4 = 0$$

2. Use Lagrange multiplier techniques to find the values of x and y that maximize
$$f(x,y) = 16 - x^2 - y^2 \qquad \text{if} \qquad x^2 - y^2 - 4y = 8$$

3. Use Lagrange multiplier techniques to find the values of x, y, and z that

minimize

$$f(x,y,z) = x^2 + y^2 + z^2 \quad \text{if} \quad 3x - 2y + z = 4$$

This problem is equivalent to the geometrical problem of finding the shortest distance from the origin to the plane whose equation is the linear equation.

4. Use Lagrange multiplier techniques to find the values of x, y, and z that maximize

$$f(x,y,z) = x^2 + y^2 + z^2$$

if $x - y + z = 0$ and $25x^2 + 4y^2 + 20z^2 = 100$. Eliminate z first using the linear relationship and then use one Lagrange multiplier. In solving this problem a set of values that minimizes f will also be obtained.

5. Use three Lagrange multipliers to find the values of x, y, z, u, and v that minimize

$$f(x,y,z,u,v) = x^2 + y^2 + z^2 + u^2 + v^2$$

if $x + y + z = 1$, $2y + u = 2$, and $x + v = 6$

6. Given the two functions

$$u = f(x) \quad \text{and} \quad v = g(y)$$

and the restriction that $x + y = c$, where c is a constant, use Lagrange multiplier techniques to determine conditions on the functions f and g that must be satisfied if $z = u + v$ is to be maximized.

7. Use the Lagrange multiplier technique with two Lagrange multipliers to find the shortest distance between the circle $x^2 + y^2 = 1$ and the line $x + y = 2$. Denote an arbitrary point on the circle by (α, β) and an arbitrary point on the line by (r,s). Then minimize the square of the distance between those two points: $f(\alpha,\beta,r,s) = (\alpha - r)^2 + (\beta - s)^2$.

8. Suppose that an individual has been allotted A dollars a month to spend on his entertainment. Assume that only three types of entertainment interest him. Let x, y, and z denote the number of times he chooses each of those respective types during a month and let p_1, p_2, and p_3 denote the costs in dollars per outing for each. His total entertainment budget can therefore be expressed by the formula

$$A = p_1x + p_2y + p_3z$$

Let $u = u(x,y,z)$ represent a utility function that expresses this individual's enjoyment as a function of any specified amounts of the three types
of entertainment.

(a) If $u = xyz$, $A = 100$, $p_1 = 5$, $p_2 = 3$, and $p_3 = 6$, find the values of x, y,
     and z that maximize u.

(b) Derive a formula, or set of equations that need to be solved, which
     can be used to determine the values of x, y, and z that maximize u
     for the general problem in which $u = u(x,y,z)$ and the p's are not
     specified numerically.

Appendix

1. Properties of logarithms.

   If $x > 0$ and $y > 0$ are any two positive numbers, then

   (i) $\log xy = \log x + \log y$

   (ii) $\log \dfrac{y}{x} = \log y - \log x$

   (iii) $\log y^x = x \log y$

   If $y = a^x$, $a > 0$, then $\log_a y = x$.

2. A brief table of logarithms.

N	$\log_{10} N$	$\log_e N$	N	$\log_{10} N$	$\log_e N$
.2	−.699	−1.609	10.0	1.000	2.303
.4	−.398	− .916	20.0	1.301	2.996
.6	−.222	− .511	30.0	1.477	3.401
.8	−.097	− .223	40.0	1.602	3.689
1.0	.000	.000	50.0	1.699	3.912
2.0	.301	.693	60.0	1.778	4.094
3.0	.477	1.099	70.0	1.845	4.248
4.0	.602	1.386	80.0	1.903	4.382
5.0	.699	1.609	90.0	1.954	4.500
6.0	.778	1.792	100.0	2.000	4.605
7.0	.845	1.946			
8.0	.903	2.079			
9.0	.954	2.197			

3. A brief table of the exponential function.

x	e^x	e^{-x}	x	e^x	e^{-x}
.00	1.000	1.000	1.0	2.718	.368
.01	1.010	.990	1.5	4.482	.223
.02	1.020	.980	2.0	7.389	.135
.03	1.030	.970	2.5	12.182	.082
.04	1.041	.961	3.0	20.086	.050
.05	1.051	.951	3.5	33.115	.030
.06	1.062	.942	4.0	54.598	.018
.07	1.072	.932	4.5	90.017	.011
.08	1.083	.923	5.0	148.41	.007
.09	1.094	.914	5.5	244.69	.004
.10	1.105	.905	6.0	403.43	.002
.20	1.221	.819	6.5	665.14	.002
.30	1.350	.741	7.0	1096.6	.001
.40	1.492	.670	7.5	1808.0	.001
.50	1.649	.607	8.0	2981.0	.000
.60	1.822	.549	8.5	4914.8	.000
.70	2.014	.497	9.0	8103.1	.000
.80	2.226	.449	9.5	13360.0	.000
.90	2.460	.407	10.0	22026.0	.000

Answers to Odd-Numbered Exercises

Numerical answers often depend on the order of operations and the extent of rounding off; hence answers may differ slightly from those given here.

CHAPTER 1

Section 3

1. (a)   (b)

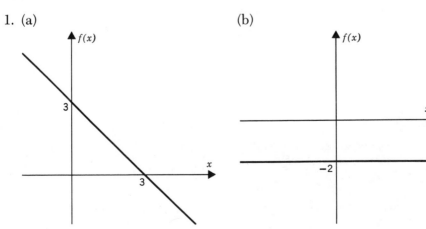

(c)

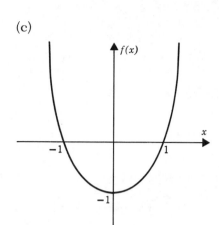

(d)

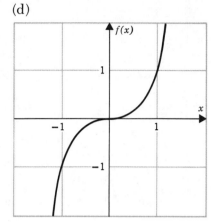

(e)

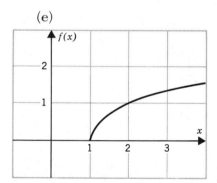

(f)

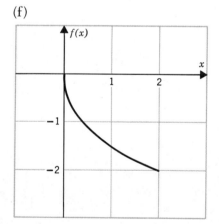

3. (a) Defined for all $x \geq \dfrac{1}{2}$ except $x = 1$.

(b) Defined for $2 < x \leq 4$.

5. (a) $y = \dfrac{1}{5} x^{3/2}$ (b)

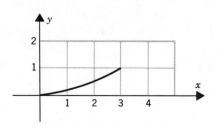

7.

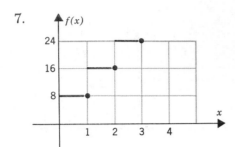

9. $W(x) = \begin{cases} rx, \ 0 \leqslant x \leqslant 40 \\ 40r + \dfrac{3}{2} r(x - 40), \ x > 40 \end{cases}$

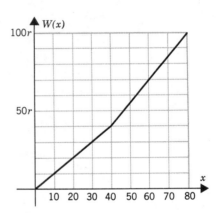

Section 4

1. (a) $\dfrac{5}{2}$ (b) $-\dfrac{1}{7}$

   (c) -20 (d) $0.$

3. (a) $-\dfrac{a}{b}$ (b) $-\dfrac{c}{b}$ (c) $x = -\dfrac{c}{a}.$

5. It cuts the x axis at $x = a$ and the y axis at $y = b.$

7. (a) $y = 3x - 5$ (b) $y = -2x - 1$

   (c) $y = \dfrac{1}{2} x - 2$ (d) $y = -\dfrac{1}{3} x + \dfrac{5}{9}.$

9. (a) $x = 3$ (b) $y = 2$ (c) $y = -3x + 2$.

11. $y = 3x + 3$.

13. $y = -\dfrac{4}{3}x$.

15. (a) $C_1(x) = 100 + 5x$
$C_2(x) = 200 + 4x$

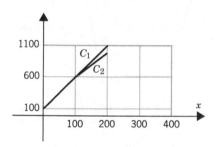

(b) Use C_1 for $x \leqslant 100$, then shift to C_2 for $x > 100$.

Section 5

1. (a)

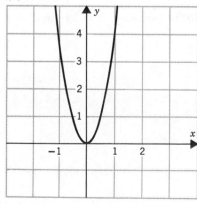

(b)

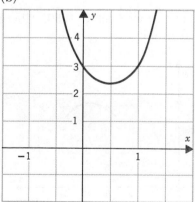

(c)

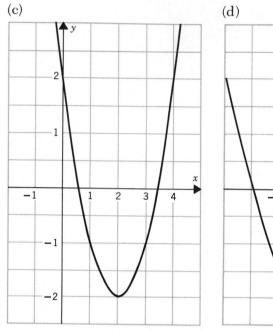

(d)

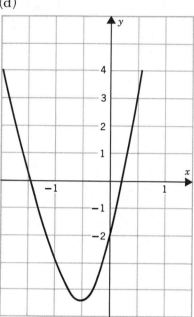

3. $y = \dfrac{1}{4} x^2 - x$.

5. $y = -x^2 + 4x - 2$.

7. $P(x) = -5x^2 + 390x - 250$. Charge \$205, because $x = 39$ produces a maximum.

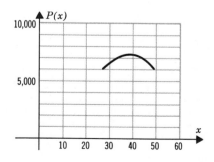

Section 6

1. (a)

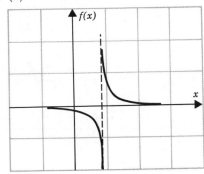

(b)

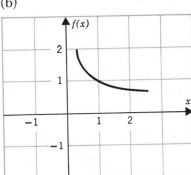

(c)

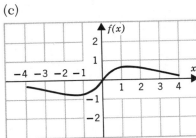

(d)

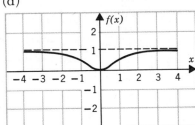

3. (a)

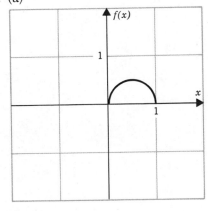

(b)

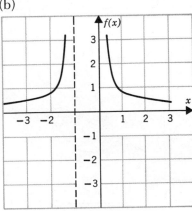

(c)

(d)

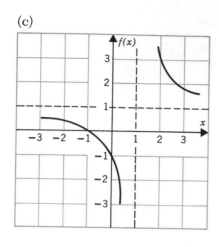 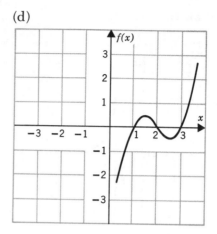

5. $E(x) = \begin{cases} 300, & 0 \leqslant x \leqslant 20 \\ 300 + 10(x-20), & 200 < x \leqslant 100 \\ 1100 + 5(x-100), & x > 100. \end{cases}$

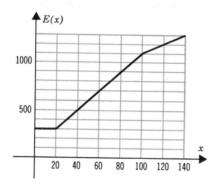

7. (a) $-\dfrac{b}{2a} = 500$ (b) $C(x) = \dfrac{x^3}{100} - 10x^2 + 3000x.$

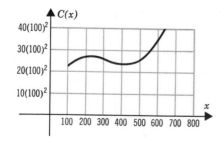

Section 7

1. (a) yes, for $x > 4$ (b) yes
   (c) no   (d) yes, except for $y = 0$.

3. $x = \dfrac{2}{y-1}$

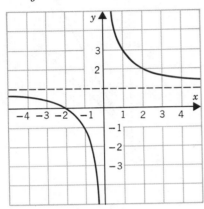
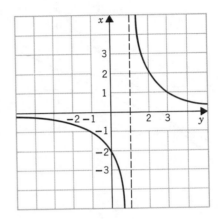

5. $x = \dfrac{1}{\sqrt[3]{y}}, \; y \neq 0$

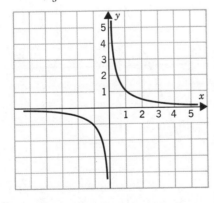
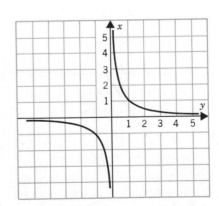

CHAPTER 2

Section 1

1. (a) $(5,-7)$ (b) $\left(\dfrac{5}{3}, \dfrac{2}{15}\right)$ (c) $(2,-3)$

(d) Inconsistent   (e) $\left(k, \frac{4}{3}k - \frac{4}{3}\right)$ (f) $\left(\frac{7}{3}, \frac{2}{3}\right)$.

3. (a)   (b)

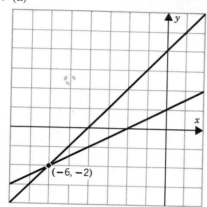

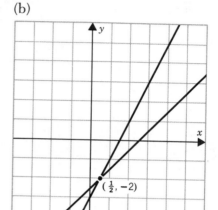

(c)   (d)

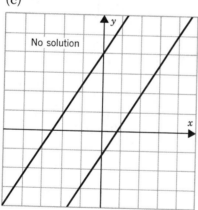

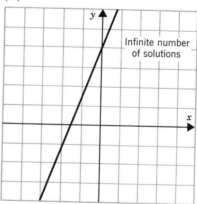

5. (a) $p = 4$ (b) $p = 6$ (c) $p = 1$
      $x = 7$ $x = 5$ $x = 8$
   (d) $p = 2$ (e) $p = 3$ (f) $p = 2$
      $x = 3$ $x = 6$ $x = 11$.

7. $a = 8$, $b = 1$, or any other pair for which $ab = 8$, except for $a = 4$ and $b = 2$.

9. $x = 16\frac{2}{3}$, $y = 33\frac{1}{3}$

Section 3

1. (a) $x = 1$, $y = 2$, $z = 3$ (b) $x = 3$, $y = 2$, $z = 2$

   (c) $x = \dfrac{6}{4} - \dfrac{k}{4}$, $y = -\dfrac{10}{4} + \dfrac{7}{4}k$, $z = k$

   (d) Inconsistent   (e) $x = 2$, $y = -1$, $z = 3$

   (f) $x = -2$, $y = 3$, $z = 4$.

3. $x = 10$, $y = 2$, $z = 2$.

5. $x = 8$, $y = 7$, $z = 5$.

Section 4

1. (a) $x = -\dfrac{17}{13}$, $y = \dfrac{2}{13}$ (b) Inconsistent

   (c) $x = \dfrac{6}{8} - \dfrac{6}{8}k$, $y = \dfrac{21}{8} + \dfrac{11}{8}k$, $z = \dfrac{43}{8} + \dfrac{13}{8}k$, $t = k$

   (d) $x = 5 - 7k$, $y = 19 - 44k$, $z = -6 + 14k$, $t = k$.

3. Pivotal reduction gives $y = 625$ and $y = 642\,\dfrac{6}{7}$; hence equations are inconsistent. Compromise by choosing, say, $y = 630$. Then $x = 240$ and $y = 630$ will satisfy the first equation but will give 198 instead of 200 for the second equation and 309 instead of 300 for the second equation. Slightly smaller values for x and y will not exceed available times.

5. No solution is possible because nonnegative values will exist only if $-5 + \dfrac{1}{4}t \geqslant 0$ and $\dfrac{75}{4} - \dfrac{29}{16}t \geqslant 0$ and these inequalities are contradictory.

CHAPTER 3

Section 1

1. (a) [4 0 −2]   (b) −1   (c) −4
   (d) 16   (e) 34.

3. If a product is zero, neither factor needs to be zero.

5. (a) $A = \begin{bmatrix} 12 & 5 & 20 & 6 & 3 \end{bmatrix}$
$B = \begin{bmatrix} 20 & 3 & 10 & 4 & 0 \end{bmatrix}$ $P = \begin{bmatrix} 10 \\ 20 \\ 8 \\ 6 \\ 5 \end{bmatrix}$
$C = \begin{bmatrix} 10 & 10 & 0 & 12 & 0 \end{bmatrix}$

(b) $AP = 4.31$, $BP = 3.64$, $CP = 3.72$
(c) $A + B + C = \begin{bmatrix} 42 & 18 & 30 & 22 & 3 \end{bmatrix}$
$(A + B + C)P = 11.67$.

7. (a) $\begin{bmatrix} 1 & 1 & -1 \\ 4 & 2 & 5 \end{bmatrix}$ (b) $\begin{bmatrix} 8 & -7 & 2 \\ 12 & -4 & 0 \end{bmatrix}$

(c) Impossible   (d) $\begin{bmatrix} -1 & 6 & 5 \\ 1 & 4 & 10 \end{bmatrix}$

(e) $\begin{bmatrix} 0 & -1 \\ 23 & 17 \end{bmatrix}$.

9. (a) $\begin{bmatrix} 3 & 1 \\ 8 & -2 \end{bmatrix}$ (b) $\begin{bmatrix} 6 & -2 \\ 8 & -5 \end{bmatrix}$

(c) $\begin{bmatrix} 2 & 10 \\ 10 & 8 \end{bmatrix}$ (d) $\begin{bmatrix} 2 & 10 \\ 10 & 8 \end{bmatrix}$

(e) $\begin{bmatrix} 8 & -5 \\ 5 & 3 \end{bmatrix}$ (f) $\begin{bmatrix} 21 & -6 \\ 24 & -3 \end{bmatrix}$.

11. $(AB)C = A(BC)$; hence, parentheses are inserted as desired.

13. Calculate $A(BC)$ and $(AB)C$ and verify that they are the same.

15. This can be shown to be true for any $m \times n$ matrix.

17. Calculate the two sides and verify that they are the same.

19. $B \neq C$; therefore, matrix factors cannot be canceled in an equality as is done in ordinary algebra.

21. $\begin{bmatrix} 29 \\ 44 \\ 32 \\ 17 \\ 19 \end{bmatrix}$.

Section 2

1. (a) $A = \begin{bmatrix} 4 & -1 \\ 1 & 2 \end{bmatrix}$, $X = \begin{bmatrix} x_1 \\ x_2 \end{bmatrix}$, $B = \begin{bmatrix} 3 \\ 5 \end{bmatrix}$

(b) $A = \begin{bmatrix} 1 & 1 & -1 \\ 2 & -1 & 1 \\ 1 & -1 & 0 \end{bmatrix}$, $X = \begin{bmatrix} x_1 \\ x_2 \\ x_3 \end{bmatrix}$, $B = \begin{bmatrix} 2 \\ 3 \\ -1 \end{bmatrix}$

(c) $A = \begin{bmatrix} 3 & 4 \\ 1 & -1 \\ -1 & 2 \\ 2 & 1 \end{bmatrix}$, $X = \begin{bmatrix} x_1 \\ x_2 \end{bmatrix}$, $B = \begin{bmatrix} 8 \\ 3 \\ 1 \\ 5 \end{bmatrix}$

(d) $A = \begin{bmatrix} a & b & c \\ b & c & d \\ c & d & e \end{bmatrix}$, $X = \begin{bmatrix} x_1 \\ x_2 \\ x_3 \end{bmatrix}$, $B = \begin{bmatrix} r \\ s \\ t \end{bmatrix}$.

3. $x = \dfrac{1}{3}, \quad y = \dfrac{4}{3}$.

5. $\begin{aligned} 3x + y + 2z &= 2 \\ -x \qquad + z &= 3 \\ 2x + y - 3z &= 4. \end{aligned}$

7. $\begin{aligned} (a_{11} - 1)x_1 + \quad a_{12}x_2 \quad + \cdots + \quad a_{1n}x_n &= 0 \\ a_{21}x_1 \quad + (a_{22} - 1)x_2 + \cdots + \quad a_{2n}x_n &= 0 \\ & \\ & \\ & \\ a_{n1}x_1 \quad + \quad a_{n2}x_2 \quad + \cdots + (a_{nn} - 1)x_n &= 0. \end{aligned}$

9. (a) $A = \begin{bmatrix} 1 & 1 & 1 \\ 2 & -1 & 1 \\ 1 & -1 & 2 \end{bmatrix}$, $X = \begin{bmatrix} r \\ s \\ t \end{bmatrix}$, $B = \begin{bmatrix} 2 \\ 3 \\ -1 \end{bmatrix}$

(b) $B = \begin{bmatrix} 1 & 2 & 1 \\ 1 & -1 & -1 \\ 1 & 1 & 2 \end{bmatrix}$, $Y = \begin{bmatrix} r & s & t \end{bmatrix}$, $C = \begin{bmatrix} 2 & 3 & -1 \end{bmatrix}$.

Section 3

1. $x = \dfrac{15}{9}, \quad y = \dfrac{10}{9}, \quad z = \dfrac{7}{9}$.

3. Inconsistent.

5. $c = 4$.

7. (a) $d - b - c + a = 0$
   (b) An infinite number.

9. (b) $x_1 = .6 - k$, $x_2 = k$, $x_3 = k$, $x_4 = .3 - k$, $x_5 = .1 - k$, $x_6 = k$.

Section 4

1. (a) $\begin{bmatrix} -\dfrac{1}{5} & \dfrac{2}{5} \\[2mm] \dfrac{3}{5} & -\dfrac{1}{5} \end{bmatrix}$

   (b) None

   (c) None

   (d) $\begin{bmatrix} 1 & 0 & 0 \\ 0 & 1 & 0 \\ 0 & 0 & 1 \end{bmatrix}$

   (e) None

   (f) $\begin{bmatrix} \dfrac{1}{2} & \dfrac{1}{2} & 0 \\[2mm] 0 & -\dfrac{1}{2} & \dfrac{1}{2} \\[2mm] -\dfrac{1}{2} & 0 & \dfrac{1}{2} \end{bmatrix}.$

3. (a) $A^{-1} = \begin{bmatrix} -1 & \dfrac{3}{4} \\[2mm] \dfrac{1}{2} & -\dfrac{1}{4} \end{bmatrix}.$

5. $X = A^{-1}B = \begin{bmatrix} \dfrac{1}{3} & \dfrac{1}{3} & 0 \\[2mm] \dfrac{5}{9} & -\dfrac{1}{9} & -\dfrac{1}{3} \\[2mm] -\dfrac{1}{9} & \dfrac{2}{9} & -\dfrac{1}{3} \end{bmatrix} \begin{bmatrix} 2 \\ 3 \\ -1 \end{bmatrix} = \begin{bmatrix} \dfrac{5}{3} \\[2mm] \dfrac{10}{9} \\[2mm] \dfrac{7}{9} \end{bmatrix}.$

7. (a) $x = 2$, $y = 1$, $z = 3$
   (b) $x = -2$, $y = -12$, $z = 18$.

9. $x = -8$, $y = 7$, $z = 1$, $t = 2$.

11. $A^{-1} = \begin{bmatrix} \dfrac{d}{\Delta} & -\dfrac{b}{\Delta} \\[2mm] -\dfrac{c}{\Delta} & \dfrac{a}{\Delta} \end{bmatrix}$,   where $\Delta = ad - bc$.

13. Multiply both sides by A^{-1}.

15. (a) $x_1 + 3x_2 + x_3 = 110$
$2x_1 + x_2 + 4x_3 = 132$
$3x_1 + x_2 + 4x_3 = 154$
   (b) $x_1 = 22, \quad x_2 = 24, \quad x_3 = 16.$

Section 5

1. $(I - A)^{-1} = \begin{bmatrix} \dfrac{75}{59} & \dfrac{30}{59} & \dfrac{55}{118} \\[2mm] \dfrac{30}{59} & \dfrac{130}{59} & \dfrac{140}{118} \\[2mm] \dfrac{55}{59} & \dfrac{140}{59} & \dfrac{355}{118} \end{bmatrix}$;   hence $X \doteq \begin{bmatrix} 8.6 \\ 17.5 \\ 30.3 \end{bmatrix}.$

3. $(I - A)^{-1} = \begin{bmatrix} \dfrac{130}{3} & 20 & \dfrac{40}{3} & 10 \\[2mm] \dfrac{195}{7} & 15 & \dfrac{65}{7} & \dfrac{50}{7} \\[2mm] \dfrac{440}{21} & 10 & \dfrac{170}{21} & \dfrac{40}{7} \\[2mm] \dfrac{75}{7} & 5 & \dfrac{25}{7} & \dfrac{30}{7} \end{bmatrix}$;   hence $X \doteq \begin{bmatrix} 900.0 \\ 610.7 \\ 478.6 \\ 246.4 \end{bmatrix}.$

5. $\begin{bmatrix} \dfrac{265}{118}\,a \\[2mm] \dfrac{460}{118}\,a \\[2mm] \dfrac{745}{118}\,a \end{bmatrix}$,   where a is the demand vector component.

7. $(I - A)^{-1} = \begin{bmatrix} 1 & 6 & 4 & 39 \\ 0 & 1 & 0 & 2 \\ 0 & 0 & 1 & 6 \\ 0 & 0 & 0 & 1 \end{bmatrix}$;   hence $X = \begin{bmatrix} 340 \\ 18 \\ 52 \\ 8 \end{bmatrix}.$

CHAPTER 4

Section 1

1. (a)

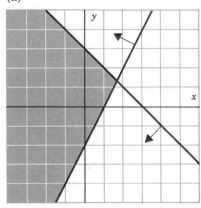

(b)

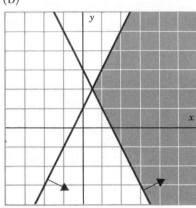

(c)

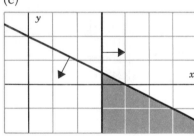

(d)

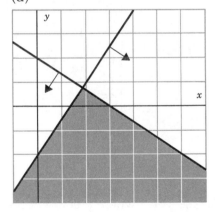

3. (a)

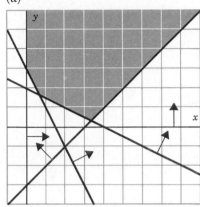

(b)

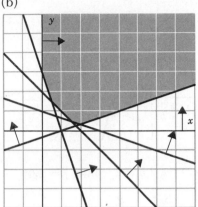

5. $\dfrac{1}{5} x + \dfrac{1}{2} y + \dfrac{2}{3} z \leqslant 2000$

$\dfrac{4}{5} x + \dfrac{1}{2} y + \dfrac{1}{3} z \leqslant 4000$

$x \geqslant 0,\ y \geqslant 0,\ z \geqslant 0.$

Section 2

1. $x = 4,\quad y = 2,\quad \max f = 22.$

3. $x = \dfrac{6}{7},\quad y = \dfrac{12}{7},\quad \min f = \dfrac{48}{7}.$

5. $x = \dfrac{5}{6},\quad y = \dfrac{4}{6},\quad \max f = \dfrac{7}{3}.$

7. $x = \dfrac{5}{3},\quad y = \dfrac{20}{3},\quad \min f = \dfrac{350}{3}.$

Section 3

1.

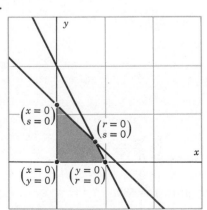

3.

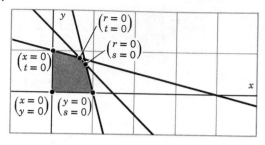

5. $x = 4$, $y = 2$,   max $f = 22$.

7. $x = \dfrac{5}{6}$, $y = \dfrac{2}{3}$,   max $f = \dfrac{7}{3}$.

9. Since that point is the corner $(x = 0,\ s = 0)$ and y has a positive value there, the value of f must be positive there. But f will assume the value zero at $(x = 0,\ y = 0)$; therefore the value of f would decrease in going to $(x = 0,\ y = 0)$, which is impossible in the simplex technique.

Section 4

1. $x = 4$, $y = 2$,   max $f = 22$.

3. $x = \dfrac{5}{6}$, $y = \dfrac{2}{3}$,   max $f = \dfrac{7}{3}$.

5. $x = 2$, $y = 2$, $z = 0$,   max $f = 8$.

7. (a) $f = 5x + 3y$
$$3x + 2y \geqslant 10$$
$$4x + 2y \geqslant 12$$
$$2x + 4y \geqslant 8$$
$$x \geqslant 0,\ y \geqslant 0$$
   (b) $x = 2$, $y = 2$,   min $f = 16$.

9. (a) $f = 3x + 5y + 4z + 7w$

$$x + y \leqslant 100, \quad x + z \geqslant 80, \quad x \geqslant 0, \quad y \geqslant 0,$$
$$z + w \leqslant 120, \quad y + w \geqslant 60, \quad z \geqslant 0, \quad w \geqslant 0.$$
   (b) $x = 40$, $y = 60$, $z = 40$, $w = 0$,   min $f = 580$.

CHAPTER 5

Section 1

1. HHHH, HHHT, HHTH, HTHH, THHH, HHTT, HTHT, HTTH, THHT, THTH, TTHH, HTTT, THTT, TTHT, TTTH, TTTT.

3. $\frac{1}{16}$.

5. Add three diagonal points RR, BB, and GG. Choose $p = \frac{1}{9}$.

7. (a) $\frac{1}{6}$ (b) $\frac{1}{3}$.

9. Yes. Yes, if the rises were larger in magnitude than the falls.

Section 2

1. (a) $\frac{2}{5}$ (b) $\frac{7}{10}$.

3. (a) $\frac{1}{16}$ (b) $\frac{1}{4}$ (c) $\frac{11}{16}$.

5. (a) $\frac{1}{10}$ (b) $\frac{2}{5}$ (c) $\frac{1}{5}$.

7. (a) .63   (b) .98   (c) .55

Section 3

1. $\frac{5}{9}$.

Section 4

1. (a) $\frac{4}{15}$ (b) $\frac{2}{9}$.

3. $\frac{2}{5}$.

5. (a) .00763   (b) .00006   (c) .98461
   (d) .01539   (e) .02313.

7. $\dfrac{16}{36} + \dfrac{6}{36} - \dfrac{2}{36} = \dfrac{20}{36}.$

9. (a) $\dfrac{247}{1700} = .145$ (b) $\dfrac{325}{1700} = .191.$

11. (a) .53   (b) .57.

Section 5

1. $\dfrac{8}{35}.$

3. $\dfrac{56}{65}.$

5. .79.

Section 6

1.

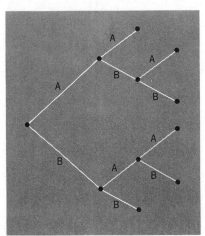

3. $\dfrac{20}{27}.$

5.

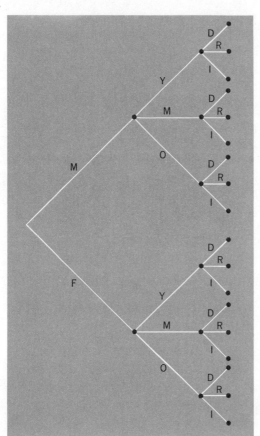

Eighteen folders
will be needed.

7. $2p^2 - 2p + 1$.

9. $\dfrac{3}{25}$.

11. $\dfrac{1}{5}$.

Section 7

1. 120.

3. $\dfrac{\dbinom{40}{10}}{\dbinom{50}{10}} \doteq .0825.$

5. .65.

7. (a) .006   (b) .90.

9. (a) .0026   (b) .01   (c) .30.

11. $\dfrac{33}{59}.$

13. $\dfrac{11}{76}.$

CHAPTER 6

Section 1

3.

x	4	5	6	7	8
$P\{x\}$	$\frac{1}{9}$	$\frac{2}{9}$	$\frac{3}{9}$	$\frac{2}{9}$	$\frac{1}{9}$

5.

x	0	1	2	3	4
$P\{x\}$	$\frac{1}{16}$	$\frac{4}{16}$	$\frac{6}{16}$	$\frac{4}{16}$	$\frac{1}{16}$

Section 2

1. $-\dfrac{2}{37}.$ Approximately a 6-cent loss.

3. 1.75.

5. $E_a = 7$, $E_b = 6$; hence, choose dice.

7. 6.

Section 3

1. $p = \dfrac{3}{7}.$

3. $p = \dfrac{2}{5}$.

5. $E_S = 1.295$, $E_B = 1.25$; hence, buy stocks.

Section 4

1. $P\{0\} = \dfrac{1}{32}$, $P\{1\} = \dfrac{5}{32}$, $P\{2\} = \dfrac{10}{32}$,

   $P\{3\} = \dfrac{10}{32}$, $P\{4\} = \dfrac{5}{32}$, $P\{5\} = \dfrac{1}{32}$.

3. $P\{x \geqslant 7\} = .88$.

5. No. Trials are not independent and p is not constant.

7. (a) .33   (b) .41.

9. .93.

11. (a) .61   (b) .014.

13. .36.

Section 5

1. (a) $p_{21}^{(2)} = \dfrac{7}{18}$, $p_{12}^{(2)} = \dfrac{7}{12}$

   (b) $p_{21}^{(2)} = \dfrac{1}{4}$, $p_{12}^{(2)} = \dfrac{1}{2}$

3. (a) $P = \begin{bmatrix} .7 & .2 & .1 \\ .4 & .4 & .2 \\ .4 & .5 & .1 \end{bmatrix}$

   (b) $p_{33}^{(2)} = .15$

   (c) $p_{13}^{(2)} = .12$
   (d) $p_{11}p_{11} = .49$.

5. $p_{ij}^{(2)} = a_{i1}a_{1j} + \cdots + a_{in}a_{nj}$; hence

   $$\sum_{j=1}^{n} p_{ij}^{(2)} = a_{i1} \sum_{j=1}^{n} a_{1j} + \cdots + a_{in} \sum_{j=1}^{n} a_{nj}$$

   $$= a_{i1} + \cdots + a_{in} = 1.$$

7. (a) [.383   .350   .267]
   (b) [.409   .348   .243].

9. $\left[\dfrac{62}{143}\ \dfrac{48}{143}\ \dfrac{33}{143}\right] = [.43\ \ .34\ \ .23]$. No difference.

11. (a) $P = \begin{array}{c} \\ A \\ B \end{array}\begin{array}{cc} A & B \\ \left[\begin{array}{cc} 1-r & r \\ s & 1-s \end{array}\right. & \left.\vphantom{\begin{array}{c}1-r\\s\end{array}}\right] \end{array}$

   (b) $x_1 = \dfrac{s}{r+s}$, $x_2 = \dfrac{r}{r+s}$

   (c) $p_n = (1-r-s)^n \left(p_0 - \dfrac{s}{r+s}\right) + \dfrac{s}{r+s}$

   (d) If $p_0 = \dfrac{s}{r+s}$, $p_n = \dfrac{s}{r+s}$ for all n; hence, the gene frequencies do not change.

   (e) $p_n \longrightarrow \dfrac{s}{r+s}$. This is the same as in (b), as it should be. Thus we have two methods for calculating the long-run probabilities here.

Section 6

1. $\dfrac{2}{5}$.

3. (a) $\dfrac{1}{4}$ (b) $\dfrac{1}{4}$.

5. $2.

7. $380.

9. (a) .45   (b) .80

   (c) .84; hence, bet the amount you wish to win each time.

Section 7

1. (a) $p_2 = .13$ and $p_3 = .217$

   (b) $p_n = 1 - \dfrac{29}{30}\left(\dfrac{9}{10}\right)^n \longrightarrow 1$

   (c) $E[T_n] = \dfrac{29}{3} - \dfrac{29}{3}\left(\dfrac{9}{10}\right)^n$

(d) $E[T_n] \longrightarrow \dfrac{29}{3}$.

3. $c = 1 - \sqrt{2/3} \doteq .18$.

5. (a) $p_5 = .69$

(b) $\left(\dfrac{3}{4}\right)^{n-1} \leqslant \dfrac{1}{2}$; hence $n = 4$ should suffice

(c) $p_n \longrightarrow 1$

(d) $E[T_n] \longrightarrow 4\dfrac{99}{100} \doteq 4$.

CHAPTER 7

Section 2

1. (a) R_1 and C_1. R wins at least 5 and C pays at most 5. Yes.
   (b) R_2 and C_2. R wins at least 1 and C pays at most 1. Yes.
   (c) R_1 and C_2. R wins at least 0 and C pays at most 1. No.
   (d) R_1 and C_1. R wins at least 1 and C pays at most 1. Yes.
   (e) R_3 and C_3. R wins at least 1 and C pays at most 1. Yes.
   (f) R_3 and C_1. R wins at least 1 and C pays at most 2. No.

3. $\begin{array}{c c} & \begin{array}{c c c} 1 & 2 & 3 \end{array} \\ \begin{array}{c} 1 \\ 2 \\ 3 \end{array} & \left[\begin{array}{r r r} 1 & -1 & -1 \\ -2 & 2 & -2 \\ -3 & -3 & 3 \end{array}\right]. \end{array}$

5. $\begin{array}{c c} & \begin{array}{c c} 1 & 2 \end{array} \\ \begin{array}{c} 1 \\ 2 \end{array} & \left[\begin{array}{r r} 2 & -3 \\ -3 & 4 \end{array}\right]. \end{array}$

7. R_1 and C_1. No.

9. R_1 and C_1. Yes.

11. R_1 (or R_2) and C_1. No.

Section 3

1. (a) $\dfrac{8}{3}$ (b) $\dfrac{4}{5}$ (c) $\dfrac{1}{6}$.

3. Expected payoff is $-\dfrac{1}{6}$. His pure strategy payoff would be -1.

Section 4

1. If a and b are of opposite signs.

3. (a) $\begin{bmatrix} 2 \\ 3 \end{bmatrix}$ because C will prefer C_1 to C_2.

   (b) [4 3] because R will prefer R_2 to R_1.

   (c) $\begin{bmatrix} 2 & -2 \\ 1 & 3 \end{bmatrix}$ because C will prefer C_1 to C_3.

   (d) $\begin{bmatrix} 1 & 0 \\ -3 & 3 \end{bmatrix}$ because R will prefer R_2 to R_3.

5. (a) $\begin{bmatrix} 5 & -1 \\ 2 & 3 \end{bmatrix}$ because R will prefer R_2 to R_3 and C will prefer C_3 to C_1.

   No value.

   (b) $\begin{bmatrix} 1 & 4 & -2 \\ 2 & 3 & 3 \\ -1 & 0 & 4 \end{bmatrix}$ because R will prefer R_3 to R_2 and C will prefer C_2 to C_3. The value is 2.

Section 5

1. (a) $p_1 = \frac{1}{7}$, $p_2 = \frac{6}{7}$, $q_1 = \frac{5}{7}$, $q_2 = \frac{2}{7}$, $v = \frac{16}{7}$.

   (b) $p_1 = 1$, $p_2 = 0$, $q_1 = 1$, $q_2 = 0$, $v = 3$.

3. $p_1 = \frac{1}{7}$, $p_2 = \frac{6}{7}$, $p_3 = 0$, $q_1 = \frac{3}{7}$, $q_2 = 0$, $q_3 = \frac{4}{7}$, $v = \frac{4}{7}$.

5. Since $\Sigma\Sigma\, a_{ij} p_i q_j > 0$ for every set of p's and q's because $a_{ij} > 0$ for all i and j, this sum must be positive for the optimal probability sets. But that sum defines v; hence, $v > 0$.

Section 6

1.

	Stone	Scissors	Paper
Stone	0	1	−1
Scissors	−1	0	1
Paper	1	−1	0

Optimal strategies are $p_1 = p_2 = p_3 = q_1 = q_2 = q_3 = \frac{1}{3}$. The value of the game is 0.

CHAPTER 8

Section 1

1. (a) $m = 1$ (b) $m = \dfrac{1}{2}$.

3. $m = 3x_0{}^2 + 3x_0 h + h^2 \longrightarrow 3x_0{}^2$; hence, $m = 3$.

5. $m = -\dfrac{1}{x_0{}^2 + x_0 h} \longrightarrow -\dfrac{1}{x_0{}^2}$; hence, $m = -\dfrac{1}{4}$.

7. $m = .305, .313, .323, .328, .332 \longrightarrow .333$ or $\dfrac{1}{3}$.

Section 2

1. $f'(1) = \lim\limits_{h \to 0} \dfrac{h}{1 + h} = 0.$

3. $f'(0) = \lim\limits_{h \to 0} \dfrac{2}{h + 1} = 2.$

5. $m = 3$; hence, $y = 3x - 2.$

7. $m = -\dfrac{1}{4}$; hence, $y = -\dfrac{1}{4} x + \dfrac{3}{4}.$

Section 3

1. (a)

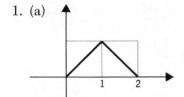

(b) $\lim\limits_{x \to 1} f(x) = 1$
(c) No
(d) Everywhere.

3. (a)

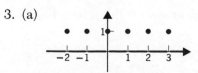

(b) $\lim_{x \to 2} f(x) = 0$

(c) No

(d) Everywhere, except at the integers.

5. At $x = 0$, 1, and 2. It is continuous at $x = 0$ but not at $x = 1$ or $x = 2$.

9. Not at $x = 1$ because the slope becomes infinite. This follows from the fact that $\lim_{h \to 0} - \sqrt{\dfrac{2}{h}} - 1$ does not exist.

11. $\lim_{h \to 0} \dfrac{1}{\sqrt{h}}$ does not exist.

Section 5

1. (a) $f' = 12x^2 - 12x^3$

   (b) $f' = 6x^5 + 4x^3 + 3x^2 + 1$

   (c) $f' = -\dfrac{4}{(2x - 1)^2}$

   (d) $f' = -\dfrac{2x}{(x^2 + 1)^2}$

   (e) $f' = 2x + 3 - \dfrac{2}{x^3}$

   (f) $f' = 70x^6 + 60x^4 - 15x^2 - 6$

   (g) $f' = -\dfrac{8}{x^3} - \dfrac{24}{x^5}$

   (h) $f' = -\dfrac{1}{(x + 1)^2}.$

3. (a) $f' = 12x^3 - 10x$

   (b) $f' = x^2 - 1$

   (c) $f' = -\dfrac{6}{x^6}$

   (d) $f' = -\dfrac{1}{(x - 1)^2}$

   (e) $f' = \dfrac{4 - 8x^2}{(2x^2 + 1)^2}$

   (f) $f' = \dfrac{x^2 + 2x}{(x + 1)^2}$

   (g) $f' = \dfrac{2 - x}{x^3}$

   (h) $f' = \dfrac{48x^2}{(x^3 + 8)^2}.$

5. $m = -\dfrac{1}{2}$; hence $y = -\dfrac{1}{2}x + 2.$

7. (a) $x = 1$, $x = -1$

   (b) $x = 0$, $x = 1$, $x = -1$

   (c) $x = 2$, $x = -2$

   (d) $x = 1.$

9. $x = 1$ and $x = 2.$

11. $y - y_0 = -\dfrac{1}{x_0{}^2}(x - x_0)$ simplifies into the answer if the relation $x_0 y_0 = 1$ is employed.

13. $f' = 9x^8 + 21x^6 + 6x^5 - 20x^4 + 12x^3 - 8x.$

Section 7

1. (a) $f = 1 + 2x^2 + x^4$; hence, $f' = 4x + 4x^3$
   (b) $f' = 2(1 + x^2)2x = 4x + 4x^3$.

3. (a) $f' = 6(x + 2)(x^2 + 4x - 3)^2$ (b) $f' = 2(x - 1)(2x^2 - x + 2)$

   (c) $f' = \dfrac{6x - 4}{(x + 1)^3}$ (d) $f' = \dfrac{2x^3(x^2 + 4)}{(x^2 + 2)^2}$

   (e) $f' = -\dfrac{6x + 5}{(x + 1)^2(2x + 1)^3}$ (f) $f' = 2(x^4 + 2x^2 + x + 1) \cdot$
                                                          $(4x^3 + 4x + 1)$

5. (a) $f' = 3(x + 3)^2$ (b) $f' = -\dfrac{4}{(x - 2)^5}$

   (c) $f' = 4x(x^2 - 1)$ (d) $f' = \dfrac{6x^2}{(1 - x^3)^3}$

   (e) $f' = \dfrac{(x^2 + 3)(3x^2 - 4x - 3)}{(x - 1)^2}$

   (f) $f' = -(x^{-2} + 4)(x^2 + 1)^{-2}(8x + 6x^{-1} + 4x^{-3})$.

7. $y = 4x - 10$.

9. $r'(x) = [f(x)]^m n[g(x)]^{n-1} g'(x) + [g(x)]^n m[f(x)]^{m-1} f'(x)$
   $= [f(x)]^{m-1} [g(x)]^{n-1} \{nf(x)g'(x) + mg(x)f'(x)\}$.

Section 8

1. (a) $y' = \dfrac{2y - x}{y - 2x}$ (b) $y' = -\dfrac{y^2}{x^2}$

   (c) $y' = \dfrac{x(y^2 - 1)}{y(1 - x^2)}$ (d) $y' = -\dfrac{y(y^2 + 3x^2)}{x(x^2 + 3y^2)}$

   (e) $y' = \dfrac{x^2 + y}{x - x^2}$ (f) $y' = \dfrac{4x^3 - 2xy^3}{4y^3 + 3x^2y^2}$.

3. $y = -4x + 9$.

5. $y' = \dfrac{x}{y}$ and $y' = \dfrac{x}{\sqrt{x^2 - 9}} = \dfrac{x}{y}$.

7. $y - y_0 = -\dfrac{x_0}{y_0}(x - x_0)$ simplifies into the answer if the relation $x_0^2 + y_0^2 = r^2$ is employed.

9. $\dfrac{dx}{dy} = \sqrt{2x - 1}$ and $\dfrac{dx}{dy} = y = \sqrt{2x - 1}$.

11. $y' = - \dfrac{mf'(x)\ [f(x)]^{m-1}}{ng'(y)\ [g(y)]^{n-1}}.$

CHAPTER 9

Section 1

1. (a) Rising for $x > \dfrac{3}{2}$

   (b) Falling for $x < \dfrac{3}{2}$, except at $x = 0$

   (c) At $(0,0)$ and $\left(\dfrac{3}{2}, -\dfrac{27}{8} \right)$

   (d)

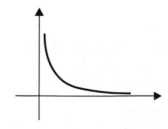

3. $\dfrac{dy}{dx} = \dfrac{b}{2\sqrt{x - c}} > 0$

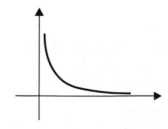

5. $\dfrac{dA}{dx} = - \dfrac{ax + 2b}{2x^2 \sqrt{ax + b}} < 0$

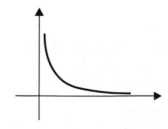

7. (a)

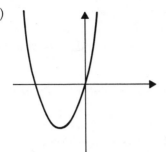

(b)

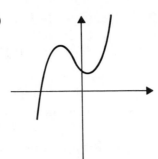

(c)

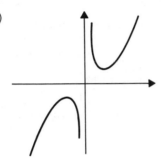

(d)

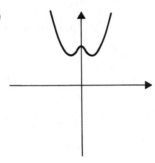

(e)

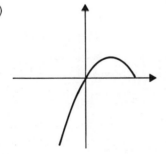

(f)

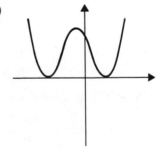

(g)

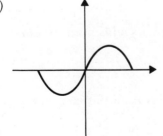

(h)

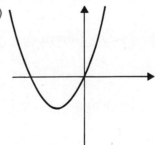

Section 2

1. (a) $f(0) = 4\frac{1}{2}$, $f(3) = 6$, $f(5) = 2$; hence, $x = 3$ yields a maximum and $x = 5$ yields a minimum.

   (b) $f(-1) = -8$, $f(1) = 0$, $f(3) = 8$; hence, $x = 3$ yields a maximum and $x = -1$ yields a minimum.

   (c) $f(-2) = 32$, $f(-1) = -5$, $f(0) = 0$, $f(2) = -32$, $f(3) = 27$; hence, $x = -2$ yields a maximum and $x = 2$ yields a minimum.

   (d) $f(-4) = -.23$, $f(-3) = -\frac{1}{4}$, $f(0) = 0$, $f(3) = \frac{1}{4}$, $f(4) = .23$; hence, $x = 3$ yields a maximum and $x = -3$ yields a minimum.

   (e) $f(0) = 2$, $f(-2 + 2\sqrt{2}) = 1.7$, $f(4) = 3\frac{1}{3}$; hence, $x = 4$ yields a maximum and $x = -2 + 2\sqrt{2}$ yields a minimum.

   (f) $f(-\sqrt{2}) = -8\sqrt{2}$, $f(-1) = -22$, $f(0) = 0$, $f(1) = 22$, $f(\sqrt{2}) = 8\sqrt{2}$; hence, $x = 1$ yields a maximum and $x = -1$ yields a minimum.

3. $x = 25$; $P(25) = 8416\frac{2}{3}$.

5. $x = 350$.

7. $x_0 p'(x_0) + p(x_0) - C'(x_0) = 0$ and $(x_0 + h)p'(x_0 + h) + p(x_0 + h) -$
$C'(x_0 + h) \begin{cases} > 0 & \text{for} & h < 0 & \text{and} & |h| \text{ small} \\ < 0 & \text{for} & h > 0 & \text{and} & |h| \text{ small.} \end{cases}$

9. Base side $= \sqrt[3]{4}$ and height $= \sqrt[3]{4}$.

11. Distance from station $= \dfrac{13}{16}$ mile.

13. (a) $x = 0$ yields a maximum and $x = 4$ yields a minimum

    (b) $x = \dfrac{1}{3}$ yields a maximum and $x = 0$ or 1 yields a minimum.

15. $x = \dfrac{a}{2}$.

17. A distance of $10\sqrt[3]{4}/(1 + \sqrt[3]{4}) \doteq 6.1$ miles from the large jet airport.

19. (a) $x = \sqrt{\dfrac{2C_2 A}{C_3}}$ (b) $T = 12\sqrt{\dfrac{2C_2}{C_3 A}}$.

Section 3

1. (a) $v = 2$ ft/m

   (b) $v = -1$ ft/m

3. (a) $-2\frac{11}{12}$ m/h

   (b) $v(1) = -4$ m/h

   (c) $t = 2$

   (d) $t > 2$.

5. (a) $\frac{5}{6}$ ft/m

   (b) $v(2) = \frac{3}{4}$ ft/m

   (c) $v\left(2\frac{1}{2}\right) = \frac{21}{25}$ ft/m; hence, close to answer in (a).

7.

9. (a) $a(t) = -2$

   (b) $a(t) = 8$

   (c) $a(t) = \frac{2}{t^3}$

   (d) $a(t) = \frac{2}{(t+1)^3}$.

11. $\dfrac{W'(t)}{W(t)} = \dfrac{b}{t}$; hence, the specific growth varies inversely with age.

13. $C'(20) = 4 + \dfrac{2}{\sqrt{401}} \doteq 4.1$.

15. $R'(x) = \dfrac{20}{(x+1)^2}$.

17. (a) $R'(x) = -.044\, x^{-1.4}$

   (b) $R'(x) = \dfrac{(38)^{5/6}}{6x^{5/6}}$.

19. (a) $R(x) = x\sqrt{100 - x^2}$

   (b) $R'(x) = \dfrac{100 - 2x^2}{\sqrt{100 - x^2}}$

   (c) $R'(x) = 0$ gives $p = \sqrt{50} \doteq 7.1$.

21. $A'(x) = \dfrac{xC'(x) - C(x)}{x^2} = 0$ gives $xC'(x) = C(x)$ for $A(x)$ to be a minimum. Average cost curve $y = \dfrac{C(x)}{x}$ and the marginal cost curve $y = C'(x)$ intersect when $\dfrac{C(x)}{x} = C'(x)$; hence, when x minimizes $A(x)$.

23. (a) $C'(x) = -\dfrac{1000}{x^2} + \dfrac{1}{10^6}$

    (b) $C'(50{,}000) = .6 \times 10^{-6}$
        $C'(100{,}000) = .9 \times 10^{-6}$
    (c) $C'(x) = 0$ gives $x^2 = 10^9$, or $x \doteq 31{,}623$ miles.

25. Elasticity $= \dfrac{2p^2}{p^2 - 10{,}000}$.

27. If $\dfrac{p}{x}\dfrac{dx}{dp} = -1$, then $\dfrac{p}{x\dfrac{dp}{dx}} = -1$, or $x\dfrac{dp}{dx} + p = 0$. But $R(x) = xp(x)$ will be maximized [for a realistic $p(x)$] when $R'(x) = x\dfrac{dp}{dx} + p = 0$.

CHAPTER 10

Section 2

1.

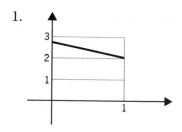

3. (a) $C(T) = .368\, C(0)$, $C(2T) = .135\, C(0)$, $C(5T) = .007\, C(0)$
    (b) $C(t) \longrightarrow 0$ as $t \longrightarrow \infty$.

5. $t = 6.3$ years.

Section 3

1. (a) $f' = \dfrac{2x}{1 + x^2}$ (b) $f' = \dfrac{x}{1 + x^2}$

(c) $f' = \dfrac{2}{x}$

(d) $f' = 1 + \log x$

(e) $f' = \dfrac{2}{1 - x^2}$

(f) $f' = \dfrac{1}{1 - x^2}$

(g) $f' = \dfrac{1}{x \log x}$

(h) $f' = \dfrac{2x + x \log (x^2 + 1)}{\sqrt{x^2 + 1}}.$

3. (a) $f' = \dfrac{1 - \log x}{x^2}$

(b) $f' = \dfrac{1}{2x \sqrt{\log x}}$

(c) $f' = \dfrac{1}{x(x + 1)}$

(d) $f' = \dfrac{1}{(x + 1) \log (x + 1)}$

(e) $f' = 3^{2x} \cdot 2 \log 3$

(f) $f' = 2x^{\log x - 1} \cdot \log x.$

5. $y = x + \log 2 - 1.$

7.

9. (a) $(0, \log 4)$ is a local minimum

(b) $\left(-\dfrac{1}{2}, \log \dfrac{3}{4}\right)$ is a local minimum

(c) $\left(\dfrac{1}{e}, -\dfrac{1}{e}\right)$ is a local minimum

(d) $\left(\dfrac{1}{\sqrt{2}}, \dfrac{1}{2} + \dfrac{1}{2} \log 2\right)$ is a local minimum.

Section 4

1. (a) $f' = 4e^{4x}$

(b) $f' = xe^x + e^x$

(c) $f' = 2xe^{x^2}$

(d) $f' = \dfrac{2e^{2x}(x - 1)}{x^3}$

(e) $f' = (2x + 2)e^{x^2 + 2x}$

(f) $f' = \dfrac{4}{(e^x + e^{-x})^2}.$

3. $y = -x + 1$.

5. (a)

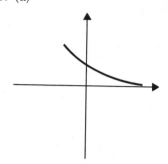

(b)

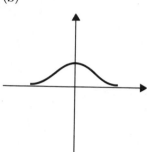

(c)

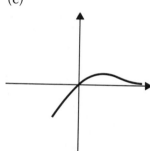

(d)

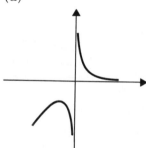

(e)

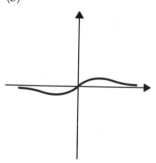

(f)

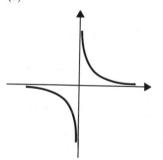

7. (a) $y' = \dfrac{y(0) \log 2}{T} 2^{t/T}$

(b) $T = \dfrac{\tau \log 2}{\log \dfrac{y(\tau)}{y(0)}}$.

9. $t = \dfrac{\log b - \log a}{b - a}$.

Section 5

1. 9932; hence, prefer 10,000 now.

3. (a) $P = 2857$ (b) $P = 2362$.

5. $t = 15.8$ years.

7. $y(6) = 20$.

9. (a) $y(t) = e^t + y(0) - 1$
   (b) $y(5) = 147 + y(0)$
   (c) $y(5) - y(4) = 93$.

11. (a) $\alpha = A$ and β any number will satisfy
    (b) $\alpha = $ fixed amount; $\beta = $ initial investment minus A
    (c) $y \longrightarrow \alpha (=A)$ as $t \longrightarrow \infty$.

13. If $y = \dfrac{ae^{at}}{c + be^{at}}$, then $\dfrac{dy}{dt} = ce^{-at} y^2$. Using the first relation in the form
    $e^{at}[a - by] = cy$ to solve for e^{-at} and substituting it into the expression
    for $\dfrac{dy}{dt}$ will reduce $\dfrac{dy}{dt}$ to $ay - by^2$.

15. $t = 8040 \log \dfrac{1}{.886} \doteq 973$; hence, $1972 - 973 = 999$, or approximately the
    year A.D. 1000.

17. (a) $\dfrac{1}{n} \sum\limits_{t=1}^{n} (C + at) = \dfrac{Cn}{n} + \dfrac{a}{n} (1 + 2 + \cdots + n) = C + a \dfrac{n+1}{2}$

    (b) $A'(x) = \dfrac{a}{2} - \dfrac{B}{x^2} - A \dfrac{xr^x \log r - r^x}{x^2}$

    (c) Since $\dfrac{r^x}{x} = \dfrac{\left(\frac{2}{3}\right)^x}{x}$ will be very small for x at all large, we may ignore
    the term involving A and solve for x in

    $$0 = 250 - \dfrac{24{,}000}{x^2}$$

    This gives $x = \sqrt{96} \doteq 10$; hence, replace every 10 years. The value of
    the ignored term for $x = 10$ is approximately 13; hence, it would have
    little effect on the solution.

CHAPTER 11

Section 1

1. $S_L = \dfrac{1}{n^2}\left[1 + 2 + \cdots + (n-1)\right] = \dfrac{1}{2} - \dfrac{1}{2n} \longrightarrow \dfrac{1}{2}$

   $S_U = \dfrac{1}{n^2}\left[1 + 2 + \cdots + n\right] = \dfrac{1}{2} + \dfrac{1}{2n} \longrightarrow \dfrac{1}{2}.$

3. (a)   (b)

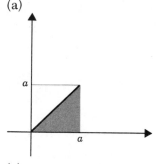 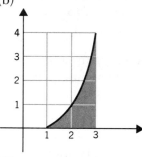

  (c)   (d)

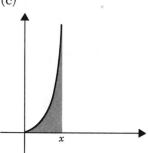 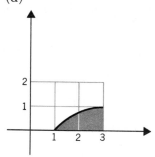

5. (a) $\displaystyle\int_{1}^{n+1} ax^b\,dx$

  (b) Integrate from, say, $x = \dfrac{1}{2}$ to $x = n+1$ to obtain $\displaystyle\int_{1/2}^{n+1} ax^b\,dx.$

Section 3

1. (a) $\dfrac{5}{3}$ (b) $\dfrac{4}{3}\sqrt{2}$ (c) 3

  (d) $15\dfrac{3}{4}$ (e) $\dfrac{4}{3}$ (f) $\dfrac{8}{3}.$

3. $A = \dfrac{4}{3}.$

5. (a) $f(x) = \dfrac{x}{2} + c$ 　　　　　　　　　　(b) $f(x) = \dfrac{x^2}{4} + c$

　　(c) $f(x) = \dfrac{2}{3} x^{3/2} + c$ 　　　　　　　(d) $f(x) = \dfrac{x^4}{4} - \dfrac{1}{x} + c$.

7. $A = 8$.

9. $A = 1 - e^{-b} \longrightarrow 1$ as $b \longrightarrow \infty$.

11. (a) $\dfrac{1}{4}$ and $\dfrac{1}{5}$ 　　　　　　　　　(b) 1 and $\dfrac{3}{2}$.

13. $s(t) = t^2 + 30t$.

15. $f(2) = 13$.

17. $F'(2) = \sqrt{5}$.

19. $x = -\dfrac{1}{2}$.

Section 4

1. $R(x) = -\dfrac{x^4}{4} + x^3 - \dfrac{x^2}{2} + 2x$.

3. (a) $R(x) = 20x - 3x^2 + \dfrac{x^3}{3} + c$,　where $c = R(0)$

　　(b) $R(x) = px$; hence $p(x) = 20 - 3x + \dfrac{x^2}{3} + \dfrac{c}{x}$.

5. C.S. $= \dfrac{1}{3}$.

7. $\log k = \displaystyle\int \dfrac{c}{t^2}\, dt = -\dfrac{c}{t} + K$; hence, $k = Ae^{-c/t}$, where A is some constant.

9. (a) $f'(x) = \dfrac{k}{x}$ implies that $f(x) = \displaystyle\int \dfrac{k}{x}\, dx = k \log x + c$

　　(b) $\dfrac{f(x+h) - f(x)}{f(x)} = k\,\dfrac{h}{x}$ is equivalent to $\dfrac{f(x+h) - f(x)}{h} = k\,\dfrac{f(x)}{x}$;

　　　　hence $f'(x) = k\dfrac{f(x)}{x}$. But $\dfrac{f'(x)}{f(x)} = \dfrac{k}{x}$ gives $\dfrac{d(\log f(x))}{dx} = \dfrac{k}{x}$; hence

　　　　$\log f(x) = \displaystyle\int \dfrac{k}{x}\, dx$. This yields $f(x) = Ax^k$.

Section 5

1. (a) $\dfrac{(2x+3)^{3/2}}{3} + c$ (b) $\dfrac{1}{3}\log(3x+4) + c$

   (c) $\dfrac{2}{3}\sqrt{3x+4} + c$ (d) $\dfrac{2}{9}(1+x^3)^{3/2} + c$

   (e) $\dfrac{2}{3}(x+1)^{3/2} - 2(x+1)^{1/2} + c$ (f) $\dfrac{1}{5}\log(x^5+1) + c$

   (g) $\dfrac{1}{22}(x^2+1)^{11} + c$ (h) $\log(x-1) - \dfrac{1}{x-1}$

   $$-\dfrac{1}{2(x-1)^2} + c.$$

3. (a) $\sqrt{3}$ (b) $\dfrac{1}{2}\log 3$

   (c) 2 (d) $\dfrac{1}{2}\log^2 2$

   (e) $3(\sqrt[3]{2}-1)$ (f) $\dfrac{1}{4}\log\dfrac{2}{17}$.

5. C.S. $= \dfrac{750}{4\log 2} - 125$.

7. $A = \dfrac{1}{2}(1 - e^{-a})$.

9. $A = 1 - \dfrac{1}{b+1} \longrightarrow 1$ as $b \longrightarrow \infty$. Finite area here whereas infinite area in Problem 8.

Section 6

1. (a) $\dfrac{xe^{4x}}{4} - \dfrac{e^{4x}}{16} + c$ (b) $\dfrac{x^2}{2}\log x - \dfrac{x^2}{4} + c$

   (c) $x\log x - x + c$ (d) $e^x(x^2 - 2x + 2) + c$

   (e) $\dfrac{x^2 e^{x^2}}{2} - \dfrac{e^{x^2}}{2} + c$ (f) $x\log^2 x - 2x\log x + 2x + c$.

3. $A = 1$.

5. C.S. $= 16\log\dfrac{4}{3} - 4$; P.S. $= 4 - 2\log 3$.

7. $6 - 2e$.

9. $\dfrac{1}{a}$ and $\dfrac{2}{a^2}$.

CHAPTER 12

Section 1

1. (a) $f_x = 3y + 6$, $f_y = 3x - 2y$

   (b) $f_x = \dfrac{x}{\sqrt{x^2 + y^2}}$, $f_y = \dfrac{y}{\sqrt{x^2 + y^2}}$

   (c) $f_x = y^2$, $f_y = 2xy - 5$

   (d) $f_x = \dfrac{x^2 + 2xy + y^2}{(x + y)^2}$, $f_y = -\dfrac{x^2 + x}{(x + y)^2}$

   (e) $f_x = e^{xy}(xy^2 + y)$, $f_y = e^{xy}(x^2y + x)$

   (f) $f_x = \dfrac{x}{x + y} + \log(x + y)$, $f_y = \dfrac{x}{x + y}$.

3. (a) $f_x = \dfrac{2y}{(x + y)^2}$, $f_y = -\dfrac{2x}{(x + y)^2}$

   (b) $f_x = \dfrac{y^3 - x^2y}{(x^2 + y^2)^2}$, $f_y = \dfrac{x^3 - y^2x}{(x^2 + y^2)^2}$

   (c) $f_x = \dfrac{2x}{x^2 + y^2}$, $f_y = \dfrac{2y}{x^2 + y^2}$

   (d) $f_x = e^y + ye^x$, $f_y = xe^y + e^x$

   (e) $f_x = \dfrac{1}{x} - ye^{-xy}$, $f_y = \dfrac{1}{y} - xe^{-xy}$

   (f) $f_x = -\dfrac{1}{x^2y} + \dfrac{8}{x^3y} + \dfrac{4}{x^2y^2}$, $f_y = -\dfrac{1}{xy^2} + \dfrac{4}{x^2y^2} + \dfrac{8}{xy^3}$

   (g) $f_x = \log y + \dfrac{y}{x}$, $f_y = \log x + \dfrac{x}{y}$

   (h) $f_x = \dfrac{2x}{x^2 + y^2 + z^2}$, $f_y = \dfrac{2y}{x^2 + y^2 + z^2}$, $f_z = \dfrac{2z}{x^2 + y^2 + z^2}$

(i) $f_x = \dfrac{-x}{(x^2 + y^2 + z^2)^{3/2}}, f_y = \dfrac{-y}{(x^2 + y^2 + z^2)^{3/2}}, f_z = \dfrac{-z}{(x^2 + y^2 + z^2)^{3/2}}$

(j) $f_x = ye^x, f_y = e^x + ze^y, f_z = e^y + e^z$.

Section 2

1. (a) $(0,0)$,   (b) No maximum or minimum.

3. If $y = 0$, $f \longrightarrow \infty$ as $x \longrightarrow \infty$; hence, no maximum is possible.
   If $x = 0$, $f \longrightarrow -\infty$ as $y \longrightarrow \infty$; hence, no minimum is possible.

Section 3

1. $\left(-\dfrac{1}{4}, 16\right)$ yields a local maximum.

3. $\left(0, \dfrac{1}{2}\right)$ and $\left(0, -\dfrac{1}{2}\right)$ are saddle points. No local maxima or minima.

5. The test gives no information at the single critical point $(0,0)$; however, it is neither a local maximizing or minimizing point.

7. $(0,0)$, $(0,3)$, $(4,0)$ are saddle points, and $\left(\dfrac{4}{3}, 1\right)$ is a local maximizing point.

Section 4

1. A square base of side 4 and a height of 2.

3. Length $= \dfrac{100}{3}$ and sides $= \dfrac{50}{3}$ each; hence, a square end and twice as long as wide.

5. $x = 500$, $y = 20\sqrt{30}$, $z = 1500 - 20\sqrt{30}$.

7. $\left.\begin{array}{l} P_x = R_1'(x) - C'(z)\dfrac{\partial z}{\partial x} = 0 \\[2mm] P_y = R_2'(y) - C'(z)\dfrac{\partial z}{\partial y} = 0 \end{array}\right\}$ give $R_1'(x) = R_2'(y)$.

Section 5

1. $x = -1$, $y = \dfrac{3}{2}$; $f\left(-1, \dfrac{3}{2}\right) = 6$ is a minimum.

3. $x = \dfrac{6}{7}$, $y = -\dfrac{4}{7}$, $z = \dfrac{2}{7}$; $f\left(\dfrac{6}{7}, -\dfrac{4}{7}, \dfrac{2}{7}\right) = \dfrac{8}{7}$ is a minimum.

5. $x = \dfrac{37}{17}$, $y = \dfrac{8}{17}$, $z = -\dfrac{38}{17}$, $u = \dfrac{18}{17}$, $v = \dfrac{65}{17}$ make $f \doteq 23.4$, which is a mini-

mum.

7. $\alpha = \dfrac{1}{\sqrt{2}}$, $\beta = \dfrac{1}{\sqrt{2}}$ make $D = \sqrt{3 - 2\sqrt{2}}$, which is approximately .4.

Index

Finite Mathematics and Calculus with Applications to Business

PAUL G. HOEL
University of California at Los Angeles

JOHN WILEY & SONS, New York • Chichester • Brisbane • Toronto

Library of Congress Cataloging in Publication Data:

Hoel, Paul Gerhard, 1905–
  Finite mathematics and calculus with applications
to business.

    1. Mathematics--1961–   I. Title.
QA37.2.H64   510   73-19505
ISBN 0-471-40430-6

Printed in the United States of America

10 9 8 7 6 5